T0232700

SCOTLAND'S PARIAH

The Life and Work of John Pinkerton, 1758–1826

PATRICK O'FLAHERTY

Scotland's Pariah

The Life and Work of John Pinkerton, 1758–1826

UNIVERSITY OF TORONTO PRESS
Toronto Buffalo London

© University of Toronto Press 2015
Toronto Buffalo London
www.utppublishing.com

ISBN 978-1-4426-4928-6 (cloth)

Library and Archives Canada Cataloguing in Publication

O'Flaherty, Patrick, 1939–, author
Scotland's pariah : the life and work of John Pinkerton,
1758–1826 / Patrick O'Flaherty.

Includes bibliographical references and index.
ISBN 978-1-4426-4928-6 (bound)

1. Pinkerton, John, 1758–1826.   2. Cartographers – Scotland – Biography.
3. Historians – Scotland – Biography.   4. Scotland – Biography.   I. Title.

GA813.6.P55O35 2014   526.092   C2014-905019-4

University of Toronto Press acknowledges the financial assistance to its
publishing program of the Canada Council for the Arts and the Ontario
Arts Council, an agency of the Government of Ontario.

 Canada Council   Conseil des Arts
for the Arts   du Canada

University of Toronto Press acknowledges the financial support of the
Government of Canada through the Canada Book Fund for its publishing
activities.

Deign on the passing world to turn thine eyes,
And pause a while from letters, to be wise;
There mark what ills the scholar's life assail,
Toil, envy, want, the patron, and the jail.

Contents

Preface

This book has its origins in the doctoral dissertation I wrote for the University of London in the 1960s. My supervisor at University College was James Sutherland, a Scot, who, while collecting material for *The Oxford Book of Literary Anecdotes*, had run into *Walpoliana*, Pinkerton's book on Horace Walpole. I'd come across Pinkerton while writing an MA thesis on ballad literature and so knew of him too. We decided he would be a good project. The examiners thought the dissertation should be published, but I knew it would have to be extensively revised before going into print. It has gathered dust in one of the offsite storage bins of the Senate House Library ever since. I turned to Pinkerton twice in the interim, once to write on his editing of the makars' poetry for *Studies in Scottish Literature* (1978) and on another occasion (2010) to read a paper to the Canadian Society for Eighteenth-Century Studies on his connection with the *Critical Review*. I have now rewritten much of the dissertation, condensed it, reorganized chapters, and brought in new material.

Pinkerton typically either gets just a mention in recent scholarship, is roundly abused, or is overlooked. There are many reasons for this and I needn't state them in a preface. He was an ornery character. Yet as I long ago read through his bibliography, through his correspondence, scattered hither and yon, and through the many attacks on him, I somehow got to enjoy him – despite his distaste for "O's and Macs." I sense his shade looking through my study window from across the Acheron, and hear him whisper "Get your book out before you have to ferry over too!"

His life story is one of high ambition thwarted, great promise only partly fulfilled, and talent too often misdirected. It is full of struggle and pathos, so much so that I think the narrative as it unfolds may appeal not just to students of history and literature but to a general readership.

I am grateful for the courtesy and help I received from librarians in recent visits to McGill University Library, the Bibliothèque et Archives nationales du Québec, Victoria University Library, the Lewis Walpole Library, the Thomas Fisher Rare Book Library, and the Queen Elizabeth II Library. I thank Humanities Editor Richard Ratzlaff, Associate Managing Editor Barbara Porter, and Copy Editor Gillian Scobie at the University of Toronto Press for skilfully guiding me through the publication process. Above all, I thank my wife, the writer Marjorie Doyle, for her support, advice, and patience as I tackled this project, especially through the trying early months of 2014 as it was nearing completion.

P.O'F.
11 Aug. 2014

Abbreviations

Bib. Nat.	Bibliothèque nationale
B.L.	British Library
B.L. Add. MS.	British Library Additional Manuscript
Bodl. Lib.	Bodleian Library
Edin. Univ. Lib.	Edinburgh University Library
LWL	Lewis Walpole Library, Yale University
Nat. Lib. Scot.	National Library of Scotland
OED	*Oxford English Dictionary*

Permissions

For permission to reprint brief extracts I am indebted to: Bloomsbury Publishing Plc for a quote from pp. 28–9 of Simon Swift, *Romanticism, Literature and Philosophy*, Continuum, 2006; Cambridge University Press for a quote from p. 33 of Jacqueline Pearson, *Women's Reading in Britain 1750–1835: A Dangerous Occupation*, 1999, and quotes from pp. 171 and 237 of Roy Porter, *The Making of Geology: Earth Science in Britain 1660–1815*, 1977; Linda Colley for quotes from p. 125 of her book *Britons: Forging the Nation*, Pimlico, 2003; Ann Douglas for a quote from p. 273 of the 2nd edition of David Douglas, *English Scholars 1660–1730*, Eyre and Spottiswoode, 1951; Anthony Edwards for a quote from his essay "Editing Dunbar: The Tradition," on p. 56 of S. Mapstone, ed., *William Dunbar, "The Nobill Poyet,"* Tuckwell Press, 2001; the Modern Language Association of America for quotes from pp. 439, 440, 441 of W.A. Craigie, "Macpherson on Pinkerton: Literary Amenities of the Eighteenth Century, *PMLA* (1927), Vol. 42; Oxford University Press for a quote from p. 135 of Alan Lang Strout, "Some Unpublished Letters of John Gibson Lockhart to John Wilson Croker," *Notes and Queries* (1944), Vol. 187 (7); Palgrave Macmillan for a quote from p. 119 of Murray Pittock, *Inventing and Resisting Britain* (1997), St. Martin's Press, a quote from p. 128 of Robin Eagles, *Francophilia in English Society, 1748–1815*, Macmillan Press, 2000, and quotes from pp. 185, 187, 198–9 of Robert Mayhew, *Enlightenment Geography: The Political Languages of British Geography, 1650–1850*, Macmillan Press, 2000; Taylor & Francis for quotes from pp. 170 and 220 of the 2nd edition of J.S. Keates, *Understanding Maps*, Longman Group Ltd., 1996, and a quote from p. 190 of Kathleen Wilson, *The Island Race: Englishness, Empire and Gender in the Eighteenth Century*, Routledge, 2003.

SCOTLAND'S PARIAH

The Life and Work of John Pinkerton, 1758–1826

1 Youth, 1758–1781

In his book *English Scholars 1660–1730* David Douglas used two citations to back up his argument about the "sudden sterility" of English medieval scholarship after 1730.

> "Ever since the time of Thomas Hearne," wrote a contributor to the *Gentleman's Magazine* in 1788, "the publication of our old historic writers has been discontinued." He was scarcely exaggerating. "The age of herculean diligence which could devour and digest whole libraries is passed away," wrote Gibbon in his old age when referring to the writers of the previous century. He was lamenting the disappearance of "those heroes whose race is now almost extinct."[1]

An irony that escaped Douglas was that the evidence he struck on to make his argument related to one of the new breed of black-letter critics and editors who appeared as the century progressed. The first quotation he used was from a series of twelve "Letters to the People of Great Britain on the Cultivation of their National History," by a writer whose *nom de plume* on the occasion was Philistor.[2] This series was a call for just the kind of attention to early historical works that Douglas said was lacking. The second was from an essay by Gibbon which, while "lamenting the disappearance" of the "heroes" who could devour "whole libraries," was emphatically pointing out that one such hero had been located. In his essay, Gibbon announced the inception of a project to edit and publish the early ("monkish") historians of England.[3] The man of "herculean diligence" who he said could carry the project through, and Philistor, were one and the same: John Pinkerton.

That a historian of Gibbon's accomplishment and reputation should have so much confidence in Pinkerton, twenty years his junior, as to nominate him for a task of this nature surely commands interest. Nor was Gibbon the only literary eminence of the day to make such a judgment. Even a cursory glance through one of Pinkerton's histories will show the absurdity of calling him "an irresponsible literary dilettante," to cite one harsh description,[4] or to dismiss him as a "miscellaneous writer" or some similar label, as is not uncommon. Oldbuck, in Walter Scott's novel *The Antiquary*, called him "the learned Pinkerton," and he deserved the epithet, as he did Gibbon's praise. This is the first attempt to publish a full biography of him and to examine his entire oeuvre as a scholar and man of letters. A man's best part is in his books, and it is to them we must turn to get the measure of him, since he left no memoir, apart from a guarded account of his sojourn in Paris in 1802–5. Reviews of his books and his own reviews of the work of contemporaries offer insight into the literary milieu in which he laboured and, for a time, had some success; they too are important elements in his biography. So are his letters, but not as much as one would think on first glance. Although many survive, a far larger number have been lost. In those that remain he was often evasive, particularly after 1797, when he was made painfully aware of the perils of laying himself open to correspondents. In any case, he was not much given to using letters for personal comment and revelation. "Writing a letter is a serious business to me," he told a collaborator who hadn't addressed points at issue.[5] He rarely went beyond his strict "usual limits"[6] in letters before his senescent years. Inquiry into his private life, and into certain parts of his literary life, thus sometimes entails guesswork.

His career provides illustrations of various historical/critical motifs, among them forgery, the position of Scottish authors in London, racism, sexual frankness in literature, the publishing industry, the role of literary vicars, the ballad revival, the interrelationship of genres, and the growth of ethnic nationalism. Some of these and other paradigms find a place in this book. But the book's focus, as in the dissertation in back of it, is on Pinkerton's life, the nature and quality of his writing, his changing intellectual interests, and his overall achievement as a scholar. The approach is circumscribed rather than broadly interpretive. The premises are: that as a player of note in the literary scene of his day he deserves a biography; and that his work merits extended treatment, not as a case-study, but for what it is. Every life can be seen as symbolic in some way or other, but it is the flesh and blood and brain of a neglected, deeply flawed, but intriguing human being, in a word "the whole man, as he looked and lived among his fellows,"[7] that the book tries to capture.

The surname Pinkerton is of Lowland Scottish origin. A place bearing that name,[8] whence the surname derives, existed as early as 1020 some six miles south of Dunbar in East Lothian.[9] In the eighteenth century the chief branch of the Pinkerton family was to be found in the West of Scotland, in the counties of Lanark and Ayr. Walter Pinkerton, described as "a worthy and honest yeoman,"[10] lived at Dalserf, a village in the valley of the Clyde in Lanark. James, one of his "numerous" children, went to Somerset and acquired a moderate fortune there as a hair-merchant.[11] In 1755 James married Mary Heron in Edinburgh.[12] She was the daughter of James Heron, an Edinburgh "physician or apothecary," and the widow of Robert Bowie, "a respectable merchant" in the city, to whom she had borne two sons.[13] Two brief letters from her survive, written in 1785 and 1790.[14] Judging from them, she was not fully literate, but despite their brevity they convey a distinct impression of her gentleness and simplicity.[15] She brought "some additional property" with her into her second marriage. James Pinkerton had become a man of substance, an investor and owner of real estate. He was said to be a hypochondriac and "of a severe and morose disposition."[16]

John, the only child of James Pinkerton and Mary Heron,[17] was born in Edinburgh on 17 February 1758.[18] Soon afterwards James moved his family from Edinburgh proper to a house owned by his wife at Grangegateside, a near-rural suburb, and it was there that the boy Pinkerton received his first formal education. He attended a school "kept by an old woman, who relieved the dryness of English grammar by a mixture of sweetmeats."[19] In "about 1764" he was sent to the Lanark Grammar School, whose history dated from 1183.[20] A grammar (or high) school was one where boys from about nine years of age "who could read English"[21] were given, first and foremost, a thorough grounding in Latin.[22] (Pinkerton's apparent earlier than usual admission was perhaps a sign of precocity.) When they finished the normal period of four or five years at such a school the pupils had a command of the language that purportedly enabled them not only to understand lectures in Latin at the universities "but to converse in it."[23] Pinkerton's teacher was one Robert Thomson. A classroom episode centred on Pinkerton has been preserved:

> Mr. Thomson one day ordered the boys to translate a part of Livy into English; when he came to young Pinkerton's *version* … he read it silently

to himself, then, to the great surprise of the boys, walked quickly out of the school, but soon returned with a volume of Hooke's Roman History,[24] in which the same part of Livy was translated. He read both aloud, and gave his decided opinion in favour of his disciple's translation, which not a little flattered boyish vanity.[25]

The school's antiquity – it was one of the oldest in Scotland – must have made an impression on Pinkerton. Later he was prompted to write a character sketch of its most celebrated pupil, the traveller and poet William Lithgow (1582–1645?), and in 1802 he made a respectful reference to his teacher in his *Modern Geography*, where he called Thomson the brother-in-law of James Thomson the poet.[26] His family's ties to Lanarkshire, not just on his father's side, may sufficiently explain why Pinkerton was sent to the "capital" grammar school there rather than to a comparable one in Edinburgh.[27]

Pinkerton is said to have been "always a shy boy, fonder of rural and solitary walks than of boisterous amusements."[28] If so, the school's surroundings offered many alluring paths to explore. A busy market town when he knew it, Lanark, about twenty-five miles southeast of Glasgow, could boast of involvement in the heroic period of Scottish history. The patriot William Wallace had figured in its past. A few miles away in the Cartland Crags overhanging the River Mouse Water was Wallace's Cave where the hero allegedly hid from the English; and in the town were the ruins of St Kentigern's Church where Wallace married Marion Braidfute, later to be murdered by English soldiers. In revenge, Wallace slew the Sheriff of Lanark, an Englishman named Keselrig whom he deemed responsible for her death. (Some of this is legend.) To add to these historical attractions, Lanark was in one of the most picturesque regions of the Lowlands. Near the town the River Clyde flowed majestically over a series of four waterfalls before passing to more level country to the northwest. Justly famous, the Falls of Clyde might inspire even the most austere mind to thoughts of poetry. Pinkerton praised the river, in his fashion:

I

While some praise the pastoral margin of Tweed,
And others the beautiful banks of the Tay,
Accept, O fair Clyde, of my dutiful lay,
Thy rural meanders no stream can exceed.

II

For oft thy wild banks in my youth did I tread
The trout and the par from thy wave to decoy;
Maria then shar'd in my innocent joy, –
But Maria is false and my pleasures are fled.[29]

Maria? Likely[30] a figment, like Chloe, Phyllis, and similar stock inhabitants of pastoral poetry. (Like Marion Braidfute too, maybe!) Other scenes and points of interest around Lanark must have been equally familiar to the youthful Pinkerton, the more so if he then possessed a trace of the inquisitiveness that later formed so striking an aspect of his character. He knew something of the region's folk traditions. Craignethan Castle, a notable ruin near Crossford downstream from Lanark, was one spot he visited. It is so ancient, he wrote about 1776, "that the country people there say it was built by the *Pechts*, which is their common way of expressing the *Picts*."[31] (Much more of them in chapter 2.) In Lanark he might have heard sung the ballad "Binnorie" which he later published in a mutilated form.[32]

Pinkerton's grammar school education lasted about six years. In the normal course of events he would now have pursued a course of study at one of Scotland's universities. It is a little surprising he didn't, given the scholarly aptitude he displayed quite early in life. But James Pinkerton had other plans in mind. He was in trade himself, and, for whatever reason, had some "dislike to university education."[33] He wanted John to study the law, and in a practical setting, intending, so his son later explained, "merely as he said to give me an insight of business."[34] John Pinkerton would inherit the estate of a rising man, and his father was trying to make him fit to manage it. The boy was obliged to stay close to home until he was old enough to begin apprenticeship in a law office. For three or four years his formal studies were abandoned, and, his situation at home being "a kind of solitary confinement," it is possible that considerable friction existed between him and his father.[35] His reading during this "confinement" was desultory but extensive. A sixteenth-century book on medals, he told Francis Douce, had been his delight at the age of fourteen.[36] He would later brag about how much he'd read before that. He was tutored privately in French and mathematics,[37] and read some standard works on early Scottish history.[38] It is unclear how much he imbibed of David Hume, Adam Ferguson, and other luminaries of the Scottish Enlightenment, and exactly when he read them. Perhaps he didn't study them deeply, just as, in our own day, Marx's *Kapital* is read by few but its central ideas remain current and its impact on the world is still felt. One thing is clear: Pinkerton was a child of the Enlightenment. He often wrote that he lived in an "enlightened age." He admired the "great and elegant writer of philosophy"[39] Hume, whose influence is felt throughout his own literary efforts, indeed in his life and beliefs. He admired the historian William Robertson. He knew Gilbert Stuart's work and that of Adam Smith. To speak generally, the spirit of freedom, inquiry, and scepticism that was abroad in Scotland in the eighteenth century he inherited and absorbed, as we will see.

In 1775, at age seventeen, there came a turning point. He was bound apprentice to William Aytoun, a prominent Writer to the Signet[40] in Edinburgh, and for five years he was trained under Aytoun's guidance to become a writer – roughly the equivalent of an English solicitor. The activities of his life now had to become more closely connected with the Edinburgh milieu, and as they did, signs of literary ambition began to appear. Rather, appear more noticeably. It is apparent that he'd caught the literary bug even earlier.

By 1775, according to William Ferguson, Edinburgh had become "one of the leading cultural centres of Europe."[41] How the cultural scene there affected Pinkerton can only be guessed at, but what stimulated him above all else appears to have been poetry, chiefly English poetry, and not just reading it but writing it. Also Scottish ballads, and not just reading them either. One of the outlets for a young poet in Edinburgh was Walter Ruddiman's *Weekly Magazine,* which had been founded in 1769 and by 1776 had a circulation of 3,000 copies. It published a selection of new verse every week. Although it is hard to distinguish his early verse from that of his contemporaries, there is reason to believe Pinkerton first published a poem in that journal.[42] At any rate, by the mid-1770s he was reading Thomas Gray and James Beattie (from whose *Minstrel* he admittedly "first derived" his taste in poetry and which he never read "without the highest delight")[43] and his ambition to be a poet appeared to be fixed. In September 1775 he wrote *Craigmillar Castle. An Elegy* and shortly afterwards sent it to Beattie, then professor of moral philosophy at Marischal College, Aberdeen, for his assessment. Beattie answered politely in December, saying that although the poem contained "many good lines" it was not "correct enough as yet to appear in public."[44] Pinkerton wrote back, asking for more detailed comment. After some delay Beattie responded, pointing to false rhymes, finding "too much hissing" in one line, and saying of another – "Dread fall their tottering basis seems to lour" – that it "is a line which I do not understand."[45] Pinkerton published the poem, anonymously and at his own expense, in June or July 1776. He dedicated it to Beattie and sent him four copies. It is an elegy lamenting the passing of the glory surrounding Craigmillar Castle at the time of Mary Queen of Scots. The castle is the setting, but Gray's country churchyard is not far off:

> The bleating flocks were pent within the fold;
> The jovial reapers long had left the vale;
> The plowman at his cheary fire-side told
> The merry story, or the mournful tale.

"Bleating flocks," "jovial reapers," "cheary fire-side" – comment is superfluous. In Gray's manner too are the annotations, which cite Scottish historians, Ovid, Shakespeare, Tasso, and Jacques du Fouilloux, the last named being brought in to support the claim that Mary "Oft bathed her snowy limbs in sparkling wine." The work is pedantic and unremarkable, yet not an incompetent production for a seventeen-year-old.

The yeoman's grandson, hair-merchant's son, was thus inaugurated into the world of letters. Pinkerton was now conning legal books and documents in Aytoun's premises, all the while doubtless thinking himself a votary of the Muse, who directed his attention to a poem first published in Edinburgh as *Hardyknute, A Fragment* in 1719. The story it told reeks of mystery. The elderly knight Hardyknute of ancient Scotland has four sons, a beautiful daughter named Fairly, and a good wife, Elenor. When the "King of Norse" attacks his country, Hardyknute and his sons ride with their armies to meet him. On their way they pass over "Lord Chattan's land" and pass by a wounded knight. At length they defeat the invaders and return home. In the original version the "fragment" ends when Hardyknute, weary with battle, sights his castle:

> Loud and chill blew the westlin wind,
> Sair beat the heavy shower,
> Mirk grew the night ere Hardyknute
> Wan near his stately tower.
> His tower that used wi' torches blaze
> To shine sae far at night,
> Seemed now as black as mourning weed –
> Nae marvel sair he sight.

Why was the castle deserted? Why was Fairly's beauty said to be "Wae-fou, I trow, to kyth and kin"? What did the wounded knight have to do with the story? As much mystery surrounded the authorship of the poem. Lady Elizabeth Ann Wardlaw of Fife, later identified as its probable author, was alleged to have said she had found it "on shreds of paper, employed for what is called the bottoms of [weaving] clues."[46] Doubts about its provenance remain.[47]

The poem appeared in Allan Ramsay's *Ever Green* (1724), Thomas Warton's *The Union* (1753), and Thomas Percy's *Reliques of Ancient English Poetry* (1765),[48] was printed separately as "the First Canto of an Epick Poem,"[49] and was sometimes compared favourably with *The Iliad* and *The Aeneid*. Gray thought it was ancient and admired it greatly.[50] Warton bewailed "the loss of a great part of a noble old Scottish poem, entitled, Hardyknute, which exhibits a striking representation of our antient martial manners."[51] In or about 1775[52] Pinkerton decided to complete it. The notorious second part of the poem, longer than the original by over 100 lines, was thus born. All is explained in it. The castle is dark on Hardyknute's return because it has been raided, his youngest son slain, and Fairly abducted; the wounded knight is Lord Drassan, Fairly's lover, whose suit for Fairly's hand has been rejected by her father. A battle ensues, Drassan dies honourably, and the guilt-ridden Hardyknute sends Fairly to a nunnery. Pinkerton prepared the completed *Hardyknute* for the press over the next year or two, adding four[53] other forged[54] pieces masquerading as ballads, together with some genuine ballads, mainly from Percy's *Reliques* and David Herd's *Ancient and Modern Scottish Songs, Heroic Ballads, etc.* (1776). His "Binnorie," as he later confessed, was half forged and half traditional.[55] Percy had similarly expanded "The Child of Elle" to "attempt a completion of the story," as he admitted in the *Reliques*.[56] Later it was discovered that the 200 lines of his version of the ballad had been puffed out from an original of thirty-nine and that only one line of the original had been preserved intact.[57] The "supplemental stanzas," Percy said, would be "easily" discovered "by their inferiority." His "Child of Elle" was as much a concoction as Pinkerton's "Binnorie." But it wasn't forgery. Pinkerton tried to pawn off his concoctions as traditional ballads. He said *Hardyknute*, earlier just a "mutilated fragment," was now given "in its original perfection" and "is certainly the most noble production in this style that ever appeared in the world." He added: "I am indebted for most of the stanzas, now recovered, to the memory of a lady in Lanarkshire."[58] One item he found, "Sir James the Rose," was neither copied from Percy or Herd nor forged. It was "from a modern edition in one sheet 12mo. after the old copy." This was an authentic text.[59] "Lord Airth's Complaint" Pinkerton claimed to have copied from a manuscript in his possession, and it seems he did.[60] Two prefatory essays were added to the mix, the first a piece, "On the Oral Tradition of Poetry," in which the text and footnotes indicate the wide range of Pinkerton's reading, quite extraordinary for someone his age.

The exact process of preparation and revision of the collection is not clear. By January 1778 Pinkerton had offered a preliminary version to James Dodsley, Percy's publisher in London. Dodsley, referring to the submission as "Hardyknute together with other ancient Scottish Poems &c.," rejected it, but said that with improvements and additions it might be publishable.[61] Nonplussed, in February 1778 Pinkerton sent him the manuscript of what would become *Scottish Tragic Ballads*. "I took the labour of writing it all with my own hand," he said, adding, "I would not have a single comma to stand but where I have placed it."[62] Dodsley again demurred, whereupon Pinkerton set the obliging Percy, with whom he had struck up an acquaintance, to work to try and get it into print.

And so he conceived a scheme that would make his name infamous in certain sectors of the scholarly world. His youth – he was nineteen when he submitted his collection to Dodsley – his inordinate desire to be noticed, and his boldness are factors that might explain the hoax, though not of course excuse it. Getting this book together was just one element in what was already becoming a busy intellectual life. It was not a solitary life. During this period he was a member of a close group of young men of literary bent, whose existence is referred to in a number of letters of his friend John Young. A brief account of Pinkerton a few years later in a London journal said he had "too long been the surly dictator of a private circle": this could refer to his group in Edinburgh, although Young's letters suggest the "circle" was hardly one presided over by a dictator.[63]

Pinkerton's reading was now voracious. He read more histories of Scotland. He was familiar not only with Macpherson's Ossian poems but with the controversy they'd aroused. At first reading, they seemed false to him, but to his subsequent mortification he came to accept their authenticity, believing them genuine "from the age of sixteen till twenty."[64] Around November 1777 he was reading Fielding's *Tom Jones*, *Joseph Andrews*, and *Amelia*, Dodsley's *Old Plays*, and the *Arabian Nights*.[65] He discovered William Drummond of Hawthornden in 1778 and wrote enthusiastically of him to Percy, who didn't need to be told who Drummond was.[66] Pinkerton was also writing prolifically. In 1776–9 he wrote most of the poems later included in his volumes *Rimes* (1781) and *Tales in Verse* (1782). In September 1778 he wrote the outline of a poem, to be called "Nature," that would cover a range of human knowledge, from mineralogy to religion – he later wrote books on both subjects – and in October he sketched out a prose work called "A Northern Tale," featuring a hermit, a ruined castle, a ghost, much bloodshed, a magician, and a maid crying for help.[67] By 1778 he had written "The Heiress of Strathern,"[68] a tragedy on a Scottish theme, had sketched the plot of an Arthurian melodrama titled "The British Princess,"[69] and had finished another tragedy called "Malvine" in which, he told Beattie, the "too much neglected rules of Aristotle I have followed with great attention so far as the genius of our drama would admit."[70] He also planned and made some headway with a collection of the "most exquisite pieces" of Latin poetry written since the Renaissance, adding those Spanish, Italian, and French poems which "are in danger of remaining unknown."[71] We might note in passing his apparent early facility with languages, though he was not strong in Spanish. His head was full of schemes. The second part of *Hardyknute* was one product of what seems to have been a creative frenzy. Nor was he finished with "completing" fragments of Scottish poetry. In 1779 the first line of the song often called "Helen of Kirconnel" appeared in print.[72] Pinkerton heard the second line somewhere, and later insisted that the third, too, was genuine.[73] The rest for certain he made up. The first stanza of the renovated poem reads thus:

> I wish I were where Helen lies
> Night & day on me she cries
> To bear her company
> O would that in her darksome bed
> My weary frame to rest were laid
> From love and anguish free![74]

He sent the song he had "lately recovered" to Percy, who was contemplating an expanded fourth edition of the *Reliques*. "But whether," Pinkerton lied, "this be the genuine old Song or only new words adapted to the air I will not pretend to say … It has never been in print so far as I can discover."[75] The "darksome bed" and "weary frame" notwithstanding, his version of the song found admirers. Walter Scott said it had "great poetical merit" and Robert Burns, in 1793, thought it better than the original.[76]

In 1779 Pinkerton was twenty-one. In the absence of a diary or of any extended comment by him on his private affairs, it is hazardous to make an early assessment of his character; yet certain elements stand out. Though studying the law, or supposedly studying it, his surviving correspondence appears to show he had no interest whatever in practising that profession. David Hume's family had forced him to study law too, and, as he famously records, "while they fancied I was poring over Voet and Vinnius," he was secretly devouring Cicero and Virgil.[77] Pinkerton was in a similar predicament, but instead of Cicero and Virgil he was taking in Milton, Gray, Beattie, Petrarch, and Copernicus.[78] Yet he must have learned something from his legal apprenticeship. How to shape an argument, maybe; how to read old script; the importance of accurate copying and citation; of disciplined study; of hard work. Which is to say, while determinedly set on a path to literary fame as poet, as he evidently was, he may also have been picking up, not just from his reading but from his quotidian labours too, certain instincts and practices of a scholar, traces of which are already apparent in his early essays and annotations. A commitment to publishing only what was factual and true – or if speculative or probable, indicating it to be so, a paramount duty of scholarship – was obviously something he didn't yet have. Yet the urge to judge him harshly must be resisted. To many in his own day and later he did indeed seem to be "a notorious fabricator,"[79] but his notoriety over this ballad issue seems out of proportion to the nature of the crime. He wasn't in the same class as William Lauder, William Henry Ireland, Thomas Chatterton, Edward Williams (Iolo Morganwg), and Macpherson. He was a minor offender. In his own "enquiry" into the subject of forgery, written after his was exposed, he said it was ridiculous to equate "literary forgery" with "penal forgery." He agreed with those who claimed that "nothing can be more innocent" than the former,

> … that the fiction of ascribing a piece to antiquity, which in fact doth not belong to it, can in no sort be more improper than the fiction of a poem or novel; that in both the delight of the reader is the only intention.[80]

Yet he came clean,[81] and early on in his career – though he could hardly do otherwise, given the way his cheat was revealed. (See chapter 2.) To get a just appraisal of his work we have to be tolerant of youthful ostentation, caprice, and error, while keeping careful watch, not just early on but throughout his career, over his equivocations and claims, and over his handling of sources.

In the aspect of his life at twenty-one that seemed to matter most to him, i.e., climbing Parnassus, he was pushy, not to say driven. He already had some literary notables as correspondents, namely Percy, Beattie, Dodsley, and the Ossian defender Hugh Blair,[82] and was using Percy to get access to Thomas Warton.[83] He liked praise – no great surprise in a young writer, or in an old one for that matter, though he seemed to need it more than most. He was given to morose anxiety about his health. A few extracts from a notebook of 1778–9[84] will illustrate this: "Never take any pepper &c at night as it excites a flow of saliva wh. freezes; but after numbing the pain stop the tooth with wax"; "3 grains Rhubarb should always be taken before a meal"; "When dreams represent the actions of real life, they show health. When visionary bad health"; "When the feet are the least damp very bad – too close understockings good"; "If the skin is red after leaving the cold bath, good Id. Always Sl. with the neck & breast bare"; 28 May 1779 plagued with legs and itching on my skn cured by Cold water pl anointing with oil"; "Rising stretch yourself often & keep your hd & ft from the morg cold."[85] His concern over his skin condition can be explained by the affliction of smallpox in childhood; his face was pockmarked, a common aftereffect of that disease.[86] He was a hypochondriac all his life. He would wear glasses in his mature years. We will hear him complaining of printed texts "hurtful to the eyes,"[87] One source says he was redheaded.[88]

In 1780 Pinkerton's apprenticeship to Aytoun was to come to an end. He could then describe himself as a writer (in the legal sense) but not as a Writer to the Signet. To get that designation he had to pass an examination and pay £100. He evidently did neither.[89] In May 1780 Aytoun died.[90] By then, Pinkerton had served all, or almost all, of his five-year term.[91] At "about the same time" James Pinkerton died.[92] He left no will, which meant that his testament would be a "testament dative," i.e., one authorized by a court. John Pinkerton was declared his father's "only Executive Dative and nearest of kin" by the "Commissaries of Edinburgh" (the Commissary Court) on 7 June 1780.[93] He soon left for London and in January 1781 was located at White Horse Court, Southwark.[94] Two aspirations of his youth were speedily fulfilled. The manuscript of *Rimes* had been prepared for the press in June 1780.[95] Pinkerton offered it to the bookseller Charles Dilly, who published it in early 1781. By June 1781 the publisher John Nichols, Percy's relative,[96] had issued *Scottish Tragic Ballads*. Both volumes were anonymous. Pinkerton had reached the London audience as both a poet and ballad "editor." Percy had even corrected the proof-sheets of the second part of *Hardyknute*.[97] Pinkerton's "imposture"[98] was out for all the world to see.

"The *Canongate*," Henry Mackenzie lamented in 1780, "is almost as long as the *Strand*, but it will not bear the comparison upon paper; and *Blackfriars-wynd* can never vie with *Drury-lane* in point of sound."[99] Not every eighteenth-century Scot would have accepted this fiat. Pinkerton surely did. But family business, it seems, brought him back to Edinburgh. James Pinkerton's testament was entered at the General Register House on 21 November 1781. It provides an inventory of debts amounting to over £1200 sterling that were "resting owing" to him at his death. At least one of these, £100 (plus penalty) owed by wigmaker Hugh Inglis of Edinburgh, was a bad debt; perhaps the others, owed by merchants in Glasgow, were as well.[100] On 26 November Pinkerton drew up, or had drawn up, a deed of factory appointing William Buchan, writer,[101] his factor in Edinburgh, authorizing him to manage his affairs "in my absence." He was directed to receive "any dividends that may arise" from the Glasgow debtors, to "pursue & uplift" the money owed by Inglis, and to "discharge all debts & Sums of money rents duties & casualties payable by the Tenants of my houses in Edr. or elsewhere."[102] Pinkerton still possessed a great deal of property. For some time the return from his investments and houses would bring him an annual income of about £300 – in the 1780s sufficient for him to "live independent," as he said.[103] His mother would continue living at Grangegateside,[104] supported by rentals collected and administered by his factor.[105] On 5 January 1782, preparing for his second trip to London, Pinkerton sold stocks to the value of £807.[106] In February he was in Markinch, in Fife, possibly visiting relatives there.[107] In late March he was again in London where he would settle in as a "gentleman of independent fortune"[108] at Knightsbridge, then on the periphery of the city.[109] He was not to return to Edinburgh for thirty years.

2 Finding His Way, 1782–1789

If sheer hard work with language and metre could make a good poet, if close study of poets of the past could make one,[1] if devotion to the Muse could, then Pinkerton would have been one. Instead, his most ambitious volume, *Rimes* (1781), features an overblown, derivative, hackneyed diction. A statement on language by Beattie in 1778 is apposite here. He said the "greatest difficulty" facing Scots who wrote in English was giving their writing "a *vernacular* cast." Scots, he said, "are obliged to study English from books, like a dead language. Accordingly, when we write, we write it like a dead language, which we understand, but cannot speak. We are slaves to the language we write, and are continually afraid of committing *gross* blunders ... In a word, *we* handle English, as a person who cannot fence handles a sword; continually afraid of hurting ourselves with it."[2] He was referring to prose, but what he said must apply to poetry too. The young poet Pinkerton had Scots on his tongue and he wrote in English, so the question arises, was he too hobbled, to the detriment of his verse, by the linguistic disconnect Beattie described?

He had to be affected by it, but to what extent is far from clear. It isn't known how much training in English he'd been given, and how much *talking* in it he did, in his home and schools as a youngster, and later. There was much to-do generally in eighteenth-century Scotland over learning English and getting "correct" pronunciations. His father had lived and worked in England; he must have been able to speak capably in English, and he would have known the advantage to a Scotsman of mastering the language. He could have insisted that his son be orally proficient in it. Pinkerton, who doubtless was a quick study, was not only well read in English literature by his late teens but a prolific writer in English, not just of poetry. He'd written plays and could write more than passable prose, as in his letters and the Advertisement to *Rimes*. His productivity suggests he didn't unduly suffer, as Beattie did, from fear of handling the medium he used. He carried a strong "accent" with him to London but there is no indication that he couldn't then converse in English with facility.

Rimes had no "vernacular cast," but not necessarily because Pinkerton couldn't give it one. It wouldn't have been appropriate. The book was a collection of odes, though he called some of them "Melodies" and "Symphonies" and also included five sonnets. An ode is normally an exaltation of someone or some thing: it requires a correspondent lifting of expression, a certain solemnity and ceremony. The language in *Rimes* was one Pinkerton deliberately chose for the genre he was using. He could write verse in quite a different style, but here he aimed for lofty refinement; he wanted the language to be exclamatory, learned, "elegant" (a word he used of the form). He was following Gray, who was a master of the ode.

Writing odes was much in vogue in Edinburgh around 1780, as a glance through the *Scots Magazine* shows. Pinkerton had studied this poetic style, experimented with it, and had many ideas about how to vary it and make it more pleasing. But his performance in writing it was dismal. It is no pleasant task to read his odes, partly because the apparatus of the form itself in the hands of eighteenth-century minor poets can grate on modern ears, but also because his specimens are particularly lead-footed, loaded are they are with adjectives, especially ones from Beattie, Gray, and the young Milton. The reader runs into "dewy," "rosy," "jocund," "sprightly," "blooming," "sacred," "tuneful," "pensive," "genial," "flowery," etc., so often as to know they serve little other function than filling out a line. Too often in *Rimes* Pinkerton wrote just as Beattie relates, as if in a dead language picked up from books; but this may reflect as much a misconception about the nature of poetry as enslavement by an unfamiliar idiom. His borrowings point to a juvenile *naiveté*. He copied not only single words but phrases: e.g., "Fancy's sacred child" from Milton, "breathing incense sweet" from Gray, and "honey from the rock" from Beattie. At one point he snatched three whole lines from Milton's "Il Penseroso." Lengthy consideration of *Rimes* is not warranted.

Yet we might look briefly at his ode "To Autumn," or, rather, a section of it. It opens with a reference to himself – "As by this ample field I stray" – and a request to Autumn to attend to the Muse, who is about to paint "the honours of thy reign." "She comes! She comes!" follows, heading a four-line description. Then this:

> All hail, thou queen of plenty, hail!
> Thine are the treasures of the vale
> That life and health to all afford,
> Best bounties of the social board:
> Thine is the orchards blushing hoard
> With balm and various nectar stored.
> Thine is the Morn so fresh and gay,
> That from her opal tower displays
> Her crimson banner's wavy blaze,
> While from the west the moon's wan ray
> Lends all the dewey landscape bright
> A double shade, a double light,
> Here gilded with the matin beam,
> There with the meek moon's silver gleam.
> Thine when o'er every dusky mead,
> The grey Mist spreads her silent sway,
> That opening to the gold of day
> The trees their pearly spangles shed,
> And smiling, thro' the twilight scene
> Reveal their robes of glittering green.[3]

Few vivid images are to be found in *Rimes*. Pinkerton seemed to think it was enough to say "wavy" or "smiling" to create a picture or feeling. But towards the middle of this passage the poet *sees* something in the world about him: shadows cast simultaneously by both the rising sun and fading moon. The line "A double shade, a double light," startling by its simplicity, stands out amidst the usual verbiage. The remainder of "To Autumn" is a letdown; we return to "tuneful" breast, "pensive" scene, and "chearful" noon.

The phenomenon of actually looking at an object recurs in the ode "To the Lark." Pinkerton again brings himself into the scene, which here is Scottish, maybe Arthur's Seat, since as he moves "upland" he can see the Forth and Bass Rock:

> And now with heedless steps I stray
> Along the woodland glade,
> And meditate my musing way

Thro' brake and warbling shade.
Of new mown hay the grateful steam
Now rises on the gale;
And chequed with many a shadowy gleam
Wide waves the grassy dale.

The "new mown hay" with the "steam" rising from it – presumably evidence of evaporation – is again striking, given the triteness surrounding it. But generally *Rimes* smells of the oilcloth, and the scenes Pinkerton presents, as in the opening long dreary piece "The Education of the Muse," mostly lack substance. Personifications (which he thought "perfectly congenial to the mind") like "Rapture" "Felicity," "Fancy," and "Wisdom" contribute to a sense of unreality. Much of *Rimes* consists of the "vacant musing" that Wordsworth spurned. As for its connection with the literary milieu of the period, it might be possible to see in a few poems signs of the approaching Romantic movement, but *Rimes* seems rather to typify the kind of arid "poetic diction" Wordsworth rebelled against and Coleridge mocked.[4]

London critics were not fooled. Perhaps they were put off by the spelling of the title – Pinkerton had a lifelong habit of such tampering – and by assurances given in the Advertisement to the volume of the "perfection" of the author's "method" in his odes, and of the harmony of "his numbers." The poems were "highly finished," said the *Critical Review*, "yet shewing no marks of true genius, are of little real value." The volume was "after all but a Lenten entertainment."[5] Edmund Cartwright in the *Monthly Review* conceded the poems showed "considerable learning," but "we meet with little of that wild and animating enthusiasm … which is the soul and characteristic of true poetry."[6] It is possible to quarrel with parts of these judgments, yet in essence they were on the mark. But while the complete failure of *Rimes* as poetry was no big event in English or Scottish literary history, to Pinkerton it was a hard personal blow. He'd pruned and polished and hoped; he told R.P. Gillies in 1813 that he'd wanted to live only long enough to publish *Rimes*, that he'd thought it a "grand poetical achievement."[7] He was mortified. In August 1781, while in Edinburgh, he addressed his critics in a poem called "The Town," terming them "dunces," "drudges of booksellers," "asses," and such like.[8] On his return to London, another edition of *Rimes* having been called for, he unwisely tacked onto his original Advertisement a denunciation of his enemies. *Rimes*, he wrote, had been written and corrected "in the fervour of youth." The poems in it were novel, but "novelty must of necessity be the companion of invention." His reviewers were trying "to crush every efflorescence of genius in the bud." He thought *Rimes* "might hold some rank in English poetry," and the testimony of "Many of the most eminent and most learned [and unnamed!] characters in Europe" had assured him his belief was justified. He would celebrate the "illiterate madmen," "garreteers," and impostors who had spurned him in a work to be called "The Assiad," which would prove "the spirits of Pope and Boileau are not yet laid."

This second edition of *Rimes*, published early in 1782, had his name on the title-page. His *Two Dithyrambic Odes* and *Tales in Verse* came out then too. He was now fair game for the critics. In July the *European Magazine* said *Rimes* contained "not one stroke of originality." There was "a total inanity or nothingness of meaning." The reviewer had advice to offer:

> Fye on it, Mr. Pinkerton, fye on it! Burn this ugly bag-pipe of thine, for the sake of *sweet poesy*, and all those pretty little things you have so prettily prattled about! Burn it, my good lad; and, for your own sake, never let the public hear another drone of it.

Pinkerton's friends, instead of granting his poems "cold assent, or concealed disapprobation," should now, the critic said, try to direct his energies into "that channel for which nature designed them." A personal assault followed, a phrase from which we have already noted. It was not entirely hostile:

> … it is not possible to conceive of a man more uncouth in his manners, less polished, and accommodating than Mr. Pinkerton. There is a sluggishness hangs about him, which he has had too much pride to correct; and having too long been the surly dictator of a private circle, who considered his solemnity as a proof of mental elevation, he is sore and petulant in all his commerce with the world, and attributes the diminution of his respect to the folly or the envy of those with whom he converses. This may be said to be a severe observation; it certainly is so, and can only be justified by its truth. He is a gentleman of benevolent dispositions, and of a strong mind; he has talents, which, by a good-natured communion with men, might have been rendered truly valuable to society, as well as pleasant to himself; instead of which he has studied the world from a darkened room, and takes all his pictures of human life, as he does the images of poetry, from the chimeras of ancient time.[9]

It is one of the few glimpses we have of Pinkerton's early days in London, though, as hinted earlier, the "private circle" could refer to his coterie in Edinburgh.[10] Much of what the *European* critic had to say can be dismissed as retaliation to the counterattack in the expanded Advertisement in *Rimes*. Yet there is a ring of truth to it. That Pinkerton could be manipulative, prideful, and petulant is confirmed by his surviving correspondence from these years. As for his being "uncouth" and lacking polish, this was echoed (though in a fit of anger) in 1808 by Francis Douce, who knew him well in the mid-1780s. Pinkerton, he wrote, was "low and vulgar in his manners."[11] It is not clear exactly what is meant by the remarks about manners – perhaps only that Pinkerton did not perform well at afternoon tea. He had views that were thought outlandish, maybe, or had too many views. He spoke "with a strong Edinburgh accent."[12] That alone would have branded him, to some, as an oafish rustic.[13] It is also possible that he at times behaved like one. Yet surliness and solemnity could have been just a front for shyness.

To turn briefly to one item in *Tales in Verse*, "The Castle of Argan." It tells the story of a warrior, Eric, who gets lost in a storm and seeks shelter in a castle "Deep in the silence of a wood." Having knocked on the "massy door" with the "pummel" of his sword without arousing anybody, he forces his way in, finds a trumpet, and blows it. An "ancient dwarf" appears, and Eric asks for "a night's abode." The dwarf tells him the castle belongs to Argan, who had tried to rape his own daughter and, having failed, murdered her. No soon had he committed the foul deed

> Than flocks of daemons swarmed around,
> By Heaven in instant wrath decreed
> A crime unheard of to confound.
> And still, since that accursed hour,
> Despair and madness have possest
> Argan's inexorable breast.[14]

The dwarf directs Eric to a peasant's cottage, and the poem ends. It has some power, and it reflects the Gothic[15] sensibility of the period. Yet raising the subject of attempted rape of a daughter would have been considered indelicate if not highly offensive by readers in the 1780s. It was typical of the young Pinkerton to take risks of this nature. It likely helped damn the entire volume. In contrast to the inflated poetizing of *Rimes*, some of *Tales in Verse* seemed to critics to be written in too plain a language. The *Critical Review* said Pinkerton didn't know how to write comic poems and that most of his tales were "dull, prosaic, and uninteresting." (Pinkerton later conceded that his "comic tales" were "totally unworthy of preservation.")[16] The reviewer pronounced even "The Castle of Argan" "very insipid."[17] Cartwright in the *Monthly* continued this vein, saying of a poem called "The Pilgrim" that there was "scarcely any prose, even the soberest and dullest, that might not, with very little trouble, be cut up into such verse, if verse it may [be] called, as this." To produce a volume like *Tales in Verse*, he said, required not poetical genius but idleness, dullness, and folly. The purpose of his strictures was to divert Pinkerton from "a practice, from which in vain he looks for celebrity."[18] Although the *Gentleman's Magazine* occasionally approved of his poems[19] and Beattie puffed them,[20] the concert of literary opinion was decidedly against them. His verse was soon forgotten. Cartwright's dismissal of *Two Dithyrambic Odes* in 1783 was the last serious notice his Muse ever received.[21] It is only to be added that the few poems he wrote in Scots are occasionally vigorous. His short lyric "Bothwell Bank" has been anthologized.[22]

In 1782 Pinkerton continued his experiments with other forms of writing, including literary criticism.[23] By July he had written a comedy called "The Philosopher" which, according to the *European Magazine*, had been "very much praised by the chosen party of his friends,"[24] and during this period too he probably dabbled at an epic in twenty books titled "Ratho king of the Orkneys."[25] But by November the "poetical incubus"[26] within him, if not suppressed, had at least been chastened, and his attention turned back to ballads. *Scottish Tragic Ballads* had been received in good faith by two of the London periodicals,[27] and Nichols said he was ready to bring out a second edition. Pinkerton proposed that a supplementary volume of ballads "of the comic kind" be added, to which he would prefix a dissertation. The ballads would be given "with a correctness not yet known in any collection of the kind."[28] Nichols liked the idea, and Pinkerton began collecting materials to fill out the new volume. It was a step forward. The "channel for which nature had designed" Pinkerton was scholarly inquiry, not poetry. He had found his way back into that channel.

The Maitland Folio Manuscript in the Pepys Library at Magdalene College, Cambridge, was one of the great unlocked repositories of Middle Scots[29] poetry. Percy had seen it and from it had taken a transcript of "Peblis to the Play," by James I of Scotland. Late in 1782 Pinkerton asked if he could borrow the copy and print it in his new ballad edition. Percy sent it in January 1783,[30] and the poem appeared in print for the first time in Pinkerton's *Select Scotish Ballads,* as his new edition was called, in the spring. His name was given as editor. By June

Pinkerton himself had gone to Cambridge, inspected the manuscript, and come away with the intention of returning to copy a selection from it for publication. He was now corresponding with the Scottish judge and scholar Sir David Dalrymple (Lord Hailes), author of the historical work *Annals of Scotland* (1776–9). Dalrymple had also edited a Middle Scots collection, derived from the important Bannatyne Manuscript in the Advocates' Library at Edinburgh.[31] "I shall proceed in the same way as your lordship has done with the Bannatyne Ms.," Pinkerton wrote, adding "I wish to prefix to my work an History of Scotish poetry; or, if it is swelled too much, to publish it by itself."[32] Dalrymple was more than thirty years older than Pinkerton and an able medievalist, though he too had been fooled by Ossian. In May 1783 he sent Pinkerton a list of sharp comments on *Select Scotish Ballads*. He'd recognized the tampering with "Binnorie" and seen through the "Hardyknute" hoax (as quite a few others had).[33] He had hard words for the second part of "Hardyknute." Pinkerton was taken aback:

> Your lordship observes … that the additions to Hardyknute are modern, because no writer near the feudal times could show himself so ignorant of the form of their castles &c. as the author seems to be. Permit me to inform your lordship, that some of the first antiquaries in England are of a very different opinion, and have asserted the antiquity of the whole poem as published by me, from the vast knowledge of feudal times that appears in it. I may safely say, for my own part, that I have studied the feudal manners and those of chivalry as much as any man in Europe, and can perceive no anachronism in the poem.[34]

The sheer gall! What is more, in *Select Scotish Ballads* Pinkerton was still toying deceitfully with readers. His new "ballad" collection contained nine more of his own poems which, he said, "have not appeared in print."[35] The boasting in the letter was, as already hinted, characteristic of him. When, in October 1783, he wrote to Nichols and offered to edit a selection of Geoffrey Chaucer's work, he was told that a scholar with a name known to the public, someone of the stature of Samuel Johnson or Richard Hurd, would be required for such a task. Pinkerton shot back, blasting both Johnson and Hurd, and adding:

> This I write to shew you how very differently a man of learning judges from the mob in point of names. I call myself a man of learning with strict modesty as I suppose I had read more before I was fourteen years of age

that would entitle me to that appellation and at present know not that man in Europe to whom I would confess an inferiority in point of genuine erudition.[36]

The proposed edition of Chaucer (intended "for the public," not, like Thomas Tyrwhitt's edition of *The Canterbury Tales* of 1775–8, "for the Antiquary") did not appear.

Pinkerton at this early stage in life did not see himself as an antiquary, a word that in the later eighteenth century had connotations even a "man of learning" such as he now thought himself did not always relish. To the "man of sensibility" the word might evoke stronger distaste. An antiquary was often thought of as a virtuoso like Johnson's Quisquilius in *Rambler* 82, wasting time collecting and pondering useless trifles from the past,[37] or else one paying inordinate attention to detail in discourse at the expense of "continuous, synthetic thought."[38] There was "nothing of the spirit of an antiquary" in his character, Pinkerton said in a note to *Select Scotish Ballads*.[39] In truth, there was a good deal of that spirit in him, and he would soon respectfully allude to antiquaries[40] and even call himself one. Yet the term scholar is a better fit for him, or historian.[41] Or man of letters. (Scientist must be permitted after 1800.) "I read every thing," he boasted to Nichols in his letter of October 1783, and it is hard to deny the wide range of his interests, as shown in his books and other writing. He was also a collector of books, manuscripts, coins, and medals; by 1784 he owned a library and cabinet. Collecting coins looks very like the practice of an antiquarian. In January 1784, he outlined to Nichols the plan of a book on medals, by which term he meant in this instance mostly coinage.[42] In June the anonymous *An Essay on Medals*[43] was published by Dodsley.[44] Conceived, written, and printed within six months, but drawing on information gathered from a hobby of years, the work was a minor classic. It was expanded into a two-volume edition in 1789, reached a third edition in 1808, and was translated into French and German. The "little essay was begun as a mere amusement and relief from idleness," Pinkerton wrote, "yet I soon found that infinite labour was required to answer my own expectations."[45] The book was a triumph of simplification, containing, as a critic noted, "*multum in parvo*, the pith and marrow of many folios."[46] For the mildly curious, for the gentleman amateur who wished to form a collection, the work was an ideal guide. It contained essays on the utility of the subject and "The various Sources of Delight and Amusement arising from it," explanations of medallist terminology, directions for forming cabinets, accounts of Greek, Roman, and British coins, and tables giving the current prices. The numismatist Richard Southgate, Assistant Keeper of Manuscripts and Medals at the British Museum, who had helped Pinkerton with the work, thought the *Essay* "the best and completest of its kind ever published."[47] The book represents a stage in Pinkerton's development – not so much his development as a historian as of him as an antiquary, although the two are closely related. An antiquary often studies objects instead of documents, and that of course was what Pinkerton did in the *Essay on Medals*. At one point in the *Essay* he commented on the comparative usefulness of medals and historical documents and said "medals alone remain as the principal proofs of historic truth."[48] That is an antiquary's boast. In 1785 he defined the "province of an Antiquary" as "the science of ancient customs, manners, buildings, dresses, coins, &c." When "applied to man and manners," he said, it was "very amusing and entertaining." It was then "an appendix to civil history."[49]

The writing in the *Essay on Medals* was lively and at times saucy, close to that of the familiar essay. His citations were by no means confined to books on medals; we note references to Pope, Addison, Smollett, Milton, Gray, and Montesquieu as well as to an impressive roster of classical authors, including Virgil. A passage on imitation in coinage reminded him of that poet, who "had not a single thing in his possession which was not stolen."[50] The material was on the whole skillfully organized. This ability to regiment data into logically devised categories was a feature of Pinkerton's work, and not just in this text. "*Arrangement* is the first quality of a good book," he noted in 1807.[51] He could attack a thorny subject with vigour, dissect it, and lay its bones bare in orderly fashion. Once done with a topic, he could leave it and move to another – not as a literary drudge or hack, but because he'd found a new interest. (The term hack can be used of him with some justification but mainly of his career much later, and even then its fitness is often doubtful. Yet a work by him on medals in 1790 – see chapter 4 – could be thought to belong in that category.)[52]

He was now ensconced in Knightsbridge and exclusively a literary man. This distinguishes him from many part-time scholars of his time. Percy was a cleric and, from 1782, a bishop; the Scot George Chalmers was a government clerk; Thomas Astle was keeper of records in the Tower of London; and Joseph Ritson, a formidable rival, was a practising lawyer. "I have no office," Pinkerton wrote in 1784, noting that he had a sufficient income to "keep a house and a girl and two servants."[53] The lease on the house he'd rented had fourteen more years to run. He thought he was well established. The "girl" could mean in this context – he was writing a bragging letter to John Young – a sexual companion, otherwise unidentified. But Pinkerton, though unmarried, may already have fathered a child, and that may explain the reference. In 1792 he had a daughter old enough to be sent to France for her education[54] and it seems unlikely she would have been younger than eight years of age. Considerable obscurity surrounds Pinkerton's relationships with the women who moved in and out of his life, either as sexual partners or acquaintances. He regarded the female sex with a condescending attitude, believing that "domestic duties and affections" should be woman's proper domain and looking upon books by women with a smile.[55] Yet he was susceptible to women, and to the pleasures they offered.[56]

The list of his male literary associates and contacts was now lengthy. Some, like Percy, Beattie, and Dalrymple, were occasional correspondents, useful but by no means close allies. Dalrymple indeed was standoffish from the start of their relationship. John Nichols was a more valuable connection; and Samuel Knight, the young elegiac poet who lived at Milton, near Cambridge, was the friend who gave him lodgings during his trip to see the Maitland Folio. While preparing the *Essay on Medals*, Pinkerton became acquainted with Southgate and the antiquaries Craven Ord and Francis Douce. The crotchety Douce was a soulmate. The memory of their early association was one of Pinkerton's few pleasant memories during his declining years. He wrote in 1822 that Douce's "recollections of our ancient intimacy give me singular satisfaction. I assure you the reciprocity is warmly felt and I shall never forget the hospitable and interesting evenings that I have passed with you. Our friendship is if I remember right of forty years standing, from Knightsbridge and Gray's Inn downwards."[57] On 24 June 1784 Pinkerton dined at Dilly's with Vicesimus Knox, Beattie, James Boswell, Johnson, and two others, six months before Johnson's death. Boswell recalled nothing of what was said except Johnson's clever remark on hackney-coaches.[58] Pinkerton evidently came away unimpressed by the great essayist and lexicographer.[59] The connection to Boswell did not blossom.[60] In August Pinkerton met Horace Walpole, who was sixty-seven, forty years his senior, and so gouty he could produce chalk-stones "with more ease and rapidity than any man in England."[61] Pinkerton struck up the friendship, offering, through Dodsley, to dedicate the *Essay on Medals* to him. Walpole declined the honour but invited Pinkerton to Strawberry Hill, his Gothic castle in Twickenham. The jealous Boswell admitted that Walpole took "prodigiously" to him.[62] To the dismay of some of

Walpole's friends,[63] Pinkerton became a frequent visitor to the "magic castle" and to Walpole's city residence in Berkeley Square. Nor did Walpole forbear to visit Pinkerton. In July 1785 he called on him twice in one week to get literary news.[64] The younger man was then the talk of the town.

Many of Pinkerton's connections with literary men were tainted by bitterness and quarrels. Sometimes it was he who was wronged; on occasion, especially to scholars seeking information, he could be considerate and helpful. More often the blame lay with him. As already noted, he was opinionated and petulant, at times deceitful. He could be a pest. And yet his conversation was at times so engaging that he could be a delightful companion. His relationship with Walpole is a case in point. Their friendship was hardly a month old before Pinkerton had given him a comedy, asked for his criticisms, and suggested he pass it on to George Colman (the elder), manager of the Little Theatre in the Haymarket, to have it staged. (He had earlier tried to reach Colman through Percy.)[65] Walpole read the play – perhaps "The Philosopher" – and made comments on it. It began, he said, with a "totally incomprehensible" song, featured dialogue that had been "too often produced on the stage," had a stock character, "a romantic old maid" named Mrs Winter, who didn't contribute "to the plot or catastrophe," and contained repartee taken from a common jest book. Nor were these his only objections.[66] In his testy response Pinkerton offered to turn Mrs Winter into a mermaid! (He had a sense of humour that came out periodically in even his soberest musings.) Walpole must have been embarrassed to send the play to Colman, though it might have been no worse than what passed in the later eighteenth century as comedy,[67] but send it he did and Colman rejected it.[68] Shortly after giving Walpole the play Pinkerton sent him the preface to a proposed history of the reign of George II, again asking for comments; a year later he suggested Walpole print an edition of Anacreon on the Strawberry Hill press.[69] This prodding and pumping continued. As shown in letters, we have a mainly one-sided view of their friendship: just a single Pinkerton letter was found for W.S. Lewis's edition of *Horace Walpole's Correspondence* (1937–83). At least twenty and probably many more are missing. Yet it is apparent that Walpole sought and enjoyed Pinkerton's company, not just in the early stages of their relationship, but for a full decade. R.W. Ketton-Cremer's judgment that Pinkerton "managed to toady Walpole with some success"[70] is unconvincing. The letters refute such a view. In 1789, when Pinkerton was severely testing his patience, certainly not toadying him, Walpole wrote that "your understanding is one of the strongest, most manly, and clearest I ever knew."[71] By the mid-1790s, Pinkerton's situation was such (as we will see) that his requests for favours from the aging Walpole became a touch onerous. "I have told you over and over" – so Walpole's last letter to him begins.[72] The earl of Orford (as he then was) was clearly fed up. Yet it was likely senescence as much as anger that caused the break.[73]

In November 1784 Pinkerton was given permission to copy from the Maitland collection[74] and went to Cambridge at once. For almost a month, set up in "a room with good light, near the library" and serenaded by a robin from a "bair linden" outside his window,[75] he took extracts from both the Maitland Folio and Quarto manuscripts.[76] According to the Scottish scholar David Macpherson, his onetime acquaintance and neighbour in London, but later, when he wrote the following note, a ferocious enemy, Pinkerton's conduct on his departure from Cambridge was, to say the least, impolite. The note was brought to light by W.A. Craigie:

> These thanks given thus publicly to the Gentlemen of Cambridge [in *Ancient Scotish Poems*, Vol. 1, xix] ill accord with the letter sent by Mr. P. accusing them of stealing his shirt. It seems he had given a very shabby old shirt to a woman to get mended for him, and by her, or his own, neglect, came away without it; on discovering the loss he had sustained, he wrote a long letter, the cream of which was that he had supposed himself among Gentlemen, and not among Newgate birds. The shirt was afterwards found, and in consequence of Mr. P.'s infamous conduct a resolution was made that in future no person shall be permitted to copy a MS. without the permission of eight trustees of the Library. Thus the literary world in general suffers for the base conduct of this dirty animal.[77]

We will get back to Macpherson later to place his words in context. His story is doubtful. (Not that there mightn't have been some fuss over a shirt!) Pinkerton had indeed irritated Peter Peckard, Master of Magdalene College, by writing a haughty letter about the delay in allowing him access to the Maitland material, and by getting Percy to intervene for him, but that was before he'd gone to Cambridge.[78] Afterwards he was on cordial terms with both Peckard and William Bywater, Librarian at the College. When *Ancient Scotish Poems* was published he sent them copies and both answered in a friendly manner, saying they wished they could have made him more comfortable during his stay.[79] The acknowledgment of their help in the preface to his book was generous and, to all appearances, sincere. The rules Macpherson mentions had been in force before Pinkerton's visit.[80]

The libel did not originate with Macpherson but was told to him by Ritson. Ritson had come to London from Stockton-on-Tees in 1775 and lambasted, in turn, Thomas Warton, Percy, Johnson, and the Shakespearian George Steevens for faulty scholarship. It was now Pinkerton's turn. In November 1784 the truth about the second part of *Hardyknute* and other forged pieces in *Select Scotish Ballads* was revealed. "Anti-Scot," i.e., Ritson, in an open letter to Pinkerton in the *Gentleman's Magazine*, said of the second part of *Hardyknute*, "Neither the lady, nor the common people of Lanarkshire, from whom you pretend to have recovered most of the stanzas, will deprive you of the honour of its procreation. The poetry is too artificial, too contemptible; the forgery too evident."[81] He said that Pinkerton was a forger of the stature of Lauder and Macpherson, that his dissertations in *Scotish Tragic Ballads* were pointless, that his own poems were "incomprehensible rhapsodies," and that he had never read Blind Harry's *Wallace*. In general, that he was ignorant and incompetent. It was the beginning of a long enmity, though the enmity was more on Ritson's side than on Pinkerton's. Ritson was the bigger hater. To him a man who could forge a ballad was a complete scoundrel.[82] His attack was diluted somewhat by the way the letter was presented in the magazine. The editor printed it only when Pinkerton gave his "concurrence,"[83] and allowed him to introduce and annotate it. This made it seem a bit silly. Nor was it entirely to Pinkerton's disadvantage to have his work attacked in the press, notoriety on a matter such as that raised by Ritson being preferable to obscurity. In any event, a few months later, in the *Gentleman's Magazine*, he responded to the charges, though not under his own name. "J. Black,"[84] allegedly of Woodbridge, said Anti-Scot's attack on Pinkerton was "illiberal." He was "very much obliged" to Pinkerton "for the high degree of pleasure which his Scotish ballads afforded me." He had always believed part two of *Hardyknute* to be "entirely" Pinkerton's, but it had given him "no less pleasure on that account." In a future edition, he hoped Pinkerton would indicate "those parts" of the ballads that were his own, and "assign the rest of them to their real authors." As "Mr. Pinkerton" was not only "a very good poet, but likewise an antiquary [*nota bene*], and a gentleman of independent fortune," J. Black proposed that he undertake "a complete edition of the Scotish poets," that being "a work for which I think he is well qualified." The poets J. Black had in mind were "such as Sir David

Lindsey, Lord Stirling, Drummond of Hawthornden, Blind Harry, &c. &c." This was so outrageous a specimen of self-advertisement as to be almost comical. How could he hope not to be detected? He likely didn't care if he were or not. The letter was a sly announcement of his literary plans. The alert reader could expect an appearance by Pinkerton as an editor of Middle Scots poetry, and a confession of his earlier transgressions over *Hardyknute* and the ballads.

These half-submerged hints would soon fully surface in his important work, *Ancient Scotish Poems*. But another book requires notice first. This was *Letters of Literature*, published in late August 1785,[85] under the pseudonym Robert Heron. (Heron, it will be recalled, was Pinkerton's mother's maiden name.) The book comprised a series of letters addressed to an imagined younger friend by someone more knowledgeable in literary matters, presumably (but not necessarily) Pinkerton himself. This epistolary method allowed for considerable variety in subject matter, digressions within letters, a certain looseness in presentation – in a word, for all the practices of a letter writer. "I use the privilege of epistolary writing," the author says, "and give you my thoughts as they rise, without studying arrangement."[86] The pieces are like personal essays, but even more conversational. "I am wandering like Montaigne, tho with a juster title to wander," he says.[87] In a familiar letter one is not under oath; and so in this work a certain latitude for overstatement, for "pert familiarity," as Pinkerton put it, might be vouchsafed the author. Pinkerton knew exactly what kinds of freedom the medium offered. He realized what he was doing might be "mistaken for arrogance and dogmatism."[88] (Mistaken hardly seems the right word, however.) His choosing to publish under a pseudonym could indicate wariness, a feeling that he might have gone beyond what the conventions of the medium permitted and needed the protection of a false name.

Letters of Literature, his single effort in literary criticism, is one of his most entertaining books. It is lively, iconoclastic, wide-ranging in subject matter, and at times funny. Walpole, who was suspected of being its author, was so delighted with it that he could hardly lay the book down before finishing it, "so admirable I found it and so full of good sense, brightly delivered."[89] The *New Review*'s critic, Henry Maty, said the book offered "an uncommon display of unaffected integrity, varied learning, much taste, a plain but nervous, and, at times, richly ornamented, style."[90] The *Gentleman's Magazine* also recommended it, though adding that it contained "opinions not conformable to our own."[91] These were early responses. As the nature of some of the contents became more generally known, the tenor of reviews changed. This was "Heron" on the language of the *Bible*:

> ... for absurd and filthy imagery, witness some parts of Ezekiel, the best of the sacred writers, the scripture yields to no composition in any language; but of sublime or beautiful style, I can from that work produce no proofs.[92]

That was part of an analysis of Biblical language, even of "Let there be light and there was light," which he said derived from "a forged addition to Longinus, not to be found in any authentic MS." It was "a common barbarism."[93] On religion generally:

> Religion is the only bond of society for the mob; and they ought not even to suspect that their superiors despise it; as they will, in that case ... imagine that their superiors have no laws, and consequently that they ought to have none. ... They must be bound in the chains of prejudice, and so led thro the road of life ...[94]

Heron said (ostensibly of Cervantes) that "no miracles are now wrought upon asses" and claimed that ecclesiastical history is the history "of human madness."[95] At this juncture[96] Pinkerton was obviously a religious sceptic, doubtless owing at least partly to the influence of Hume. Yet in his writing he never denied the existence of God; what he thought fit to question was organized religion and literal readings of scripture. The distinction was too subtle for many vigilant Christians of the day. The young Pinkerton seemed unaware of how dangerous to his career as a writer even a suspicion of infidelity could be.

In *Letters of Literature* it wasn't just the heterodox pronouncements and ironies that gave offence. The author was an outspoken whig[97] in politics, attacking "tory principles of government which have prevailed thro this pitiful and miserable reign, and have made it one blot in the British Annals."[98] He berated contemporary writers such as Gibbon (whose conclusion to the *Decline and Fall of the Roman Empire* was "worthy of an old woman") and "the nonsense-reading-and-expounding" George Steevens; he was left "disgusted" by James Thomson's *The Seasons*; and he once again airily dismissed Virgil as a mere imitator.[99] He also wished to "improve" the English language and gave an example of his desired orthography by rewriting the opening of Addison's famous *Spectator* essay No. 159.[100] But it may be that the real sources of readers' antipathy to the work were two passages of a sexual nature, the first a remark on Johnson and Steevens' 1778 edition of Shakespeare. Pinkerton's notes on that text occupied three lengthy letters. In discussing *Hamlet* he came to the prince's line to Ophelia in Act III, "Do you think I meant country matters?" which was glossed in the edition as follows in a footnote:

> *Do you think I meant country matters*?] I think we must read, *Do you think I meant country* manners? Do you imagine that I meant to sit in your lap, with such rough gallantry as clowns use to their lasses? JOHNSON.[101]

Pinkerton responded:

> 'Country matters.' The commentator is so chaste, that he seems not to know that both of these words are dissyllables. Tho I would be sorry to claim the praise of Agnolo Poliziano,[102] of finding obscenities where the meaning was possibly innocent, yet such *matters* should either not be understood, or understood aright.[103]

A few pages after this reflection, in a letter on luxury, he turned to male homosexuality among the ancient Greeks. He didn't use the h-word but spent a paragraph explaining its ubiquity. The learned men of Greece, he said, naming Pindar, Aeschylus, Euripides, and Sophocles, "we know for certainty, were given up entirely to a vice too black to mention." And "Socrates himself had different catamites beside Alcibiades; tho his passion for the last was the most notorious." Those who say such love was Platonic "only shew their total ignorance of Grecian learning, and of Grecian manners." Grecian manners "did not regard this vice as of the slightest moment. Sophocles and Euripides practised it we know in the open fields around Athens." The "passion was looked upon as equally innocent with legitimate love."[104] In broaching this subject, as well as in the note on *Hamlet*, Pinkerton stepped outside the limits of what was tolerated in the polite literary discourse of his day.[105] A twenty-first-century reader can only view his words in a different light.

The book caused a ruckus. Pinkerton was soon spotted as the real author. The *European Magazine* in August denounced the book in a long review and followed it up with three more instalments in succeeding issues. By October it recognized that "A most horrid poetaster, one Pinkerton, talks exactly like our author."[106] The *Gentleman's Magazine* backtracked in September, saying that the book's only novelty lay in deviations from "great and established truths."[107] The *Edinburgh*

Magazine, also in September, said the author was like a monkey which "clambers upon the back of poor Virgil, and amidst a thousand puerilities and monkey tricks, bespatters him with his nauseous criticism."[108] The *Critical* and *Monthly* reviews were similarly ill disposed.[109] A queue of correspondents, mainly literary vicars writing to the *Gentleman's Magazine*, continued the onslaught, carrying it through 1786 and into 1787.[110] Among them were: John Duncombe; Anna Seward, the "Swan of Lichfield"; John Aikin; Thomas Townson; Joseph Towers; and William Pettman. Pettman's response was book-length.[111] Pinkerton defended himself in a series of letters to the *Gentleman's Magazine*. As "Eusebes" he pointed to the "misquotations and misrepresentations" of his book in the *European Magazine*.[112] As *"The Author of Letters of Literature"* he denied suggestions in the *Critical Review* that he had modelled his book on William Jackson's *Thirty Letters on Various Subjects* (1783),[113] and that he had resorted to puffing his own works.[114] He seemed especially outraged by the latter suggestion, loftily asserting "upon the whole faith and veracity of a man who hath a much more precious character than that of author to maintain" – the gentleman pose again – that "I never have used, nor shall use, any such infamous arts." As "Vindex" he wrote two letters[115] censuring attackers and reaffirming belief in the opinions expressed in the book. He was "not so insane as to wish to attract enmity," he said to those claiming he merely wanted to "attract notice"; he was "an advocate for the cause of truth and science." In the second "Vindex" letter, in December 1786, he struck back at the strictures in the *English Review* of the previous January which he thought had been written by Gilbert Stuart, the historian alleged to be the review's "despotic ruler,"[116] who had died in the interim. It had been an especially savage review. The "real author of these letters," it said, "is some disappointed, disgusted, and damned poetaster, who wishes to revenge his quarrel with the public, against those celebrated authors who have been their favourites in all ages. The weakest white wines make the sharpest vinegar … and the man who has deserved the gallows is the fittest person to fill the office of hangman." The acrimony in the book sprang from "moral demerit."[117] Pinkerton in response told of the "viper of disappointment" that gnawed at Stuart's mind and hinted at his immorality. This brought an enraged reply from Joseph Towers, who said Stuart had not written the review, but instead had sent in a favourable one that had been rejected by the magazine's proprietor. (That being Edinburgh-born John Murray.) Was it "brave or manly," Towers asked, that Pinkerton waited until Stuart was dead to make his charges?[118] Good question!

Pinkerton took some pains to answer, writing two more letters to the *Gentleman's Magazine* on the matter.[119]

The book and ensuing controversy further blackened Pinkerton's already tarnished reputation. In 1787 he planned a second volume of *Letters* and, it appears, would have maintained or clarified some of the provocative opinions in the first. Two titles in the new volume were to be "Sublime and beautiful of scripture agt. Louth" and "Virgil's fame."[120] Eventually he recanted. In 1799 he disowned the book as containing juvenile and crude ideas long since abandoned[121] – a belated and pointless effort at damage control. The book was by then gathering dust but its effect had been to create or extend the line-up of critics and prudes who shuddered on hearing the name Pinkerton. As late as 1847 his effort to amend English orthography could still provoke Thomas de Quincey to brand him a "horrid barbarian" and "monster."[122] The poet William Cowper took a shot at him as well. Heron, he wrote, is like "him of old/ Who set th' Ephesian dome on fire":

> And for traducing Virgil's name
> Shalt share his merited reward;
> A perpetuity of fame,
> That rots, and stinks, and is abhorr'd[123]

– a condemnation more suitable for a mass murderer or serial rapist than for a young critic stretching his wings and showing off. Pinkerton didn't traduce Virgil; he questioned his originality, as others did before and after him, while acknowledging that the poet "deserves all his fame."[124]

It remains to ask if the book made any mark in the history of its genre, literary criticism. The answer must be, only a slight one. The book rehashes some of the perennial topics of eighteenth-century discourse on writing. In answering the question "*In what quality does the perpetual and universal excellence of poetry consist?*" we find him pondering the importance of "good sense," "imaginary invention," "style," and "truth of nature," and coming up with no very original perceptions. Yet a strain of new thinking can be detected. Amidst the opinions in *Letters of Literature* are calls for freedom from the rules imposed on poetry by prior critics. Of course he was not the first to adopt this line but he pursued it with vigour, as if he were. Poetry "is a faculty, not an art," he wrote, not "an exertion of the mind to be circumscribed by rules." "Poetry knows no rules. The code of laws which Genius prescribes to his subjects, will ever rest in their own bosoms."[125] He also raised an "impious hand"[126] against that favourite eighteenth-century poetic form, the imitation, which he dismissed – wrongly, but in a good phrase – as "only a decent and allowed plagiarism."[127] In general, he distrusted the dictates and models from the past. *Letters of Literature* has been cited as illustrating the "tendency of the time to get rid of the dead weight of neo-classicism."[128] It more than illustrated that tendency; it drove it forward. And who can say what hidden effects such a brazen work might have had? "Be free" he said to artists; and he told writers they would produce nothing of lasting fame unless they had "a soul free as the mountain winds, and large as the universe."[129] "A soul free as the mountain winds": that is a touch Wordsworthian. "Large as the universe" has a whiff of the Lake Country too.

Pinkerton's *Ancient Scotish Poems, never before in Print. But now published from the MS. Collections of Sir Richard Maitland … Comprising Pieces written from about 1420 till 1586* was published on 15 December 1785.[130] It was his first major work of scholarship and his most important contribution to the study of Middle Scots literature. How qualified was he for the task of editing the makars'[131] poetry? W.A. Craigie, the lexicographer who, in addition to his work on the *OED* edited various texts, including, as we've noted, the Maitland Folio, said that by 1785 Pinkerton "had clearly made extensive studies in the older literature of Scotland, so far as it was then accessible."[132] To Percy he seemed "the only person in the kingdom" capable of editing the manuscript.[133] This was a stretch – Ritson was surely capable of doing it, as were Dalrymple and some others. But there can be little doubting Pinkerton's capacities. He certainly had no qualms himself. He had by 1785 "read almost the whole of ancient Scotish poetry," he asserted. The word "almost" may even hint at a dawning humility, in keeping with his admission that he'd written the second part of *Hardyknute* and other pieces among the ballads he'd published, the alleged motive in passing them off as ancient being "to give pleasure to the public."[134]

Two items in *Ancient Scotish Poems* had appeared in print before. "Christis Kirk on the Green," which Pinkerton attributed to James I of Scotland, is in the appendix.[135] It had been printed previously in a number of versions, including his own in *Select Scotish Ballads*, based on the text in John Callander's *Two Ancient Scottish Poems* (1782). Pinkerton included it so that out of the Maitland and Bannatyne versions a "standard edition of this celebrated poem" could be made. *The Twa Merrit*

Wemen and the Wedo, Dunbar's longest poem, which has been called "the greatest, and grimmest, satire" in Scottish literature,"[136] had been issued in a Chepman and Myllar imprint in 1508. (Walter Chepman and Androw Myllar were Scotland's first printers.) Scholars did not know of the existence of that and other Chepman imprints until 1788.[137] It is in the first volume of *Ancient Scotish Poems*, following the one with pride of place, Gawin Douglas's "King Hart." (If it is by Douglas.)[138] "The Freiris of Berwick" is the third work. Pinkerton supposed it to be by Dunbar.[139] "For vivid comic narrative," says one critic, "this poem has no equal in Scots until 'Tam o' Shanter'."[140] Twenty "Poemes be Dunbar" follow. Among them are the well-known "Of James Dog," "Aganis the solistaris in court," "Dunbar's Complaint," and "On the warlds insta-bilitie." The "Poemes be various authors" completing the first volume include "Adveyce to a Courtier" by Quintyne Schaw, who is mentioned in Dunbar's "Lament for the Makaris," three by Alexander Arbuthnot, two by John Maitland, two by Alexander Montgomerie, and a sonnet by James VI. The second volume begins with twenty-nine "Poemes be unknawin makars": of these probably the best known are "The Bankis of Helicon," later ascribed to Montgomerie,[141] and "Song of Absence," which Pinkerton suspected was by James I of Scotland. The poems end with twenty-six by Sir Richard Maitland and a few in praise of him and John Maitland. One of the fragments towards the end is Dunbar's "Of Sir Thomas Norray." Concluding Volume 2 are explanatory notes on "particular passages" in the poems and an appendix, together compris-ing about 150 pages of commentary and additional texts.

Showing editorial restraint, Pinkerton did not print Dunbar's "Goldyn Targe" and his exquisite "Lament for the Makaris," both having Mait-land transcripts. These were in the Bannatyne Manuscript too, and had been in Dalrymple's *Ancient Scottish Poems* and Ramsay's *Ever Green*. He was conscious of adding to or improving what been done by his pre-decessors, and indeed his editorial work can only be properly evaluated in relation to what previous scholars had accomplished or attempted. Knowledge and editorial practice in this field advanced incrementally, through republication of unavailable or badly edited printed texts, dis-covery of new texts and facts, conjecture, argument, correction of error, and recognition of the value of the literature being uncovered. Pinker-ton's contributions within this continuum may be briefly summarized: 1. He brought a great deal of little known, unknown, or poorly edited Middle Scots poetry into the open, as it were, not just in *Ancient Scotish Poems* but in two other books, to be discussed later. 2. He offered his selection to readers, not "as effusions of nature, shewing the first efforts of ancient genius, and exhibiting the customs and opinions of remote ages" – Percy's apologia in the *Reliques*[142] – but as pieces with genuine poetic merit, as determined by himself. He championed the makars and the language they wrote in. Thus John Barbour's poetry "is as smooth as that of Chaucer, with great descriptive, and expressive, powers"; the poems of James I of Scotland "have superlative merit"; Dunbar's work has "humour, description, allegory, great poetical genius, and a vast wealth of words."[143] These were three of the seven early Scottish poets he considered classics, the others being Drummond of Hawthornden, Douglas, Sir David Lyndsay, and Blind Harry. (Leaving out Robert Henryson, whose work would not get the attention it deserved until the nineteenth century.) 3. His introductory essay "A List of all the Scotish poets; with brief remarks," together with the extensive annotation ending Volume 2, formed a significant piece of scholarship. It was not so much a history of Middle Scots literature as a call to action, combining inquiry into the lives of the poets, discussion of the texts in print, brief assessments of each writer's work, and a plan for "new editions" of the seven poets, "in which labour much remains to be done." He noted hesitantly: "Perhaps the editor may in time give new editions of the whole of these poets." He hoped "the public will so encourage the design, that the printer will have no occasion to desist."[144] His plans extended beyond new editions of the poets. In *Ancient Scotish Poems* he announced that he intended to publish no fewer than thirteen separate works relating to early Scottish literature, including a "general Glossary of the Scotish language," an edition of the sixteenth-century *Complaynt of Scotland* ("the only classic work in old Scotish prose"),[145] a copy of which he'd "discovered" in the British Museum, and the lives of the Scottish saints. 4. As editor, he aimed at presenting texts that were accurate, yet intelligible. Throughout *Ancient Scotish Poems* he had been "so very tender of every particle" of the texts "that he believes the most rigid antiquary will not censure him. … Where in one or two places, a word, or line, was palpably lost, the editor has supplied them; but every the most minute supplement, or alteration of an evidently wrong word, tho it be but a *That* for an *And*, or the like, is always put in brackets [thus]. And the reader may depend upon finding thro-out a *literal* transcript of the MS. save in these very rare instances, and as far as human fallibility would permit."[146]

This last claim requires discussion. Pinkerton's idea of what constituted a literal text was different from, say, that of Ritson, to whom an edition was almost an exact reproduction in print of the original. Pinkerton later ridiculed this approach in his review of Ritson's *Scotish Song* (1794)[147] and elsewhere. In *Ancient Scotish Poems* he said that presenting early works "in their original orthography" meant "not one in a hundred of the peasantry could read them."[148] He edited his source in the following ways: he used his own punctuation, capitalization, and italics; he did not copy minutiae, always writing "for" for "ffor," using "y" for yoghs and "th" for thorns, and expanding contractions; he sometimes regularized the metre of the original by inserting words in parentheses; in some cases he omitted a (to him) unnecessary word that made the metre of a line irregular, without noting the omission; he occasionally corrected a perceived grammatical (or other) error, as when he wrote "is" for "ar" or "quhilk" for "quhill"; and he composed lines and inserted them in parentheses to make up for lacunae. He made what divisions he liked in the poems, preferring a sequence of "cantos" or "books" to the unorganized mass of lines in the manuscript. The result of such interventions is that Pinkerton's texts are quite readable. We may compare him with Craigie:

(Craigie)
Servit this quene dame plesance all at richt
first hie apporte bewtie and humilnes
with mony vtheris madinis fair and bricht
Reuth and gud fame fredome and gentilnes
Constance patience raddour and meiknes
Conni*ng* kyndnes heyndnes and honestie
Mirth lustheid lyking and nobilnes
blis and blythnes plesance and pure pietie[149]

(Pinkerton)
Servit this Quene Dame Plesance, all at richt,
First *Hie Apporte, Bewtie,* and *Humilnes;*
With mony utheris madinis, fair and bricht,
Reuth, and *Gud Fame, Freedome,* and *Gentilnes;*
Constance, Patience, Raddour, and *Meiknes,*
Couning, Kyndnes, Heyndnes, and *Honestie,*
Mirth, Lustheid, Lyking, and *Nobilnes,*
Blis, and *Blythnes,* [*Gudenes*] and pure *Pitie.*[150]

He substituted *Gudenes* for plesance in the last line (he couldn't see how plesance could serve plesance). Otherwise the transcript is reasonably faithful. He didn't just copy out what he saw at Cambridge. He tried to make sense of it; he was thinking his way through it; and he was doubtless working fast, which may account for many of the blunders, omissions, and changes Craigie points to in the notes to his own edition. It is well to remember that Pinkerton was the first to edit a substantial portion of the two manuscripts, neither one plain sailing. And he copied it himself, as Craigie did not. A more recent scholar has said that "Given the speed with which Pinkerton must have had to work, the level of accuracy is quite creditable … his intrusions seem, by the standards of his age, quite restrained."[151]

5. One further contribution: he provided a glossary of Middle Scots terms. Writing twenty years after the publication of *Ancient Scotish Poems*, George Chalmers said he found twenty "mistakes" in that glossary. Of these it appears ten were genuine mistakes; in seven cases Pinkerton was right and Chalmers wrong, in three cases both were wrong.[152] Pinkerton's mistakes reflected, not ignorance or indolence, but rather the state of Middle Scots studies in his time.

Of the 165 poems in the Folio he printed 71; from the Quarto's 96 poems he chose 28. In his appendix he listed the contents of both manuscripts and explained his omissions. Dunbar's "Vanitas Vanitatum" is *"A moral poem which is very dull"*; his "Confession" is so dull not even *"a patient monk"* could listen to it; his "On Christ's Passion" is *"as stupid as need be."*[153] In general, he rejected poems of a religious character except those painting the monks' lechery. Three of Sir Richard Maitland's poems are tossed out peremptorily: one is *"A pious and pitiful performance"*; the second *"Another leaden lump of godliness"*; the third *"Another psalm!"* One poem is dismissed in this fashion: *"The point of it is that Christ had a woman for mother, but no man for father. It is subscribed* quod Dunbar in prays of woman; *but I dare say he is innocent of it."*[154] (He is guilty of it.) It is no doubt remarkable that so many "pious lines" in the Maitland manuscripts are bad poetry while the lusty *The Twa Merrit Wemen* and "The Freiris of Berwick" are *"Tales* equal to any that Chaucer has written," but the view can be defended, with the proviso that some of the pious lines, certainly Dunbar's, can pass muster too. Pinkerton's quick dismissal of items with a moralistic turn was bound to irritate his critics, as was his opinion of publishing verse containing "immodesty." While that term, he said, has been applied to a "too free revelation of amorous affairs," he had no objection to raise to such revelation. It has "the palliation of ever delighting every mind that is not callous to nature's best and finest sensibilities." Examples of "the most virtuous authors of every period and country" could be given "who have amused themselves with describing in writing the effect of this most important and elegant of the propensities of nature ... this species of writing has no shade of immorality."[155] We find him objecting strongly to the "bashfulness" of one of his contemporaries, saying "If the generation of man be a matter of shame and infamy, it follows that man is the child of shame and infamy."[156] He went so far as to insist that immodesty was an "essential" feature of *The Twa Merrit Wemen*, and claimed that "castrating a book, and putting asterisks" tended "to raise far worse ideas in the guessing reader than those omitted."[157] Printing *The Twa Merrit Wemen* and "The Freiris of Berwick" was a daring move (though he omitted eight lines from the latter).[158] His attitude may be contrasted with that of Percy, who promised that in the *Reliques* "great care has been taken to admit nothing immoral and indecent,"[159] and Dalrymple's, which was that nobody would "ever venture to publish" Lyndsay's *Ane Pleasant Satyre of the Thrie Estatis* because it was "loose and indecent beyond credibility."[160] "I wish that the indecent poems had been omitted," Dalrymple wrote to Pinkerton on receipt of *Ancient Scotish Poems*.[161] In 1802 the editor John Leyden condemned Pinkerton's "shameless defence of obscenity," and in 1830 Walter Scott repeated the charge.[162]

The deliberate omission from "The Frieris of Berwick" brings us back to David Macpherson, who quoted Ritson as saying that "where a passage uncommonly difficult" occurred in the Maitland manuscripts Pinkerton "got over the difficulty by omitting it."[163] Pinkerton omitted thirty-one lines from the poems he transcribed, including the eight from "The Freiris of Berwick." The other twenty-three were practically illegible and Craigie himself had to supply many of them from the Reidpath Manuscript, a seventeenth-century transcript of parts of the Folio that was unavailable to Pinkerton, or to admit his readings were uncertain. Except in the case of the eight ribald lines and two other instances, Pinkerton signified his omissions by a series of asterisks. He did not avoid a text because it was hard to read in the source. *The Twa Merrit Wemen*, where the original was in a mutilated condition, he printed as best he could, and two of the serious omissions occurred in the exceptionally difficult "Of Sir Thomas Norray." Since the text of Dunbar "is both a nightmare and a challenge,"[164] Pinkerton, the first editor of many of his poems, might be allowed some leeway.

Though Scots could see that the editor of *Ancient Scotish Poems* was a patriot of sorts ("I glory in Scotland as my native country"), much in the work could unsettle and displease them. This patriot pronounced that good taste in poetry and history was "astonishingly wanting in Scotland to this hour," said Allan Ramsay's *Gentle Shepherd* was a "dunghill" and "hyper-monster," promised to write "a shilling pamphlet" to prove Mary Queen of Scots guilty of murdering Darnley, and, while admitting, in his second prefatory dissertation, "An Essay on the origin of Scotish poetry," that as a poet James Macpherson was "the Homer of the Celtic tongue," emphatically joined the list of Ossian sceptics and gave good reasons for his doubts.[165] He also had words for Bible lovers, to add to what he'd said in *Letters of Literature*. He now asserted, among other critical remarks, that scripture had no validity as a source of history, since "The silliest rabbi might inform us that [it] is allegorical, from beginning to end," and that it was wrong to suppose that the numerous races of men "are all from one parent."[166] Yet comments like these may not explain all the resistance he stirred up. Anyone making an entrance into a scholarly field with as much éclat as Pinkerton's into Middle Scots studies could expect a degree of hostility from rivals; and jealousy, as much as offended national pride, devotion to Ossian, or piety was a factor in how his work was received. Lord Hailes, doubtless upset by what was said of scripture, but just as likely motivated by a sense of injured merit, now wrote him off, saying pointedly that he no longer wished to be mentioned as a correspondent in any future publication.[167] When he was later asked to help with an edition of *The Bruce*, he refused. (But he had agreed to help with it before reading *Ancient Scotish Poems*.) William Tytler, who had found James I of Scotland's "The Kingis Quair" and published it in James's *Poetical Remains* (1783), took umbrage too. He anonymously sent Pinkerton over three pages of disparaging remarks on *Ancient Scotish Poems*, and asked "Why are you so illiberal as to suppose that learning worth &c &c are confined to one party?"[168] Pinkerton did not "confine" learning to himself alone. He acknowledged the work of others, including Tytler who, he said, had "done so well" in his edition that James I's works "hardly need to be republished."[169] David Herd, George Paton, David Macpherson, and still more Scottish literary figures joined the list of opponents and resenters, and would soon have stronger reasons for castigating him.

Nor did the general public support his work in Scottish literature. In the two years following the appearance of *Ancient Scotish Poems*, a mere forty copies were sold in Edinburgh. In that book Pinkerton printed a prospectus for a work to be called *Vitae Antiquae Sanctorum qui habitaverunt in ea parte Britanniae nunc vocata Scotia* – lives of the Scottish saints – and asked for 100 subscriptions of 10/ each. He explained that he couldn't bear the cost himself.[170] After two years only thirty-six subscribers had appeared, of whom few were Scots. He would get that title in print, but his plans to proceed with such large projects as editions of the *Complaynt of Scotland* and Andrew of Wyntoun's *Orygynale Cronykil*, also announced in *Ancient Scotish Poems*, were eventually dropped. (The extracts from Wyntoun in the appendix to Volume 2, from the Cotton Library manuscript in the British Museum, were the first lengthy pieces of his poem to be published.)[171] This falling off in interest in editing Middle Scots literature took time. During the period immediately after the publication of *Ancient Scotish Poems* his search for texts went on, and indeed the hunt picked up throughout the scholarly world, inspired in part by what he'd written in his "List of all the Scotish poets." By 1787 he had succeeded in establishing connections in Scotland that helped him, not without difficulty, get access to the treasures in the Advocates' Library. This working from a distance to locate and have the early Scottish documents he wanted copied was a great disadvantage. Why he made no effort to travel north and make his own arrangements is a good question, especially since he owned property in Edinburgh that by 1786 required repairs, and had a mother there who also needed looking after.[172] But he was immovable. John Davidson, a Writer to the Signet who knew Scottish legal history, Adam de Cardonnel, the first curator of the Society of Antiquaries of Scotland, his factor Buchan, and the "daft" 11th Earl of Buchan, David Steuart Erskine, managed to get transcripts of manuscripts and rare books for him and kept him informed of scholarly developments in the north. The most important of these assistants was Lord Buchan who, though at times a bewildering correspondent, was an eager student of Scottish poetry and history. Pinkerton approached him late in 1785, mainly for his help in securing a transcript of Barbour's *The Bruce* from the Advocates' Library copy.[173] Their exchanges were at first testy but they resolved their differences and became friends.[174] Percy too helped in the hunt, despite what he termed his "difference in opinion" with Pinkerton on "some essential points … of great importance to the welfare of society"[175] – alluding to the remarks of a religious nature in *Ancient Scotish Poems* that ruffled Lord Hailes. The bishop could tolerate what the layman Hailes couldn't.

Pinkerton's interests were never entirely focused on editing Scottish texts. One of the introductory essays in *Ancient Scotish Poems*, the "Essay on the Origin of Scotish Poetry," is essentially a disquisition on Scottish prehistory. It was, he told Percy late in 1785, merely the beginning of "one of the most laborious tasks ever attempted":

> It is some years since I formed the design of writing the History of Scotland from the earliest accounts till the reign of Mary: to be comprised in forty books, forming two volumes 4to. But the earliest part, from the beginning till Malcolm III. 1054, is so overwhelmed in fiction, that I find it absolutely necessary to dig a foundation, and clear away rubbish, ere I venture to build an edifice. This I mean to do by publishing first in 8vo, *An Enquiry into the History of Scotland prior to the reign of Malcolm III. or year 1054.* So far as I have gone, I find that it is to the most violent and pitiful national prejudices alone that we are indebted for the obscurity of our early history.

Compared with this, "all my other labours are a jest," he wrote. His book would set the early history of Scotland "upon the firm basis of ancient authorities, that nothing can shake." It will convince all men of science, and as for others, "Let them put up with the dreams of the father of Ossian, and other followers of prejudice."[176] A few notebooks[177] and letters remain to tell of the patient toil that occupied much of his time during the next four years.

The scholar often lives mostly in seclusion, with his books and ideas. Family, friendships, fixing the chimney, the daily news – the stuff of ordinary domestic life – may be secondary. Or lower still. Such was likely the case with Pinkerton, with this difference: he also wanted public recognition, and fast. But the bent of his mind was towards research in thorny areas of knowledge, chiefly at this stage history, and a certain kind of history. It is possible that he became so continuously immersed in study and writing that his life in the day-to-day world outside his library was ruinously affected, beyond what his irascibility and egotism might have caused anyway. Not only that. Prolonged pondering of historical fact and argument, of the past, can of course lead to enlightened understanding of the present. It can also have the opposite effect. It can lead to infatuation with one's own ideas. So it was to a degree with Pinkerton: he got stuck in his discoveries. There are moments lying ahead when patience with him will be sorely tested, far more than it has already been. There is much to extenuate, not just in his writing but in the manner he conducted his personal life, where it is occasionally difficult to offer much sympathy. Yet his intellectual qualities – his learning, intelligence, free-thinking, boldness – place him so far apart from the stuffy piousness of so many of his rivals and contemporaries that, despite his errors and shortcomings, he cannot be shrugged off as a nuisance or "miscellaneous" scribbler but demands that attention be paid to him as a serious writer.

While finding his way through the maze of authorities and documents on the early history of Scotland, he still found time for other literary pursuits. In 1786, he made some progress with "An Easy Introduction to the Greek Language,"[178] and in that year too he published a jest-book entitled *The Treasury of Wit* in two volumes under the pseudonym "H. Bennet, M.A." This work, certainly Pinkerton's,[179] need not detain us: it is "a lounging book for the drawing room," comprising apophthegms, anecdotes, and jokes copied from other books of the kind. "A Discourse on Wit and Humour" prefixed to the second volume is shrewd and learned, but somewhat anomalous in a work in such a hackneyed genre. Around this time he probably also wrote part of a "Gothic Tale" where he tried, with little success, to put his knowledge of Pictish names and Scandinavian mythology to creative use.[180] In 1786 he met the Icelandic-born Danish scholar Grímur Thorkelín and translated into English for him a ninth-century Icelandic fragment dealing with the Danish invasions of Northumberland.[181] He acted as reviewer for the *Gentleman's Magazine,* but it is not clear just how many reviews he wrote.[182] These were merely distractions from his main enterprise. He had planned to publish his *Enquiry into the History of Scotland* in 1787, but a number of factors caused delay. One was that to make his book convincing he needed documents from Ireland. In 1786 he began corresponding with Charles Vallancey and Charles O'Conor (the elder), scholars in Irish prehistory, asking for transcripts and translations. He told O'Conor of one piece he especially craved, "A literal copy of the famous metrical Chronicle of Albany written in Irish under Malcolm III." He had come upon references to it in the early Irish historians. "By this chronicle as the most ancient we must abide," he said, "and as it differs from our common chronicles it is of all others the most important."[183] This was the eleventh-century *Albanic Duan* – sometimes called Alban Duan or Duan Albanach – a poem[184] in the Irish language that, translated, was to be a choice item in Pinkerton's *Enquiry.* O'Conor answered, saying he had a copy of the poem, and that both the original text and a translation would be remitted after he recovered from a "languid state, bound by rheumatic pain." In the same letter he asked "how could the language of the third century in your country be preserved pure to this day in the Highlands of Scotland? How could the poems of Ossian be preserved by oral tradition through a period of 1,500 years?"[185] – as if

Pinkerton, being a Scot and a historian, was somehow implicated in the Macpherson fraud, as many in his day thought it was, and had to answer for it.

Another reason for putting off publication of the *Enquiry* was that he found it necessary to prepare the world for his ideas about Scotland by issuing an introduction to the early history of Europe. In June 1787 his *Dissertation on the Origin and Progress of the Scythians or Goths* was published by George Nicol.[186] It was an attempt to simplify an immensely complicated subject, one beyond the reach of eighteenth-century inquirers, owing largely to the fact that archaeology and etymology[187] were in their infancy. The subject was the settlement patterns of the various early inhabitants of Europe named by ancient writers, who were copiously cited by Pinkerton to support his theories. He postulated four "grand races": Celts, Iberi, Sarmatae, and, most important, the Scythians, whom he also identified as Goths or Getae. These Scythians were, he said, a distinct race originating in Asia and using one language, the Scythic or Gothic. Prior to the Assyrian empire they ruled all Asia. The "Scythian Empire was the first of which any memory has reached us."[188] With them "commences the faintest dawn of history: beyond, altho the period may amount to myriads of ages, there is nothing but profound darkness."[189] By 2000 BC they had moved over the Caucasus Mountains and settled on the east, north, and west of the Euxine (the Black Sea). This was the "Parent Country of the European Scythians." In time they left this home on the Euxine and advanced westward in six stages, populating as they went Thrace, Illyria, Greece, Italy, Germany, and Scandinavia, driving the Celts before them until they were "pent up in the extremity of Gaul."[190] Britain and Gaul were the "final receptacle of almost all the Celts."[191] Inhabiting the area from the Danube to the Baltic Sea were the Basternae, a Scythian tribe, two of whose peoples were the Peukini and Sitones. The Sitones advanced into Scandinavia and "it is most likely that a part of the Peukini"[192] went with them, or even went there before them. Pinkerton inferred "from every ground of cool probability" that the Peukini ultimately "emerged under the name of *Picti*, the *Pehtar*, or *Peohtar*, or *Pihtar*, of the Saxon Chronicle … and *Pehts* of ancient Scotish poets, and modern natives of Scotland, and the north of England."[193] Thus he linked the Scythians (or Goths) with the Picts of Scotland. Many of his ideas about the Peukini and "Piks" (as he now christened the Picts) were given, Pinkerton confessed, "as mere conjectures," and in his forthcoming *Enquiry* he would not insist on connecting the two peoples.[194]

While Pinkerton acknowledged the presence of an "author of nature" in the *Dissertation*, its contents, including a citation from the Roman historian Jornandes containing the highlighted Latin words "VAGINA NATIONUM,"[195] could only have rattled the godly. Some of his remarks about Scythian antiquity drove the age of man well back from the widely accepted date of Creation as laid down in the seventeenth century by the Irish Protestant bishop James Ussher, i.e., 4004 BC. What was more, he gave his "full assent" to the view of the Flood pronounced by the "latest and best Natural Philosophers," which was that it was "impossible," being contrary to "the immutable laws of nature."[196] The common opinion among nations that a flood had occurred, he said, "arose from the shells found even on the tops of mountains." Nor was this all. He once again discarded the idea of humankind's descent from two first parents. Just as the "author of nature" formed varieties in the plant and animal kingdoms, so he "also gave various races of men as inhabitants of several countries. A Tartar, a Negro, an American, &c. &c. differ as much from a German, as a bull-dog, or lap-dog, or shepherd's cur, from a pointer." The differences were "radical, and such as no climate or chance could produce."[197]

We pass by these insights, merely noting that Pinkerton would pay dearly for them in the story stretching ahead. As for the historical validity of his reworking of European prehistory, we need not delve into it at length. It was learned, ingenious, and, on almost every major point, dead wrong. The Scythians weren't Goths, nor were the Getae. Pliny, in the first century AD, said that "The name 'Scythian' has extended, in every direction, even to the Sarmatae and the Germans," and added: "this ancient appellation is now only given to those who dwell beyond those nations, and live unknown to nearly all the rest of the world."[198] Pinkerton was drawing conclusions from the often imprecise and conflicting comment of ancient authorities. The Scythian advance from the southeast to northwest, the primordial Scythian empire, the Scythian colonies throughout Europe – all are illusions. The Peukini/Pict identification is also fantasy.

The most startling feature of the *Dissertation*, and surely the most obnoxious, was the view of Celts advanced in it. These were, Pinkerton said, always "a weak and brutish people"; they "have been savages since the world began, and will be for ever savages while a separate blood."[199] Anyone who doubted that could travel to the Celtic parts of Wales, Ireland, and Scotland "and look at them, for they are just as they were, incapable of industry or civilization, even after half their blood is Gothic, and remain, as marked by the ancients, fond of lyes, and enemies of truth."[200] Their intellect is marked by "idiotic credulity"; their language reflects the "dark understanding in the people who use it"; their personal manners are nasty and filthy, as they were in ancient times when they washed their bodies and cleaned their teeth with urine.[201]

What motives, what feelings, prompted the studious antiquary to introduce such an offensive strain into his book? Walter Scott thought it "to be in a great manner assumed, for the sake of attracting attention."[202] There could be some truth in this, yet it is not a fully convincing explanation. Possibly it was an overreaction to the prolonged Ossian controversy and the mass of argument *pro* and *con.* it engendered that he, as a historian, had to claw his way through to get a clearer picture of early Scotland. He had, as noted, succumbed to Ossian as a youngster and even defended the bard in a poem that caused him some embarrassment:

> Envy in vain
> Shall seek to dim the light of thy name.
> When the eagle from his rock
> Descries the crows dark children of the wood,
> He degrades not his pride
> By the base encounter;
> But rising in the blaze of noon,
> Leaves his foes in the regions of darkness.
> Such shall be thy praise
> Thou Son of the Mighty!

That was in *Rimes* in 1781.[203] When the volume was reissued in 1786 a note said: "*Cancel the piece called* The Harp of Ossian, *a production of early youth. Fingal was an Irish hero; the poems ascribed to Ossian are quite modern.*"[204] Between these two dates, in his second prefatory essay in *Ancient Scottish Poems* – where there is no evidence of anti-Celtic feeling – he decided where he stood on the boasted antiquity of the Ossian material. The evidence, he said there, indicated that nine-tenths of Macpherson's work was "his own," while the remaining tenth had been "so much changed by him, that all might be regarded as of his own composition."[205] This change of heart and mind had been hard-won. By 1787 it had hardened into something close to rage:

> It is to the lyes of our Celtic neighbours, that we are indebted for the fables of English history down to within these thirty years, and the almost total

perdition of the history of Scotland and Ireland. Geoffrey of Monmouth, most of the Irish historians, and the Highland bards, and senachies of Scotland, shew that falsehood is the natural product of the Celtic mind: and the case is the same to this day. No reprobation can be too severe for these frontless impostors: and to say that a writer is a Celt, is to say, that he is a stranger to truth, modesty, and morality.[206]

It is tempting to picture a waspish, solitary, mortified hypochondriac, stomping around in his book-filled porch, brooding over past error, feeling ashamed, not just for himself but for Scotland, filled as it was with believers. The language he later used to describe being taken in by Ossian, his "abhorrence of being made a dupe,"[207] makes such a scene almost believable. The racist element in the *Dissertation* could then be accounted for, at least partly, as an emotional counterblast, showing, to the world and maybe to himself, that he was free of Ossian at last. When the Irish antiquary Thomas Campbell dined with him "more than once" late in 1787 he found Pinkerton to be "what few bookish men are, very communicative," and except for "the prejudices which overpower him ... much to my taste."[208] *Overpower*: a striking word.

The source of the distemper could lie deeper than we know, in parts of his human experience that the documents do not disclose. It probably had in it some trace of the prejudice that Lowland Scots inherited from their ancestors, who endured the Highlanders' periodic raids and uprisings – one as late as 1745 – and responded with fear and loathing.[209] Pinkerton, to repeat, was from a Lowland family, and had lived at an impressionable age in a Lowland area, namely Lanarkshire, not far from the Highlands, where it would have been hard to avoid catching the virus. He seems to have been conscious, through his life, of his affinity with the Lowlanders, who were, he pointedly stated, "in all ages, a distinct people from those of the western Highlands."[210] It is well to remember too that racial prejudice, often directed at Irish, Welsh, and Scottish people, and not just at the Highlanders but at all Scots, with the former bearing the brunt of it, was an element in the intellectual and social life of the day in England. Pinkerton no doubt encountered it, and might have felt that the view of Highlanders held by many in London, spread about by John Wilkes and his followers, as disorderly, alien, and rapacious "vermin,"[211] damaged his own standing, even though he wasn't one of them. He made that specific point twice in *The Treasury of Wit*, once quoting Smollett as saying "That the Scotch Highlanders, by their rank ignorance, and impudence, disgusted every man of sense against the country that gave them birth," and again, in his own voice, claiming that the Highlanders were "a people different from the Scotch as our Welch are from us,[212] and as noted for stupidity as the other Scotch are for acuteness."[213] However they originated, and however deeply felt, his opinions about Celts were nevertheless shameful, and it was foolishly naive for him to express them so openly. Perhaps he was thinking of himself when he later wrote "The greatest minds often act, and speak, meanly; the wisest, foolishly,"[214] the truth of which maxim it takes no great hunt through history to verify. In expiation, the biographer can only say that the prejudice in time moderated and then almost[215] entirely disappeared from his writings. He tried to account for it too, as we will see, though never in a completely satisfactory manner.

Back to his books and the British Museum he went, working on the *Enquiry into the History of Scotland preceding the Reign of Malcom* [sic] *III*, the expanded second edition of his *Essay on Medals*, and *Vitae Antiquae Sanctorum*, all three appearing early in 1789 and constituting a remarkable scholarly performance for a single year. The greatly expanded *Essay* appeared in February, this time with a leaven of asperity that had been missing in the original edition, provoking more chastisement from reviewers.[216] The *Enquiry* came out in May. It had "loaded me so much for many years," he explained in March, "that I am at present almost sick of composition."[217] *Vitae Antiquae Sanctorum* was reviewed in the *Gentleman's Magazine* in July; it appeared in May or June.[218] The *Enquiry* got him involved in controversy straightaway. A notable feature of the book was its display of authentic early documents, one being the *Albanic Duan*. Even Ritson was impressed with that item. The *Duan*, he said, was the "most curious piece of genuine Irish poetry that I know."[219] Pinkerton had got his hands on it by nagging his Irish correspondents to send it to him. In 1786 O'Conor sent a transcript of the poem to Vallancey, asking him to remit it to Pinkerton.[220] For some reason Vallancey sent the transcript to John Lorimer, a doctor friend of his in London with an interest in Scottish history, and said: "You will be pleased to *communicate* the enclosed to Mr. Pinkerton."[221] Lorimer did send it to Pinkerton, who was an acquaintance if not a friend,[222] but before he did he took his own inaccurate transcript and forwarded it to a correspondent in Scotland. It found its way to John Gillies, who later became Royal Historiographer for Scotland. A month after the appearance of Pinkerton's *Enquiry* Gillies published a translation of the poem in the *Gentleman's Magazine*, with an explanatory letter. Pinkerton's name was not mentioned. This enraged him. The *Enquiry* was not selling well,[223] and here was someone undermining his claim to the most intriguing item in it. In July he wrote to the *Gentleman's Magazine*, exposing Lorimer's "breach of confidence."[224] A hot exchange ensued.[225]

The *Enquiry into the History of Scotland*, though drastically injured by Pinkerton's continued Celt-bashing, was nevertheless a significant contribution to scholarship. As with *Ancient Scotish Poems*, it has to be judged in context, and in this case that context is extensive. Early Scottish history had been much written about by 1789. The outstanding work of earlier decades was Thomas Innes's *Critical Essay on the Ancient Inhabitants of the Northern Parts of Britain or Scotland* (1729) which demolished the arguments of Hector Boethius, John of Fordun,[226] and others who claimed great antiquity for Scots under a line of forty kings.[227] The Scots, Innes claimed, came from Ireland in the third century, when Scotland was already peopled by the Caledonians[228] or Picts, and had no kings prior to Fergus, whose rule began at the end of the fifth or the beginning of the sixth century. Studying Scottish history, he argued, meant looking into the interrelation of the different races, the Picts, Scots, and Britons. Innes seemed to have swept away much of the rubbish cluttering early Scottish history, but in historical writing, not just in Scotland but everywhere, old rubbish tends to wash back in, often bringing new rubbish with it. Some considerable revision of his work was attempted, chiefly by those who wished to affirm the primacy of the Goidelic Celts or Gaels and, patriotically, to reassert the antiquity of Scotland. In direct opposition to Innes, they held that the Scots had colonized Ireland. William Maitland in his *History and Antiquities of Scotland* (1757) said that the earliest names for Ireland, i.e., Ierne, Hibernia, and Scotia, belonged properly to the Scots, and that the Irish, "as they are descended from the Scots," and "having no actions of their own worthy of commemoration," had pirated the name Scotia "to rob us of the great an noble feats performed by our brave ancestors, against the mighty Cumri, or Welsh, Picts, Saxons, and Danes."[229] Maitland's ideas were restated by Walter Goodall in 1759 in his recasting of Fordun's *Scotichronicon*, and further publicized by the Highlanders John and James Macpherson. John Macpherson in his *Critical Dissertations* (1768) argued that the "Ancient Caledonians" were Gaelic Celts from Gaul and the ancestors of the Picts and Scots. They were also, he said, the ancestors of the Irish. "The languages of the Caledonians and Scots were the same," and "the Pictish and Irish tongues were so likewise." The "oldest inhabitants of Ireland were colonies from the Western parts of modern Scotland."[230] The idea that the Irish had colonized Scotland had originated when the

Irish were "a degenerate race of men."[231] James Macpherson copied his kinsman John. In his *Introduction to the History of Great Britain and Ireland* (1771), which had gone through three editions by 1773, he blamed medieval monks for perpetrating the lie about Irish colonization of Scotland and reiterated the notion that Picts and Scots were descendants of the ancient Caledonian Gaels. The image of early Scotland drawn by these writers was appealing to Scottish nationalists, fitted well with the picture drawn in the Ossian poems, and offered an uncomplicated narrative. It was challenged by John Whitaker in his *Genuine History of the Britons Asserted* (1772). He pointed out some of James Macpherson's misrepresentations of sources and argued that the Scots were of Irish origin, but his use of Ossian to support his own claims made him a less than impressive opponent. The Macpherson version of Scottish history was influential, leading even Gibbon astray.[232]

Among the mysteries that remained was: what happened to the Picts?[233] They were alluded to by Roman writers from the third century onwards. Bede, writing in the eighth century, noted that "it is said" they had come to northern Ireland from Scythia, by which term he likely meant Scandinavia, whence "according to tradition, [they] crossed into Britain" and settled "in the north of the island." "At the present time," he said, there were "in Britain five languages and four nations – English, British, Scots, and Picts," the fifth language being Latin.[234] When, in 843, Kenneth I became the first king of all Scotland, he had, according to some medieval historians, vanquished the Picts and extirpated them.[235] The twelfth-century English historian Henry of Huntingdon said the Picts and had "entirely disappeared, and their language is extinct, so that the accounts given of this people by ancient writers seem almost fabulous."[236] That Kenneth had destroyed the Picts was the view commonly held by historians until the beginning of the eighteenth century. In 1710, however, Sir Robert Sibbald, in his *History of Fife and Kinross*, said the people there were descendants of the Picts who, far from being extirpated by Kenneth, had been "incorporated with us under our Kings."[237] In 1729 Innes put the matter beyond doubt, or seemed to. The Picts, he said, were "the most ancient and most valiant inhabitants of Britain." Their "dominions far exceeded those of the Scots, as did also the number and strength of their nation." The idea that Kenneth wiped them out was "false and fabulous." The "royal family and present inhabitants of Scotland are … as well the descendants and progeny of the ancient Caledonians or Picts, as they are of those Scots who came in from Ireland."[238] Innes's claim for the antiquity of the Picts' monarchy was based on a crucial fourteenth-century manuscript catalogue of their kings (the *Pictish Chronicle*) that he had discovered in Paris. It was published in an appendix to his *Critical Essay*.

Innes had answers to questions about the Picts' identity. They were Celts, they originated, like the southern Britons, in Gaul, and they spoke a Celtic dialect more akin to the Brittonic (as in Wales) than to the Goidelic branch. But another school of thought had already begun. Sibbald said "it appeareth to be very clear … that the *Picts* were of a *Gothish* Extract, and came from *Norway* and the places upon the *Baltic*."[239] This idea, because it seemed to be supported not just by Bede's comment on the Picts' origin in Scythia but by the historian Tacitus in his first-century AD life of Agricola,[240] was discussed in Innes's *Critical Essay* and dismissed as improbable.[241] In 1750, however, it was again voiced, this time in Sir John Clerke's *Dissertatio de Monumentis Quibusdam Romanis* (written in 1731). Clerke was convinced the Picts spoke a Germanic language. But Maitland and the two Macphersons made little of the Picts. To them the Picts were Gaels, had come to Scotland from Gaul, and played no big part in Scottish history. They sneered at Innes's list of Pictish kings. Maitland judged it a forgery. John Macpherson said the list was no less dubious than the "forty Scottish Kings whom Innes had been at so much pains to erase."[242] James Macpherson similarly ridiculed Innes's "great Pictish Monarchs," noting they had "sunk away into their original nonentity."[243] The index to his *Introduction* made no mention of the Picts at all.

Pinkerton now enters the scene. His *Enquiry* was lengthy, learned, well documented, and densely argued – hard going for the common reader of history. We may summarize his achievement as follows. 1. In Pinkerton, the Macphersons ran into someone every summarizer and romancer of history fears to confront, namely a scholar. In denouncing and dismissing their historical efforts, tearing into Ossian as a sidelight,[244] Pinkerton also in a way reinstated Innes to his proper place in Scottish historiography, not, indeed, by agreeing with everything he wrote, but by accepting much of it, and building on the foundation his forerunner had laid. 2. He brought the Picts back to where Innes had left them, and out of the Gaelic pit where the Macphersons had tried to dump them, dismissing their opinions as "absolutely ignorant and false, and contradictory to all authorities and facts."[245] He devoted 300 pages to the Picts; they were central to his book. He scoffed at the idea that they were Gaels. He denied that they were Britons. He admitted that some Briton place names existed in the east of Scotland, but explained that away by arguing that the Britons had been in possession of the territory before the Picts' arrival and so had left some place names behind them. His belief, as already noted, was that the Picts were Gothic immigrants from Scandinavia, arriving in Scotland c. 200 BC. He dealt at length with the lists of Pictish kings found in Innes's appendix and elsewhere, and located and analysed supporting data confirming the kings' existence in Irish chronicles. He "reposed" his mind "on the truth, that we have as complete evidence of our Pikish kings, of the kings of all Scotland, from the Fifth to the Ninth century, as human history affords in such cases."[246] Not only did Kenneth not extirpate the Picts, "The probability is clearly that he was a Pikish prince."[247] (Innes had said that Kenneth had some hereditary claim to the Pictish throne.)[248] The Pictish language, he said, was the Gothic of Scandinavia. It was "a lost language," never a written language. He thought half of the words still used in the remote Lowlands derived from it. 3. Here we must cite the nineteenth-century Scottish historian William F. Skene, who said there were essentially two sets of chronicles from which to attempt a reconstruction of early Scottish history. The first group consisted of five documents: the *Pictish Chronicle*, the *Synchronisms* of Flann Mainistrech,[249] the additions to Nennius's *Historia Britonnum*,[250] the *Albanic Duan*, and the *Annals* of Tigernach:[251]

> They are entirely consistent and in perfect harmony with each other. The same chronology runs throughout the whole, and they stand apart, and far above all other chronicles in authority, – first, from their superior antiquity; secondly, because they emerge from the native races themselves, whose early annals they profess to give; and thirdly, because they were compiled before any of those controversies, whether secular or ecclesiastic, arose, which … led to the falsification of records and to the perversion of history.[252]

The second group comprised documents written between the years 1165 and 1327. They present lists of Pictish and Dalriadic[253] kings and "may possibly contain, in their statements, a germ of historic truth." But they were compiled for the sole reason of combating English claims to the Scottish throne, and were therefore not to be trusted by the historian:

> Upon these chronicles, however, the early history of Scotland has been based by all the more recent historians of Scotland who have entered upon that portion of the history at all, from the ponderous *Caledonia* of George

Chalmers down to the latest history of Scotland. The only historian who has estimated correctly the value and superior claims of the earlier documents, and saw somewhat of their true bearing upon the early history, was John Pinkerton, but they were to a very limited extent available to him. He obtained a correct copy of the Pictish Chronicle, but the Synchronisms of Flann Mainistrech were unknown to him. Of the Irish additions to Nennius, he had an imperfect and incorrect extract, and their meaning was perverted by a bad translation. The Albanic Duan he possessed, but unfortunately he altered the order of the stanzas, and the position of the two kings Dungal and Alpin, and placed the stanza containing them immediately before that in which Kenneth mac Alpin appears, from an idea that one of the leading differences between it and the later chronicles arose from a mistake of the transcriber, – an idea which the Synchronisms of Flann would have corrected, if he had possessed them, and thus prevented him from missing the full bearing of the Duan. The great work of Doctor O'Connor,[254] containing the Annals of Tighernach, Innisfallen, Ulster, and the Four Masters, had not been published, and he only knew what the Annals of Tighernach contained through an inaccurate transcript of the Annals of Ulster[255] which usually repeat his statements, in the British Museum, and a translation published by Johnstone in his "Antiquitates Celto-Normanicae."[256] Still it is remarkable how near to the truth he came …[257]

"Near to the truth" was a generous assessment, since in many details Pinkerton was in error, led there in part, as Skene said, by his "unreasoning … prejudice against everything Celtic." This prejudice was in operation throughout the *Enquiry*; we find it even in his glance at "circles" such as Stonehenge which were, he claimed, neither Celtic nor Druidic but Gothic "Courts of Justice."[258] Nor was his handling of authorities immaculate, his most serious blunder being (again) that he placed trust in the forged *De Situ Britanniae* of Richard of Cirencester.[259] Yet his instincts were often solid; and it was in the pursuit, study, and differentiation of historical sources, mulling over language and chronology, teasing meaning out of fragmentary evidence, in short, the down-and-dirty hardscrabble labour of scholarship, that much of his contribution lies. He was the first Scottish historian to make use of the *Albanic Duan* and the *Annals of Ulster* and printed them (extracts of the latter) along with other documents in his appendixes.[260] He was the first to make extensive use of the lives of the Scottish saints, in which, he noted, "tho the miracles be fiction, the geography and history are always real."[261] His *Vitae Antiquae Sanctorum*, containing two such lives never before printed,[262] was a handy compendium for later historians.

A major source of error in the *Enquiry* was Pinkerton's wrong notion of the extent of Pictish territory. He thought that as the Romans left Britain the Picts took possession of the Lowland regions of Scotland, indeed beyond – of all the country down to the Humber, as he put it. The permanent home of the Picts was in fact[263] north of the Antonine Wall (the Roman wall between the firths of Forth and Clyde) and they evidently never made any full-scale settlement south of it. The misconception led him to the belief that the language of the Lowlanders was descended from the Pictish and reflected its "Gothic," i.e., Teutonic, character. Instead, Lowland language is descended mainly from that of the Northumbrian Angles. The character of Lowland speech, cited by him as one of the main supports for his conviction that the Picts were Goths, had no bearing on the Pictish question at all.

That question – just who were the Picts? – puzzled many scholars in the controversy the *Enquiry* stirred up. The nineteenth-century historian John Hill Burton said the controversy "leaves nothing but a melancholy record of wasted labour and defeated ambition."[264] But that is to misunderstand what historical inquiry is. Wasted labour and defeated ambition are two of the costs of driving that inquiry forward. It is out of them that understanding emerges. The *Dissertation on the Scythians or Goths* and *Enquiry* did much to stimulate scholarly interest in early British history, not just Scottish. Celts and their defenders, lathered with abuse throughout these works, were at first stunned into silence but soon struck back. In 1791 William Webb produced *An Analysis of the History and Antiquities of Ireland* and devoted sixty pages to attacking the *Dissertation*. Some of the abuse was lively, but he was no match for Pinkerton. The Welsh antiquary (and forger) Iono Morganwg was "riled ... beyond endurance" and amassed "a collection of furious snippets insulting" Pinkerton, but refrained from going into print until 1794 when he termed his enemy "*diabolus*" and "*vermicule*."[265] The Welsh would keep up the argument with Pinkerton, though the chief responders would be Scots. John Lanne Buchanan brought out his flimsy *Defence of the Scots Highlanders* in 1794. (They needed defending: Pinkerton called them "stupid, indolent, foolish, fawning, slavish.")[266] James Tytler flayed the *Enquiry* in his *Dissertation on the Origin and Antiquity of the Scottish Nation* (1795). He preferred Boethius to "Pinkerton the Great," and ridiculed the "scraps of old insignificant authors" Pinkerton and his "English and

Irish associates" were using to write Scottish history.[267] It must have seemed to many other patriots that the *Enquiry* into the history of their country was poor compensation for the orderly narratives of Boethius and his followers. To Robert Macfarlan, whose essay against Pinkerton was prefaced to his translation of George Buchanan's *Dialogue concerning the Rights of the Crown of Scotland* (1799), such "a ridiculous scrap" as the *Albanic Duan* could serve no purpose "but that of laughter and contempt." The *Pictish Chronicle* he judged to be a feast fit only "for the gross palates of Gothick book-worms."[268] Pinkerton thus touched off anti-Gothic as well as pro-Celtic sentiment. In 1803 the "Picto-Gothicists"[269] got a drubbing when the anonymous author of *A Vindication of the Celts* (later identified as a vicar, later Archdeacon, William Coxe, author of mostly travel literature),[270] challenged the central argument of the *Dissertation on the Scythians or Goths*, arguing that Scythians "could not have peopled even ancient Scythia, much less Thrace, Illyricum, Greece, and Asia Minor, 1500, and all Europe, except the North-western extremity of Gaul, 500 years before our aera. The whole fabric, therefore, of this system, is without foundation."[271] It was a hard stroke, but not a fatal one. (See chapter 6.)

The *Enquiry* continued to provoke argument well into the nineteenth century. Pinkerton's idea that the Picts were Goths was upheld in Malcolm Laing's antiquarian work, James Sibbald's *Chronicles of Scottish Poetry* (1802), David Irving's *Lives of the Scotish Poets* (1804), and John Jamieson's *Etymological Dictionary of the Scotish Language* (1808). It was attacked in Chalmers' *Caledonia* and Ritson's posthumous *Annals of the Caledonians* (1828). Not that Ritson was altogether displeased with the *Enquiry*. In *Scotish Song* (1794) he wrote of Pinkerton's book that no one "can refrain from lamenting that a discussion so curious and important, and in the course of which the enquirer has evinced uncommon industry and singular acuteness, should be degraded by groundless assertion, absurd prejudice, scurrilous language, and diabolical malignity."[272] From an enemy as fierce as Ritson, that amounted to praise. Apart from its occasional reappearance in opaque contemporary academic studies, where Pinkerton typically gets scant and scathing notice,[273] the Gothic/Celtic fracas now lives only[274] in the pages of Scott's novel *The Antiquary* (1816), where an argument breaks out between Jonathan Oldbuck and Sir Arthur Wardour while Oldbuck's mysterious young friend Lovel looks on:

> "Why, man, there was once a people called the Piks –"
> "More properly *Picts*," interrupted the Baronet.

"I say the *Pikar, Pihar, Piochtar, Piaghter,* or *Peughtar,*"[275] vociferated Old-
   buck; "they spoke a Gothic dialect –"
"Genuine Celtic," again asseverated the knight.
"Gothic! Gothic! I'll go to death upon it," counter-asseverated the squire.
"Why, gentlemen," said Lovel, "I conceive that it a dispute which may be
   easily settled by philologists, if there are any remains of the language."
"There is but one word,"[276] said the Baronet, "but, in spite of Mr. Oldbuck's
   pertinacity, it is decisive of the question."
"Yes, in my favour," said Oldbuck: "Mr. Lovel, you shall be the judge – I
   have the learned Pinkerton on my side."
"I, on mine, the indefatigable and erudite Chalmers."
"Gordon[277] comes into my opinion."
"Sir Robert Sibbald holds mine."
"Innes is with me," vociferated Oldbuck.
"Ritson has no doubt!" shouted the Baronet.[278]

It is tempting to make light of it, but to Pinkerton the fate of the *Enquiry*,
which sold poorly both in London and (understandably) in Scotland,
was no joke. Nor was much of the abuse directed at him. In November
1789, he defended himself against the harsh strictures of the *Analytical
Review* on the *Enquiry*. He drew up his answers in legal fashion, his training under Aytoun coming in handy.

   That to extract two or three warm pages from a thousand cool ones, in
   order to give an estimate of a book, is unjust; and yet it unfortunately
   so happens, that the worst pages in the Remonstrant's book are given as
   specimens. That great labour is apt to beget peevishness for the time. That
   if the Critic had gone through half the Remonstrant's toil, he might perhaps have been peevish too …

Had he been too severe on Highlanders?

   [He] has not a more earnest wish, than that the Highlanders were as much
   superior to the Lowlanders in the virtues of civilization as they are inferior. That his anger is that of patriot, who hates to see idleness and savage
   manners prevail so long among a people, and oppose industry, and the
   other virtues, which do honour to human nature. That [he] was angry to
   find, that late writers gave the whole history of Scotland to the Highlanders; as an Englishman would be justly angry to be told, that there was no
   history of England but that of Wales.

Was he too critical of other historians? He "is young, and of warm feelings."[279]

Peevishness brought on by toil, patriotism, and youth: so he explained himself. It would hardly suffice as an apology to the Celts. He would have to make other amends.

To add to his troubles, in 1789 his life was disturbed by financial and domestic difficulties. The nature of these disturbances is not clear, but they were sufficiently serious, when added to his vexation over the reception of the *Enquiry*, to make him think of leaving England and beginning life anew in Switzerland.[280] From this point onwards, the profit motive entered more fully into his activities as an author. In 1789 he moved from his leased house in Knightsbridge to Kentish Town,[281] bringing with him the woman who had borne him at least two children since 1784 but who now may no longer have pleased him.

3 The Great Work, 1790–1797

As 1790 dawned John Pinkerton was thirty-one years old. He had already published twelve books, three of which had gone into second editions. The *Essay on Medals*, *Ancient Scotish Poems*, and *Enquiry into the History of Scotland* had been notable contributions in knotty areas of knowledge, though the *Enquiry* was marred by racist views. Despite its manifest flaws, the *Dissertation on the Scythians or Goths* was also impressive. Having risked censure by forging ballads, challenging orthodoxies, and generally bulling his way onto the literary scene, Pinkerton now had enemies galore, more than he knew he had. He had some friends, not many, and a well-deserved reputation for cantankerousness and eccentricity. What his exact or even approximate total earnings were from book sales is not known, but it is apparent that he had worked hard for little financial return. And if he'd expected thanks for his scholarly labours from a grateful public in Scotland, he didn't get it, or ever would. The *Enquiry* had stirred up "violent disgust" there, as Lord Buchan plainly told him.[1] What he'd gained by 1790 was, simply, that he was no longer a tyro in literature but an established figure, known not just for pugnacity and effrontery but for his learning. He was sometimes a clumsy prose stylist, as many scholars were, yet he could write fluently, with force and clarity, and with "an inveterate and indelible tinge of mannerism" that was identifiably his own.[2] The disappointments he'd experienced, while they appear to have made him more circumspect and secretive, didn't stop his pen. In 1790 he published four more books. But the underlying purpose of his literary life for the next seven years was centred on an ambition outside the scope of those books. He had a larger view in mind, namely to produce a work of history that would lift his name out of the imbroglios of the 1780s and onto a higher level of literary achievement. He needed success, not infamy; he would try again to get it.

Three of his new books were published anonymously. One was on religion. Despite his occasional inflammatory sentences about the Bible and other sensitive topics, Pinkerton, as we have noted, was no atheist. At times he would let on that he was just a Protestant – his forebears were Presbyterian – with mildly sceptical views.[3] In truth, he was much given to brooding about religion and once even drafted a sermon against pride and fanaticism.[4] His pamphlet *The Spirit of all Religions* (1790),[5] also issued under the title *A New System of Religion*, was an attempt at theology. In it he dismissed all religions so far created by man as the products of pride and ignorance and in their place proposed what he thought, or said, was a new doctrine. There were, he wrote, "numerous … scales of spiritual existence, between the deity and man." He envisioned three main divisions above man in this spiritual scale: first, the Deity, the all-powerful, all-good "First Cause"; second, the Gods, created by the Deity and presiding over worlds of their own creation, who are "the spirits next to himself in power and dignity"; and third, the "genii" or "daemons." The Gods are "created pure intelligences, for good, and not for evil"; the genii are not all good and may even "partake of matter" to a degree imperceptible to man. It is probable that the genii "intermix in human affairs, and produce those singular events, which are by some imputed to fortune, and by others to a special providence." Evil springs, not from the Deity, but from these "inferior agents" who are "neither all good nor all powerful."[6] He argued also for the existence of a "moral instinct" or conscience in man and for the "probability" of the soul's immortality.

It is hard to say how seriously Pinkerton believed in these notions about gods and genii, though he would refer to "rational polytheism" in a later book[7] and, still later, had *The Spirit of all Religions* reissued in a new and expanded Latin edition – of which more in chapter 7 – and boasted that the system espoused in it represented "an epoch in the progress of the modern mind."[8] It was no such "epoch"; it was at bottom a version of the old idea of a great scale (or chain) of being. There was no shortage of intrusive demigods, spirits, angels, etc. between man and the godhead in systems of religion before Pinkerton hit on the idea. Soame Jenyns indeed had advanced just such a notion in his *Free Inquiry into the Nature and Origin of Evil* in 1757. To the biographer, the most intriguing element in Pinkerton's pamphlet was the low opinion it presented of the philosophical and moral stature of mankind. Man is "dust," "but a particle of littleness," "the shadow of nothing," ignorant, prideful, deceived by his senses and reason, narrow-minded, credulous, blasphemous, fanatical, insignificant, foolish to think that he can unravel the workings of the universe. He is twice labelled "a worm."[9] The pamphlet is an example of what is derisively called worm theology, that is, a system of belief that emphasizes the degraded, fallen condition of humanity by comparison with the glory of God. It is unclear where Pinkerton found this theology. Perhaps it came straight out of Job 25:6, where both man and the son of man (i.e., Jesus) are referred to as worms. He was young to have accepted or even entertained such a conclusion about the nature of human life. It is a gloomy thought to suppose that it was a reflection on his own life to date.

Why he kept his authorship of the tract a close secret is another question. Perhaps he feared ridicule. Or else some of the revelations in it were too intimate to let out. Towards the end he made statements that may shed more light on his inner life – what truly motivated him – than anything he wrote in his larger works or letters. In a chapter on man's "duties" we learn that "Our more immediate motive is the good opinion of others; and a regard to the order of society." And in a chapter on man's "happiness," these remarks: "Happiness consists in the active pursuit of some great object"; and "The improvement of the mind is the most important of human pursuits."[10] That Pinkerton hankered for "the good opinion of others" while writing books that clashed with the views of so many readers seems especially poignant.

The comment about "a regard to the order of society" is also noteworthy and, to a degree, unexpected. Pinkerton was stirred by what had happened in 1789 in Paris. His anonymous *A New Tale of a Tub* (1790)[11] was a satire attacking Tories and defending the French Revolution. The book recounted British history by means of a fable whose characters are "James Tory," "Will Whig," and a newcomer, "Jack Common-good." Jack's ideas are advanced ones; perhaps we hear in him the views of a yeoman's grandson. "Can that country be wise, or happy, where a soldier is a title of honour, and a mechanic a term of contempt?" Jack asks, adding "I know no real nobles except the mechanic and the farmer."[12] Jack visits "Frankland" which, "long opprest by bad management, had all of a sudden become one of the freest estates in the country." The "tenants" there

> … had seized upon the management themselves; and made a new code of
> rules, to determine all questions by … The public good was to be the only

law. Tithes were abolished; and the imprudent power and claims of the richer annihilated. They would allow of no nobility, but that of merit; and all men were to be equal as God made them.

Some heads had been broken upon the occasion. But what is the death of a few men, to the good of thousands ... Never had such a revolution been accomplished with as little bloodshed.[13]

Returning to his own estate, Jack eliminates rotten boroughs, institutes other parliamentary reforms, introduces laws against bribery, repeals some foolish laws, improves the coinage, and makes six other unspecified changes.

"Bliss was it in that dawn to be alive." Pinkerton felt the thrill too, and he remained sympathetic to the Revolution long after 1790. He was drawn to France, as so many Scots had been through their history. He would in time make an extended visit to Paris and write a book about it; and he spent his last years there. Yet his attraction to the ideas behind the Revolution was only partial. Like many other observers, he worried that democracy, if not "wisely adjusted," might slip into mob rule, which would be the worst form of tyranny. Republicanism was an experiment. America, he thought, was a federation of republics, not yet a full republic. "Perhaps it may be one republic, but it is not yet tried." (It would indeed be "tried"!) As to whether Britain should follow its neighbour, he was doubtful. "Let France try the experiment," he wrote, while "we look on. ... Let them try. Let us mend our own ways, and wait."[14] His interest in politics extended far beyond what he wrote in *A New Tale of a Tub*. It was lifelong. In 1817 he explained that he had tried to steer an honest course between political extremes. Looking back, he saw himself as "neither a jacobin nor a slave."[15] A "regard to the order of society" was never far from his mind.

A third book from 1790 was *The Medallic History of England to the Revolution. With Forty Plates.*[16] When Walpole learned, in 1788, that the *Medallic History* was underway he said he was sorry to learn that Pinkerton was "going to be the editor of *another's* work" rather than "composing himself."[17] The letter from Pinkerton that provoked the remark is lost, but in it he had clearly described himself as editing an already existing text, and in the preface to the book "the Publishers" take credit for the work. The book was intended as an update and completion of Thomas Snelling's *Thirty-three Plates of English Medals* (1776), the additional material being "more than a third" of the plates and the descriptions of the medals. Snelling, whose name does not appear on the title-page, had supplied no descriptions. Pinkerton's task was to locate some of the added medals (as the letter from Walpole shows him doing) and, though this is not stated anywhere, to write the descriptions. The preface noted that Charles Combe, the physician and medallist, had a "capital Collection of Prints and Drawings of English Medals" which was drawn on for "many" of the "articles," meaning images, in the book. He also provided "several explanations" in the descriptions. (Combe had helped Pinkerton in preparing the second edition of *An Essay on Medals*.) Did Pinkerton approach the publishers to suggest the book be done, or did they approach him to work on a book they had decided to reissue? That isn't clear. His name appears neither on the title-page nor, surprisingly, anywhere in the book. Perhaps he didn't wish to appear as an author in such a humble role. It is more probable that the publishers wanted to make use of his knowledge of medals without incurring the risk that his notoriety might injure sales or stir up controversy. The book was elaborate and expensive to produce. It was, the *Monthly Review* noted, "a very beautiful volume for a gentleman's library."[18]

The forty plates each contained, on average, about ten medals, with obverse and reverse exposed, the reverse only if embossed (the word medal here being defined narrowly, to exclude coins). Each medal was to be described. So there was a lot of writing to be done, and of a certain kind, succinct (normally), fact-choked, with little opportunity for opinion or embellishment. Each item was identified, as in "A small medal of Henry VIII," and marked as a medal proper, or as a jetton or counter. The obverse details were then noted, including the (generally) Latin legend, often abbreviated, with translation, plus detail on the exergue if there was one. The reverse of the medal was similarly treated. A comment could be provided on a number of themes: e.g., the meaning of the images, the source of the medal, historical or biographical background, correction of error, or the medal's quality. Pinkerton, on internal evidence undeniably the describer, did his dry piece of hack work competently, even, rarely, with humour. His knowledge of English and Scottish history, and of literature, was on display throughout.

In Pinkerton's three-volume edition of Barbour's *The Bruce*, his fourth publication from 1790, his name reappears. The title-page asserted this to be *"The First Genuine Edition"* of the poem, as it surely was.[19] It would not be replaced as the accepted text of the work until 1820. The source was a fifteenth-century manuscript in the Advocates' Library, transcribed for Pinkerton and at his cost. Pinkerton took some pains to get an accurate text, insisting that Lord Buchan examine the transcript, check it against the original, and correct any mistakes. Buchan made a formal attestation that he'd carried out "a very minute investigation & comparison,"[20] which was inserted in the edition. Still, error crept in. Pinkerton divided the poem into twenty books, prefacing a lengthy commentary and adding notes and a glossary. He knew he risked being "accused of nationality" by saying he preferred *The Bruce* to the work of Dante and Petrarch; but say it he did. (*Accused* of it: he was wary of being earmarked as a Scottish nationalist.) The reader will find in it "few of the graces of fine poetry, little of the Attic dress of the Muse," he wrote, "but here are life, and spirit, and ease, and plain sense, and pictures of real manners, and perpetual incident, and entertainment." He again admonished Scots to make their early poetry available in careful editions, noting that the manuscript containing *The Bruce* also had in it Blind Harry's *Wallace* which, though "a mere wild romance," might be published "for the sake of the language, and manners." He left this "to some gentlemen residing in Scotland, and curious in such matters."[21] If the latter part of Wyntoun's *Cronykil* were printed, he said, it would form, with *The Bruce*, "a chain of memoirs" for the history of Scotland down to "the commencement of our memoirs in Scotish prose."[22] He had ideas about how the gaps in prose might be filled, projecting "two large octavo volumes," but giving no hint that he would undertake such a work. His suggestion about Harry's *Wallace* was taken up at once by the Morisons, publishers in Perth, who issued it in an edition dated 1790. They acknowledged a debt to Pinkerton for "the arguments which they have prefixed to the Books" and printed his plan for dividing the poem into twelve books rather than eleven, the number in the manuscript.[23] Yet his subsequent response to the book in the *Critical Review* (see below) was far from laudatory.

The notes to *The Bruce* reveal that Pinkerton's attention had turned back to the history of Scotland. He had begun research on a new work in the field by February 1790;[24] by July, amassing materials ("I have not yet put pen to paper in it")[25] took up much of his time. He read at the British Museum, going through the Caligula MSS and other documents. As Isaac Disraeli recalled in 1838, he was one of the few regulars there:

> I passed two years in agreeable researches at the British Museum, which then (1790) was so rare a circumstance, that it had been difficult to have made up a jury of all the spirits of study which haunted the reading-room. I often sate between the Abbé de la Rue[26] and Pinkerton, between Norman antiquity and Scottish history. There we were, little attended to, musing

in silence and oblivion; for sometimes we had to wait a day or two till the volumes, so eagerly demanded, slowly appeared.[27]

Pinkerton got help from friends and correspondents, especially the Earl of Buchan. It was on Buchan's advice that he decided to end his history with James V, thereby avoiding the controversies surrounding Mary Queen of Scots. Had he included Mary, he would have become more of a mainstream historian to the Scots; Hume, Robertson, and Gilbert Stuart had all dealt with her. Omitting her no doubt affected sales of his book too, since owing to the pathos and mystery of her life she still drew a crowd. Pinkerton was to offer a more remote and drabber subject: the first seven Stuart kings of Scotland, perhaps beginning with a synopsis of earlier centuries. Dalrymple was approached for assistance – he had been collecting material on the statute law of Scotland between James I and VI – but he curtly demurred.[28] He explained to Buchan, who had interceded, that Pinkerton had connected one large work, i.e., the *Enquiry*, with his *Annals of Scotland* at one end, and now planned a "larger work" to connect with the other; and "Your Lordship will judge whether I can be very willing to go down to Posterity as an intermediate link of Mr Pinkerton's Historical chain." What was more, if Pinkerton were to say in his preface that he owed "even a hint to me," it would suggest that "I read with patience what he has said of a great & respectable Body of my Countrymen." So "I must not" help. Were Pinkerton to propose a collection of original papers, he would subscribe. But "*so far and no further.*"[29]

Pinkerton had undertaken no ordinary work. It was, he told his good friend Thorkelín in 1791, when some of the writing was completed, to be "rather philosophical and political, than antiquarian."[30] It wasn't to be a "dissertation" like his *Enquiry*[31] or a compilation like Dalrymple's *Annals*. He would not depend just on secondary materials. "I want original authorities," he said in 1792.[32] He would then mould this material into the dignified narration that was now, to him, history. In the *Enquiry* he claimed that history was "a science";[33] in the new work fine writing and philosophy were to be added. He would produce a book that would not only repair his injured reputation but give him an enduring fame. This was the "great object" he had in mind. He would become the historian of Scotland.

Becoming the historian of Scotland did not take up every hour of this full-time writer's day. In 1791 he began writing for the *Critical Review* where, at the beginning of the year, the proprietor, Archibald Hamilton, had begun a new series of his monthly periodical. That meant changes were afoot, and Pinkerton was shortly taken on as a contributor. The publisher Dilly said he was a "monthly" contributor.[34] How this arrangement was worked out – *what* it was exactly – is unknown, since the affairs of journals such as the *Monthly* and *Critical* reviews were conducted in secrecy, and no marked file of the latter or collection of papers relating to it has come to light. Nor did Hamilton leave a memoir. Accordingly, we know a great deal about the minor *English Review*[35] and hardly anything about the more important *Critical*, apart from what is disclosed by the journal itself and commentary in letters and other contemporary publications. Any papers dealing with the *Critical*'s operations in the 1790s must have been destroyed by the "tremendous fire" that swept through Falcon Court, Fleet Street, where its premises were located, in February 1803.[36] That Pinkerton started writing for the journal in 1791 is certain. There was no sign of his hand before then. Grímur Thorkelín's political news sent to Pinkerton from Denmark in August, 1791, was incorporated with very little change in the *Critical*'s "Review of Public Affairs,"[37] which appeared regularly at four-month intervals from May of that year. (On arriving in England in 1786, Thorkelín had introduced himself at once to Pinkerton and by 1790 was a close friend.)[38] In October Pinkerton thanked Thorkelín for the political "information," saying "I shall make use of it in print," and asking for Danish literary news. "I told you my design of communicating an account of foreign literature to one of our journals," he added.[39] He was referring to the "Retrospect of Foreign Literature," which appeared first in the *Critical* in September 1791, and continued again at four-month intervals. It is likely that Pinkerton compiled, if not all, then much or most of the contents of these two series from 1791 until his first break with the journal, which occurred about four years after he began writing for it.[40] The articles are mainly dry collections of facts – another instance of his hack work – yet the whiggish (and other) views embodied in them correspond with Pinkerton's in his later books, and the style is often recognizably his as well. However, it may have been that in the conducting of a large magazine, and over time, others had a share in writing them.[41] The clear change in the overall political outlook of the journal in the new series from "moderate conservative opinion" to reformist views, noticed by Derek Roper,[42] can partly, if not wholly, be attributed to Pinkerton's influence.

Of more significance to the biographer are his book reviews. He could now give his opinions of books and writers with the protection of anonymity. Pinkerton was not slow to take advantage of his hiding place, and it is no hard task to identify many of his contributions, given what we know of his character, his friendships, his enmities, and his literary style. His first appearance in the journal was in the April 1791 issue where he reviewed three books: Thorkelín's *Sketch of the Character of his Royal Highness the Prince of Denmark* (1791), Dalrymple's translation of the third-century *Address* by Tertullian to Scapula (1790), and the Morison edition of Blind Harry's *Wallace*. He praised the first one, quite extravagantly and undeservedly.[43] The second book he slighted. It was a clear case of hitting back at someone who'd snubbed him, but it wasn't just that. Pinkerton continued to acknowledge the merit of the *Annals of Scotland* long after the snub.[44] In a 1786 book Dalrymple had taken issue with Gibbon's treatment of Christianity in the fifteenth and sixteenth chapters of the *Decline and Fall of the Roman Empire*. That had rubbed Pinkerton the wrong way. In the preface to the *Address* Dalrymple returned to Gibbon, saying that in drawing up his notes he had found "strange inaccuracies" in the *Decline and Fall* and that in the first volume of the work, independently of the two famous chapters, there was a "wide field for literary and historical criticism." The *Critical* reviewer responded:

> In the historical and juridical antiquities of his own country, we are always happy to meet with lord Hailes: and we can hardly conceive what motive induces him to abandon a province in which he is eminent, for one in which he makes but an indifferent figure. Perhaps he thinks it especially incumbent on him, as a layman, to undertake the defence of Christianity against that legion of lay infidels which now assails it. But we must beg leave to remind his lordship that the defenders of Christianity are already many, and strong; that the fugitive infantry which he arranges will do no execution on the foe; and that the worst kind of enmity is a weak defence.[45]

The review of *Wallace* was mostly hostile. In 1789 Pinkerton had told Lord Buchan that he had "dropt all thoughts of Wallace,"[46] and that he hoped the Morisons of Perth could be persuaded to publish it from the manuscript in the Advocates' Library. Buchan then had the poem transcribed and edited. He ignored two of Pinkerton's suggestions. He had found a portrait of Wallace and decided to insert it. "It is surely a mistake," Pinkerton said. "Even in France, paintings of monarchs can hardly be got of the fourteenth century."[47] He'd also warned Buchan not to retain textual minutiae.[48] In his review Pinkerton reprimanded the editor on a number of fronts, including the use of the portrait and the retention of the old orthography. But he complimented the Morisons on "their attention to the ancient literature of their country, which has been too much neglected."[49]

The personal element in his reviews soon became noticeable. One of them from 1792 illustrates how sensitive he could be over any slight to his reputation. By then he had been honoured by three learned societies. He was one of the elected "correspondent members" of the Literary and Antiquarian Society of Perth,[50] and an honorary member of the Royal Society of Icelandic Literature (Copenhagen) and the Royal Society of Sciences (Trondheim).[51] Late in 1791 Lord Buchan, who had founded the Society of Antiquaries of Scotland in 1780, asked him if he wished to join that body; if so, he would propose him. Pinkerton was flattered but he warned Buchan not to propose his name "if not certain of success."[52] For whatever reason, Buchan did not submit his name. Perhaps he had been forewarned that the application would be fruitless. But he had "been forced by the hand of power" to resign from the Society in 1790[53] and it may be that he was in no position in 1792 to recommend anyone for membership. The thin-skinned Pinkerton, who must have watched with disbelief as such honours and a job offer at the British Museum were handed out to the lightweight Thorkelín,[54] was undoubtedly insulted. In August he set upon the first volume of the Society's *Transactions* (1792) in a review extending over two numbers and tore it to shreds. He mocked the pretended "superior dignity" of the Society, condemned it for not paying proper regard to its own manuscript collections, and ridiculed the Scotticisms and ignorance displayed in the essays by its members. One of them, Alexander Geddes, in a "Dissertation on the Scotto-Saxon dialect," knew not "the first principles of antiquarian or of historical reasoning":

> When he asserts … that the names of the Pictish kings seems [sic] to be Celtic, he must have been ignorant that the author of the late Enquiry into the History of Scotland has demonstrated them to be Gothic, has explained the meaning of many of them in that language, and has even produced some of the same *identic* names from Scandinavian monuments.

The "more learned members" should contribute to future *Transactions*, Pinkerton concluded, lest this work be interpreted by foreigners as "a test of national erudition."[55]

That such a lengthy piece on an inconsequential book should have been accepted by the *Critical*'s proprietor was surely a sign of Pinkerton's status among the contributors. So was the fact that he was allowed to puff his own work. This puffing became more conspicuous in subsequent reviews, one of which, in 1792, was almost autobiographical. This was the review of Webb's *Analysis of the History and Antiquities of Ireland*, already alluded to as one of the attempted rebuttals of the *Dissertation on the Scythians or Goths*. It "is likely," said the reviewer, that "Mr. Pinkerton" would now feel that certain pages in his *Dissertation* were "ridiculously warm," and it is "not to be supposed" that because he displayed "some improper warmth of youth in a polemical work" he would therefore "stain the page of history with passion, to the perdition of his reputation and labour." No indeed – a hint is thrown out that history with a capital H is coming, sober and elevated. And as for Webb's censuring the author's "hasty expressions" about the Celts,

> … he forgets that the attack on the Celts was but a retort courteous to Mr. Macpherson, who had extoled them, at the expence of all other nations, and had expressed his particular contempt for our Gothic ancestors. When we consider the insolence, the anger, and the perversions of the authors whom the Dissertator on the Scythians has attacked, we the less wonder at his occasional warmth, though we must blame it even in a youthful writer.[56]

Having summarized the case between Pinkerton and Webb, Pinkerton left "the verdict to the public."

Joseph Ritson's inoffensive *Pieces of Ancient Popular Poetry* appeared in 1791 and the *Critical* reviewed it savagely in January 1792.[57] This was almost certainly another of Pinkerton's reviews. For once Ritson was staggered. He wrote to the editor and denounced the journal's "partiality, insolence, and malice," terming *Critical* reviewers "a pack of snarling curs." The letter was not noticed in the *Critical*'s "Answers to Correspondents" – Pinkerton filed it among his papers,[58] probably having had it passed to him by Hamilton. No owner of a magazine wants to have his writers called a pack of snarling curs; perhaps Hamilton wanted to placate this enemy. If so, that may explain the treatment given Ritson's next book, his edition of *Ancient Songs and Ballads* (1792), later in the year in a review, unquestionably by Pinkerton, that ruptured one of his early friendships. *Ancient Songs* contained an essay entitled "Observations on the Ancient English Minstrels" in which Percy's observations on the same subject in the *Reliques* were "criticised, and perhaps confuted," to used Ritson's description of what he'd done.[59] Percy had maintained

that the minstrels were musicians who sang verses of their own composition; Ritson said the term meant, simply, musician without the added
connotation of poet. It was a moot scholarly point, but it was a sore one.
Percy's feelings towards Ritson were bitter. Ritson had even questioned
the existence of the Folio Manuscript, which Percy owned. It was one
of the sources of ballads in the *Reliques*. In July 1792, the same month
Ancient Songs appeared, Percy arrived in London from his see at Dromore, Ireland, and met with Pinkerton, who had drawn up a paper on
his behalf in opposition to Ritson.[60] That document has not been found,
but Pinkerton suggested, either orally or in the paper he'd written, that
the Folio Manuscript be put on public display to prove its existence.
Percy refused, since "this was the very end to which Mr. Ritson has
been driving."[61] He may have known of Pinkerton's connection with the
Critical[62] and could have suggested using the journal to expose Ritson's
error about the Manuscript. The next step Pinkerton took was decisive.
In November he reviewed *Ancient Songs* in the *Critical* and not only
described the volume as "curious, entertaining, and instructive" but
sided with Ritson on the minstrel question. The French and English
minstrels, he wrote, "appear to have been solely musicians." Percy
"has confounded orders totally distinct; has used the word minstrel
in a sense completely new and improper."[63] It was not until two years
later that Percy learned what the review contained. He sent Pinkerton
a list of objections to the points raised in it and proposed he insert it in
the *Critical*. Pinkerton replied that he was "thoroughly convinced, that
Minstrel only implied musician, and *was never used for a bard, maker,
or poet*."[64] He added that Percy should correct his essay on the lines
Ritson had indicated. The friendship was over.[65]

In 1792 Pinkerton made his last appearance as editor of early
Scottish poetry, and once again his and Ritson's paths crossed. *Scotish Poems, reprinted from Scarce Editions* was issued in December,[66]
in three volumes. The first volume had three texts, starting with *The
thrie Tales of the thrie Priestis of Peblis*, an anonymous work from the
fifteenth century reprinted from the 1603 Charteris quarto belonging
to the scholar Richard Gough. The tales were "more moral than face-
tious," Pinkerton wrote, their chief merit lying "in a *naif* delineation
of ancient manners."[67] He thought they were likely a satire on James
III's government. Following *The thrie Tales* came Gawin Douglas's *The
Palice of Honour* and Sir David Lyndsay's *The Historie of Squire Meldrum*,
both reprints from Charteris quartos (1579, 1594). (Pinkerton had been
forestalled on *The Palice of Honour* by the Morisons of Perth.)[68] The

entire second volume was taken up by Lyndsay's *Ane Pleasant Satyre of the Thrie Estatis*, printed from a transcript of the version in the Bannatyne Manuscript. Allan Ramsay had intended to publish the work, but didn't. Dalrymple wouldn't print it, recoiling from its indelicacies. Pinkerton was full of praise for it, saying it contributed more to the Reformation in Scotland than all the sermons of John Knox, had much native humour and good poetry, and as the first specimen of Scottish drama "claims a distinguished notice in Scotish literature." He thought the date of the play was 1552 – too late. He dismissed any comparison of *Ane Pleasant Satyre* with English mystery and morality plays, saying it belonged to "a mixt class," mingling the ideal personifications, virtues, and vices of the moralities with features of "the regular drama."[69] In the "Preliminaries" to his work Pinkerton added several details to his life of Lyndsay in his 1785 "List," including a letter by the poet he found in the British Museum.

In the third volume were: the sixteenth-century comedy *Philotus*; the metrical romance *Gologros and Gawaine* (called by Pinkerton *Gawan and Gologros*), copied for him from the recently discovered Chepman and Myllar imprints of 1508; a number of "ballads," as he chose to call them, from the same source;[70] and three previously unpublished items, Sir Richard Holland's "The Howlat," a long fifteenth-century political allegory from the Bannatyne Manuscript, Henryson's "The Bloody Serk," from the same source, and "The Awntyre of Gawaine" (he called it "Sir Gawan and Sir Galaron of Galloway"), another metrical romance. Pinkerton said the last item, a choice find, was "copied many years ago, by a learned friend, from a MS. belonging to Mr. Baynes of Gray's-inn, who was a noted collector of romances of chivalry."[71] Therein lies a tale that requires a brief recounting.

On the publication of *Scotish Poems* Ritson rushed into print to explain where Pinkerton got the "Awntyre of Gawaine" – from him! Not directly, but "surreptitiously." Owing to the "goodness" of the late John Baynes, Ritson had "possession" of the manuscript. Pinkerton, through a "learned friend," had applied to him for permission to publish it, but he refused, "having myself an intention to publish it." Ritson was "well convinced" that nothing "could have given Mr. Baynes more uneasiness than an idea that either his MS would be printed, or even his name mentioned, by Mr. Pinkerton." The "learned friend," who had obtained a copy of the manuscript before Ritson did – there had been a dispute over ownership by one of Baynes' relations – accepted what he'd been told and "assured me it should not be printed." But

Pinkerton, "to whom my refusal was communicated," went ahead and printed it.[72] Thus Ritson. On Ritson's death in 1803 the "learned friend" in the story, who was Francis Douce, bought the manuscript and wrote in it his account of Pinkerton's conduct. Before the manuscript fell into Ritson's hands, Pinkerton had borrowed Douce's transcript and promised that it wouldn't be printed "without leave from the future proprietors." Douce told Ritson, once the manuscript became his property, of the transcript and promise, and Ritson asked him to tell Pinkerton that he would never consent to have the poem published by anyone but himself. Douce did that, and Pinkerton put it in *Scotish Poems* "in direct violation of his promise."[73] Thus Douce. The two accounts are not identical, but they are the same in essence. (In another document Ritson claimed that Douce's copy was "surreptitiously obtained" as well.)[74] It would be good to have Pinkerton's views of the transaction but he evidently left none. We might ask how Ritson could justify saying he was "well convinced" nothing would give Baynes, if he were alive, more uneasiness than to have his name even mentioned by Pinkerton – he might have spoken for himself rather than put that cheap shot into a dead man's mouth. But that slur wasn't answered either. Pinkerton stands convicted.

Aside from that blot, his performance as editor in *Scotish Poems* was creditable. The book added considerably to the rich store of early Scottish literature on hand for readers and critics to enjoy and ponder, and the "Preliminaries" expanded his earlier commentary in the "List of all the Scotish Poets" of 1785 and its supplement in the *Bruce*. Many of his findings and speculations through this extended commentary have been challenged by later investigators, but others have stood up. Some editors have acknowledged his work, others have not. While "much of his work is quite correct," W.W. Skeat remarked, his edition of the *Bruce* should be avoided.[75] Similar comment is not hard to find. Slighting precursors happens all the time in scholarship. No big deal. Pinkerton's texts in *Scotish Poems*, while generally readable and solid, have been replaced by better ones, just as the better ones were knocked aside by still better ones. As the episode over the "Awntyre of Gawaine" illustrates, he was capable of taking risks to get ahead of competitors and push texts into print. His version of Lyndsay's *Ane Pleasant Satyre* shows him scrambling too. He ought to have waited until he could find the 1602 Charteris edition of the play, the superior authority. He'd borrowed a copy from the Edinburgh custom house clerk and collector George Paton in 1787, but Paton demanded its return before he could transcribe it. Pinkerton didn't get his hands on another copy until the second volume of *Scotish Poems* had passed through the press. His Bannatyne text of *Ane Pleasant Satyre* is a confused concoction.[76] Only one of his versions, that of *The thrie Tales of the thrie Priestis of Peblis,* has been found to have independent textual value. He had a perfect copy of the 1603 Charteris quarto. T.D. Robb, in preparing the 1920 Scottish Text Society edition, had only the Douce copy, which was missing 194 lines. Robb supplied them from *Scotish Poems.* He found that Pinkerton, "except that he took liberties with the lettering – printing u for v, v for u, and y for z – kept very close to his authority."[77]

Pinkerton's contribution to the late eighteenth-century revival of interest in early Scottish poetry is many-sided, but perhaps the main elements to be emphasized are, first, the sheer bulk of the makars' poetry he made available and defended in his editions, and, second, the 1785 "List of all the Scotish Poets," including the publishing plan outlined in it (and subsequently added to). He pointed out what he thought needed doing, and in the years ahead the need was partially answered. His name is to be directly linked with editions of the makars issued by the Morisons in cheap, pocket-sized editions between 1786 and 1790, including the selected poems of Dunbar (1788). He encouraged their enterprise, and urged Lord Buchan to give them material and advice; they responded. Other texts were soon in preparation. The decade and a half after *Scotish Poems* appeared saw the publication of Wyntoun's *Cronykil* (1795), which followed guidelines on content that Pinkerton had suggested – and to which we will shortly return – John Graham Dalyell's *Scottish Poems, of the Sixteenth Century* (1801), Sibbald's *Chronicles of Scottish Poetry*, Walter Scott's *Sir Tristrem* (1804), and Chalmers' *Poetical Works of Sir David Lyndsay* (1806). He was not, of course, the father of all good things; and he has to share credit with various scholars and editors, including Ritson, for what happened. He was not averse to seeing his own work replaced. When John Jamieson brought out a new *Bruce* in 1820, he wrote that Pinkerton admitted to him that his text of 1790 had been "in many respects inaccurate." With "laudable candour," Jamieson said, Pinkerton had "urged [him] to undertake a new edition."[78]

To return to 1792 and an episode in Pinkerton's private life. The woman he had been living with in Knightsbridge had been introduced to those who visited him as his wife.[79] She was part of his "family" when he moved to Kentish Town.[80] Thorkelín found that his bookish friend had a wife and two daughters and seemed happy with his domestic situation. The Icelander grew fond of the two children and after his return to Copenhagen in 1791 mentioned them and "Mrs. Pinkerton" in his letters. Thus in August 1791 he sent his "best compliments to Mrs. Pinkerton and the two young ladies."[81] Pinkerton answered in October, assuring Thorkelín of their tender feelings towards him: "Mrs. Pinkerton desires her best compliments, and my two little girls offer their love to you. Sophia encloses a kiss for her distant sweetheart."[82] In November he told Thorkelín that "Mrs. Pinkerton has made me a present of a fine little boy."[83] Thorkelín wrote in his next letter of "the inward joy which I sincerely feel on account of the increase of your family, and the prosperous state which your family enjoys, and which, I pray Heaven, you may long share with the best partner, and those blooming roseate pledges of conjugal affection which surround you."[84] These scraps of evidence suggest that Pinkerton at this time, though evidently unmarried,[85] enjoyed at least periodically a tranquil home life. He told Douce in 1823 that he was "of all men the most domestic,"[86] and although this somewhat strains belief it is likely he disliked family upheavals as much as any man. One now came his way, however. (More lie ahead, as we will see.) In 1792 he left his partner, children, and home in Kentish Town. Only a few signs of this crisis are in his surviving correspondence. He told Thorkelín in May that his health "has been rather bad" and that he'd been "occupied with sending my daughter to France for her education, and with other matters."[87] He did not specify the "other matters" and went on to discuss literary news in his usual way. The letter made no mention of "Mrs. Pinkerton." In September, in another letter to Thorkelín, he mentioned his "eldest daughter" and added, "I have found Kentish Town too damp a situation, and have left my house there."[88] Thorkelín should write to him in care of publisher George Nicol. Pinkerton now left London and stayed away, apparently, for four months. He spent part of this time in Hampshire, courting another woman. (See below.) In January 1793, he was back in London "for a month or two," staying at an address on Tottenham Court Road. He wrote to Lord Buchan and said he'd been "so exceedingly occupied with domestic and literary affairs for these some months past, that I have almost omitted all epistolary correspondence."[89] He was in good spirits. The crisis had passed.

What happened to the three children and the woman who lived with him in Kentish Town? David Macpherson had an answer. In *Ancient Scotish Poems* Pinkerton wrote that where ribald poetry "has led one to debauchery, it has induced twenty to marriage."[90] Macpherson glossed:

It had not then had that effect on Mr. P. who was living in concubinage with a woman, whom after passing for his wife for about 14 years he repeatedly turned out with an infant of three or four months old to pass the night in the fields behind my house, and at last dismissed with her three children to make way for a marriage with a woman who kept a school in Kentish-town and taught those very children.[91]

Macpherson lived in Kentish Town with his wife and child and met Pinkerton there.[92] Much of his statement is corroborated by other evidence. In a letter to the lawyer and editor Macvey Napier in 1818 Pinkerton said he did "dismiss" a woman he'd lived with, and it was probably this woman of Kentish Town he then referred to.

A former woman I left on account of her bad conduct and drunkenness which led her to insult my friends particularly Mrs. Kemble (mother of Mrs. Siddons) by whose advice I became awake and shook off the disgraceful yoke. I have nothing on my conscience having always been the dupe and never the deceiver.[93]

Macpherson did not write his comment until 1799 or later, by which time he'd developed, as will be explained, a savage hatred of Pinkerton. It is possible that this latterly acquired enmity was so deep that it led him to distort what he saw to give vent to his malice or somehow to injure the object of it. Yet he could have looked out upon some hard scene or scenes behind his house; the breakup of a marriage or close conjugal partnership, as those who have lived through it will testify, can be ugly, and the drunkenness[94] of one or both participants can make it uglier. To an observer of sensibility the process may be appalling. And so Macpherson might have misinterpreted the scenes he witnessed. On the other hand, maybe he described what he saw accurately. If Pinkerton "turned out" the woman and infant – "repeatedly"! – in the fields to spend the night, he was guilty of abhorrent cruelty. Comments on his children in his letters, infrequent as they are, appear to display normal parental affection. A more revealing statement in a work of 1806, while it might not bear directly upon his own situation, sheds light on his thinking about family life. It "is of far more importance," he wrote, "to preserve the lives of adults, than to multiply marriages and children; and innumerable good parents have been brought to the grave, by incessant efforts and anxious cares to provide for their families." He went on: "Those who write against the utility of foundling hospitals are … utter strangers to the sensations which the birth of a child gives to a poor parent, and the tranquillity restored by its death."[95]

This first Pinkerton family now leaves the scene. If he contributed anything towards their livelihood in the years ahead, he kept the matter a close secret. No doubt he tried to keep the breakup a secret too, but the gossip seems to have reached Ritson, likely via Macpherson. Ritson was soon writing to a friend about the "criminal parts" of his adversary's "literary and moral character."[96]

In 1793 a new "Mrs. Pinkerton" enters the picture. Henrietta Maria Burgess was the eldest daughter of a grocer, seemingly a successful one, in Odiham, Hampshire, "a man of excellent understanding and sincere piety." Her mother's "connexions were highly respectable."[97] Henrietta was an accomplished and intelligent woman, could play the piano, and had "some money."[98] Pinkerton went to Odiham late in 1792 and arranged the marriage. The arrangement included payment to him of at least £300.[99] On 14 August 1793[100] the ceremony was performed by Thomas Burgess, the youngest of Henrietta's three brothers, a priest in the Church of England, an author, and a decent, erudite man. Henrietta's two other brothers were William, the eldest, who had inherited property from his maternal grandmother, and John, who had a business in London and from it "acquired a considerable fortune."[101] Henrietta, in short, came from a rising, respectable, middle-class family. The couple returned to London and settled in a cottage near the Flask Tavern, in Flask Walk, Hampstead, where Pinkerton had been living on his own for some months. Hampstead in 1793 was a village outside London and a fashionable resort for artists and actors. To get to London one took the Hampstead stage,[102] but the distance was not too great to cover on foot. Pinkerton could walk to the British Museum in just under an hour. His cottage was small, carpeted, with portraits on the parlour wall of Horace Walpole, George Buchanan, and (after 1797) himself.[103] He had a library, a servant, and an expensive piano at which Henrietta would sometimes entertain him with "an exquisite performance of Scotish music."[104] One feature of his home life was unpleasant. His wife brought her sister Rhoda to live with them – exactly when is not clear. He did not take to her, nor she, it appears, to him.[105] But for this – quite a but – his life was happy. His "finances" had risen, his "life and spirits," he said in April 1794, "have never in my life been so good as within these two years."[106] One element that made it so, apart from his marriage to a "quiet, good wife,"[107] was the testimonial he received from the historian of the Roman Empire.

As already noted, in 1788 Pinkerton, under the name Philistor, had written a series of articles in the *Gentleman's Magazine* called "Letters *to the* People of Great Britain *on the Cultivation of their* National History." There he pointed to the large number of works by early British historians that remained in manuscripts and corrupt editions and that hardly anyone had shown any interest in publishing for nearly a century. What was needed, he said, was support from patrons, since to publish "any ancient English historian, with illustrations, would not at present pay for the printing." The fact was that "Nobody reads such books." The study was "too masculine for our trifling times."[108] He went on to contrast the neglect of the subject in England with the interest shown to such works in other European countries; to point to specific desiderata in the field; and to suggest that a society of national history be created and allotted a revenue by the government to enable it to "publish original writers, in an elegant manner, worthy of a great and wealthy nation."[109] He reiterated this theme in the *Critical Review*. "The original monuments of our history," he wrote in 1791, "are much neglected."[110] In 1792 he said it was a "disgrace" that the ancient Brehon laws of Ireland hadn't been published, though the British Museum had a manuscript of them "of considerable antiquity," and that the *Annals* of Ulster, Inisfallen, and Tighernac remained in manuscript.[111] Suddenly Pinkerton's prodding seemed to bear fruit. He was given a chance to become the British Muratori.

Unknown to Pinkerton, Gibbon had also conceived a scheme like that proposed in the Philistor papers. While writing the *Decline and Fall* he had been struck by the attention paid by the French and Italians to their early historians, and had suggested to his countrymen that a work like Martin Bouquet's *Recueil des historiens des Gaules et de la France* (1738–52) "might provoke our emulation."[112] Privately he had misgivings about finding an editor capable of conducting such a work. Pinkerton had come to his notice after the publication of *Letters of Literature*. Gibbon followed his career with some interest, read the *Dissertation on the Scythians or Goths*, was pleased with the *Enquiry*, and became a subscriber for *Vitae Antiquae Sanctorum*.[113] Pinkerton's learning impressed him. On his return to England from the continent in 1793, Gibbon suggested to George Nicol that now was the time to publish a collection of the early English historians and that Pinkerton was a suitable person to be its editor. A meeting was arranged for mid-July at Nicol's bookshop, but Gibbon was, ominously, unwell and had to call it off. The two now exchanged letters. Pinkerton suggested that Gibbon's name should appear as superintendent of the collection; his name alone "would insure success to the work."[114] He seemed to think of himself as doing the labour while Gibbon got the credit. In his reply Gibbon praised Pinkerton's "independent spirit," "love of historic truth," and "Herculean industry." "The best judges have acknowledged your merit," he said, "and your rising fame will gradually extinguish the early prejudices and personal animosities which you have been, perhaps, too careless of provoking." ("Too true. – *Peccavi!*" Pinkerton commented to Lord Buchan later.)[115] Gibbon would not edit the work: that was Pinkerton's province. Nor was he willing to follow its progress by "regular inspection or frequent correspondence." But he would use all his influence to promote the reputation and sale of the work, and would write the prospectus once Pinkerton supplied the materials.[116]

By October, Pinkerton had written an "Address to the Eminent, the Learned, and the lovers of the early litterature and history of England, concerning an intended publication, to be intituled Rerum Anglicarum Scriptores," which he sent to Gibbon to be made into "an animated and beautiful creation."[117] In late November the two met for the first time. They subsequently conferred and decided that Pinkerton was to be sole editor, Gibbon was to write a preface to the whole work and subsidiary ones for each volume as well as the prospectus, and Thomas Cadell rather than Nicol was to be the publisher. Pinkerton began to tell his friends about the plan and solicit subscriptions:

> Mr. Gibbon, I need hardly tell you, is now in England; and has projected and undertaken a scheme worthy of his talents and industry, namely a splendid republication of all the ancient English Historians, down to the year 1500, in one chronological body, with prefaces, notes, indexes, maps, and other illustrations. But as he goes to Lausanne in the spring, he only issues a prospectus in January, in which he delegates the execution of his plan to me; and has been pleased to recommend your humble servant, as the only person capable of answering his ideas as a proper editor, and has expressed himself of me in terms which modesty forbids my mentioning. I have had repeated conferences with him upon the subject; and hope his name will ensure the success of a plan, in which if I fail it will not be for want of love of the subject, or of industry.
>
> It is supposed the publication will amount to twelve volumes folio, at £2. 12. 6 a large volume, on the plan of Bouquets French historians. Thus the common paper will be 30 Gs. the large 50 Gs. Subscribers for the large

are to form a committee for managing the funds, as special patrons of the work.[118]

This was in December 1793 to the influential Thomas Astle, who assured Pinkerton of his assistance.[119] At dinner at Dilly's a day later, with Boswell (whose snooty comment was that Pinkerton "behaved himself very well and moreover went away early"),[120] the poet Samuel Rogers, and others, he met the lawyer Charles Butler who soon after promised his cooperation in the "great work."[121] Sir Joseph Banks was enthusiastic; Douce must have approved.[122]

By December Gibbon was becoming more animated about the project. On his return to Lausanne he planned to read the old British historians in chronological order to become conversant with the subject – it was to be "the last and most favourite occupation of his life."[123] The prospectus (or address)[124] he wrote reflects an eagerness above that of a mere patron. The limits of the work were now settled: it would include the writings of English historians from 500 to 1500. The Welsh, Scottish, and Irish would be excluded, except for certain documents relating to English history. He had long contemplated such a design, Gibbon wrote, but was prevented from executing it by the lack of an editor, since the "age of Herculean diligence, which could devour and digest whole libraries, is passed away." But "the man is at length found." All the "fiery particles" of Pinkerton's nature, Gibbon said, "have been discharged." His "juvenile sallies" against Virgil, his other improprieties, all attributable to his youthful "ignorance of the world," his love of paradox, and "the warmth of his temper," have been thrust aside. There remains "a pure and solid substance, endowed with many active and useful energies." His mind is "replete with a variety of knowledge," and he is blessed with "a spirit of criticism, acute, discerning, and suspicious." As soon as he was told by Nicol "of my wishes and my choice" (of him as editor) he responded "with the generous ardour of a volunteer, conscious of his strength, desirous of exercise, and careless of reward." If it is objected that the work is beyond the powers of a single man, "I must answer, that Mr. Pinkerton seems one of the children of those heroes, whose race is almost extinct; that hard assiduous study is the sole amusement of his independent leisure; that his warm inclination will be quickened by the sense of a duty resting solely on himself; and that he is now in the vigour of age and health."[125] Pinkerton, he added, would always be willing to receive "the advice of judicious counsellors."

The prospectus was to be published on 20 January 1794. But Gibbon, suffering from "a dropsy, with which, and a rupture, he had been long troubled,"[126] died on January 16, and his plan for the publication of the English medieval historians died with him. G.M. Young said the blow was "the cruelest which fate has ever dealt to English scholarship."[127] It was certainly a blow to Pinkerton who, at the height of his powers, contented at home, and full of ambition, might well have pulled it off. He considered getting a similar collection of Scottish documents together, perhaps to prove that he was capable of the larger task, but it didn't get beyond the planning stage.[128] Gibbon's praise, which was made public in 1796,[129] was to be a solace to him in the many disheartening years lying ahead. No doubt the praise was exaggerated. Yet it stands in sharp contrast to the sneers of Ritson, David Macpherson, Chalmers, and others. It could well be that Gibbon, who was able to view Pinkerton through eyes undimmed by moral righteousness or professional jealousy, made a more accurate assessment.

As this drama and the marital one were being played out, Pinkerton was still preoccupied with his new history. In 1790 Lord Buchan had suggested to him that to help sell the book he should include portraits of Scottish noblemen. Pinkerton didn't warm to that idea but he fancied there was a need for a work devoted solely to Scottish portraits, or at least a chance to make money from one, owing to the current "passion for portraits"[130] among the public. He tried without success to get George Nicol interested in such a book. In 1794 he met the young bookseller Isaac Herbert, who liked the idea, and the prospectus for *Iconographia Scotica or Portraits of Illustrious Persons of Scotland* appeared in May. The work, supported by subscription, was published in parts and numbers, the first of which came out in December. It was not completed until 1797. Pinkerton's task, as in the *Medallic History*, was to collect items for engraving and add what he termed in a letter "short and heraldic" notes.[131] The book was poorly executed. It is tempting to say it was a bad idea to start with, since Pinkerton was in Hampstead and a large number of the portraits he needed were in Scotland where he had few reliable supporters and friends. Once again, distance from his sources created difficulty. Pinkerton did have contacts in Aberdeen, and the drawings he got from there were of good quality. (One result being that Aberdeen Scots were over-represented in the book.) He erred in depending for many items on Buchan, who was eager to help but whose submissions tended to be careless and incomplete, usually head and shoulder drawings with the head set eerily above faintly sketched shoulders and arms. Buchan aimed at catching "the soul" of his subject, perhaps leaving the periwig unfinished. This was not a contemptible objective – given how hard it was to get true portraits of early Scottish figures, the "soul" might be all that was available to be caught. Yet he was "too erratic, fanciful, and credulous to be a safe guide."[132] To be fair to Buchan, his view was that Pinkerton, not he, was the bungler.[133] Herbert's engravings[134] were shoddy – Walpole termed some of them "scrapings."[135] Judged by portraits in the age of photography, they all seem scrapings. Yet when we consider the tortured process through which Pinkerton's images passed – sometimes copied, prior to engraving, from a copy of a painting, or the copy of a copy – the clarity of some of them might be thought impressive. Most mortifying of all to Pinkerton was the inclusion of an idealized Mary Magdalen for Mary Queen of Scots.[136] He protested that it was not authentic, but Herbert had received the drawing directly from Buchan and in deference to him included it in what was supposed to be a collection of genuine portraits.[137] Pinkerton became disillusioned with the work and eventually demanded that his name not appear on the title-page. Herbert "impudently did"[138] insert it. A year after the book appeared Pinkerton disowned it in a letter to the *Monthly Magazine*, where he said Herbert alone had been responsible for the faulty plates.[139]

The book had other faults. Chronology wasn't followed. The selection of portraits other than those of royalty seems arbitrary. Among the forty-seven plates are several bishops, two professors of divinity (Henry Scougal and Robert Boyd), a plenipotentiary (Alexander Erskine), three more Erskines (two Johns and another Alexander), a "gentleman" of whom "no memorials have yet arisen" (Sir Conrad Ruthven), an Englishwoman (Alicia Stuart), and one "beauty" (Frances Stuart). What Pinkerton mainly had in mind was to locate and print genuine pictures of Scottish kings and their queens, and of Mary Queen of Scots and other high-ranking historical notables like Darnley and Lennox. But the kings and Mary were primary – he still felt the need to destroy the old imaginary line of forty kings found in Boethius and elsewhere. The emphasis on kings and Mary – about one-third of the portraits – explains the somewhat incongruous inclusion of six plates of seals at the end of the book. They were important because on them were images of some of the earliest kings for whom he couldn't locate proper portraits. He probably also sought out pictures of the Scottish painter George Jameson, the source of a number of his portraits, and of John Lesley, the sixteenth-century historian he admired. Other images were chosen simply because he could get them, and because they were "illustrious." To the twenty-first-century reader the selection is questionable because of what was left out, i.e., representatives of most of the Scottish people. Pinkerton explained to Buchan that "in a monarchy, the kings attract the chief notice: in short, the chief actors in history will ever form the first class of illustrious men." He added, "no philosophy can make a shrub an oak."[140] It is a little peculiar to find him, who prated about the nobility of the mechanic and the farmer in *A New Tale of a Tub*, and later opined that a "free and vigorous yeomanry may well be regarded as the chief glory of any state,"[141] engaged in a work like this. Nor was it his only effort in such an enterprise, as we will see.

His remarks on the images in *Iconographia Scotica* require no lengthy analysis. He did not comment on every one, and only with extreme terseness on some. His book, he said, was "rather to be regarded as an account of portraits, than of persons,"[142] which meant he normally condensed the biographies, often giving little more than a succession of dates, and focused on the images, their meaning, source, questions arising from them, costume in them, etc. He didn't wish to repeat information in standard sources such as Dalrymple's *Annals* or Robertson's *History of Scotland*. His statement about Darnley, whose fate "is too well known, to need any recapitulation here," is typical. When he got to Jameson he referred readers to Walpole's *Anecdotes of Painting* for an account of his work, adding that not much more additional information could be found. Much of what he wrote is dry fact, though reflecting an admirable instinct for clarity and accuracy. His personality occasionally came through and, all too rarely, a hint of irony and humour, as when he said of Henry Scougal, a divine who had written a book of practical piety, that "being of an amorous complexion, he sometimes loved God, and sometimes loved women." His accounts of the Stuart kings are weighted with the knowledge he was picking up from work on his forthcoming history of Scotland. Yet he was sometimes stumped by the material before him. On a symbol consisting of oak leaves in one "Compartment" of a painting in Kensington Palace he wrote: "Its meaning must be left to some future antiquary."

At one point in the tripartite dealings over *Iconographia Scotica* Buchan told Herbert of the antipathy towards Pinkerton in Scotland, and Herbert brought it up, no doubt out of worry about sales there. The exasperated Pinkerton asked Buchan not to let Herbert know of "any prejudices which fools may have against me" since "he can form only one idea, that I have committed murder or robbery."[143] The exchange reflects his concern over his reputation, even more an issue with him now that he had a big new book in progress, one designed to show a different persona from that seen in earlier works. And if, as is likely, the planned collaboration with Gibbon had done something to assuage the hostilities he'd aroused among the literati, he didn't want a loose tongue to threaten it.

This heightened sensitivity was evident in the 1794 reissue of his *Enquiry*, where he was at some pains to retreat from the harsh anti-Celtic strain in the first edition, to the extent that it was feasible to do so, and attributed it to the "temporary feelings" of "warmth and impatience" engendered by "long and dry labour."[144] It was apparent also in his *Critical* reviewing. Here we must note a comment he made to Joseph Banks in 1800. "Now I," he said, "for five years (1794–1799) neither wrote one word in that Review, nor ever saw it."[145] He made a similar remark to William Godwin in 1799.[146] Maybe by "1794–1799" he meant the period between those years but not including them, and the figure of five would almost naturally come to mind in a hasty correspondence or exchange. Also, on both occasions it was in his interest to stretch the truth. (See chapter 4.) Whatever the explanation is, he did write reviews for the *Critical* both in 1794 (at least one of which was printed in 1795) and in 1799. In 1794 he reviewed Buchanan's *Defence of the Scots Highlanders*, the attack on the *Enquiry* and *Dissertation on the Scythians or Goths*, already referred to. When it reached Pinkerton he told Lord Buchan he could find only "one typographical error pointed out, *Scytharum* for *Scythicum*."[147] The *Critical* reviewer in October said "we leave Mr. Pinkerton to fight his own battles"; yet Buchanan had pointed out only one error in the *Dissertation*, and that is "perhaps typographic, *Scytharum* for *Scythicum*, in a quotation from Ovid." He went on:

> While in Pinkerton's *thousand pages* we must reprobate *twenty* or *thirty sentences*, as disgustingly acrimonious, and only to be vindicated from that degree of ill humour which long and dry labour is apt to produce, we cannot shew equal lenity to Mr. Buchanan's violent language. In a work of industry and utility, a candid reader will pass such defects with a sigh, or a smile; but in a light and superficial publication, acrimony cannot be pardoned.[148]

"Disgustingly acrimonious" was perhaps bending over a little too far to hide his hand. Ritson, who termed Buchanan "as very a Celt as his antagonist could possibly wish for,"[149] published *Scotish Song* in 1794.

It contained more vitriolic language than Buchanan's dull imagination could ever have mustered. At one point he suggested that Pinkerton had been engendered by a devil.[150] Pinkerton hit back in the *Critical Review*, damning the book as trivial, acrimonious, inaccurate, and dull. The "repeated attacks, in texts and notes," he wrote, "on Pinkerton's youthful productions, the Select Scotish Ballads, &c. we pass in silence, for they are too *savoury* for our pages." Yet "no error is pointed out, but only general invectives."[151] Ritson's "savoury" behaviour might have signalled an incipient mental disorder. Pinkerton had become his *bête noire*. Ritson watched him like a hawk, hunting for error and misjudgments to lay at his charge, as when he found to his delight that Pinkerton had been one of the signatories to the "Certificate of Belief" in the fake Ireland Shakespeare manuscripts.[152] By May 1795 Ritson had become aware from what he termed "the falsehood, impudence, and scurrility" of the *Critical* that Pinkerton was one of "its principal contributors."[153]

The *Scotish Song* review, which appeared in January 1795, had probably been written in 1794, and a couple of other reviews by Pinkerton may have spilled over in this fashion. When Ritson spotted him he was off the payroll, to return in 1799 for another long stretch. As far as can be determined from a study of the *Critical Review*'s contents, it appears he did not write for it from February 1795 to April 1799 (inclusive) – four years and three months. Why the break with the *Critical* occurred is unclear. It could be that writing his history demanded so much time that he had little left over for reviews. That is not quite convincing, since it isn't often that we find Pinkerton at this stage turning away from a source of income. His departure might have been due to changes at the *Review*. Archibald Hamilton the elder, the journal's founder, died in March 1793. His son Archibald, a printer, had died the year before, leaving two sons, Archibald and Samuel, also printers. It was this third Archibald, or he and Samuel together, who presumably ran the *Critical Review* in 1794–5. Amidst these upheavals and, perhaps, economies, Pinkerton's services could have been seen as dispensable. No evidence of a quarrel has come to light. The fact that he resumed writing for the review in 1799 suggests that there had been no quarrel, and indeed his two letters from Samuel Hamilton in 1802 are friendly, even intimate.[154]

Writing his history was slow and painful. In November 1791 he had predicted it would take him three years to finish it.[155] Two years later he had only "about a quarter" finished and knew the book wouldn't be completed until 1795 or 1796.[156] In April 1794, though he had worked "every day nay hour" through the previous year, he wrote that in "about two years my book will be ready for the press."[157] He simply couldn't proceed any faster. He told Lord Buchan that "the chronological defects of our history" in the period he covered were "incredible." Simply getting events in proper order took "extreme" labour. He found letters, charters, and other records that "move events from one to ten years sooner or later" than the dates assigned by early writers.[158] He could find few documents relating to James II's reign[159] and was forced to go further afield than usual to try to get authentic data for it. (It would be an inferior part of his work.) To add to these and similar problems of content, he paid so much attention to style that his work was perhaps doubled. In normal circumstances, in reviews or dissertations, he wrote quickly and often concisely. More, he thought, was demanded of a historian. "The style I pay great attention to," he wrote in 1793, "but I study the ancient simplicity of narrative, mingled, I trust, with some degree of grandeur and grace, more than the modern turgidity."[160] The style "I have polished as highly as I could," he told Bradfute, the Edinburgh bookseller.[161] Polishing to achieve "grandeur and grace" was no easy matter for Pinkerton. Whether he should have attempted it is doubtful since style can't help but seem forced and concocted if it is not the natural expression of a writer. He sought assistance and advice on style from Walpole, Percy, Beattie, Astle, and (possibly) John Douglas, Bishop of Salisbury,[162] and in the end decided Gibbon would be his model.[163] At length he abandoned the idea, never firmly held, of including an introductory account of Scottish history before the Stuarts.[164] On 23 July 1795 he offered Dilly the manuscript:

> After the labour of many years, I have at last compleated for the press my History of Scotland under the House of Stuart, or from the year 1371 to 1542. It will form one large volume 4to, begins where Sir David Dalrymple's Annals end and closes at the accession of Mary, of whose reign there are several good histories by Robertson, Stuart, etc., so that a work embracing her period could pretend to little novelty. This, on the contrary, being chiefly drawn up from original letters and papers, will be found entirely new; and I will pledge my character and judgment that it is the most elaborate and exact history of any country ever published. But it must speak for itself, and I shall only add that the plan and style have been warmly approved by Lord Orford, Mr. Gibbon,[165] Bishop Percy and other eminent judges who have seen it.
>
> I have as yet shewn it to no bookseller and should wish to part at once with the copyright. I have no doubt that it will be a book in constant

demand, as long as history is read, as it has nothing of party or of a temporary nature; and I have exerted any little abilities I have to unite the research and accuracy of an antiquary, with the philosophy, animation and clearness of arrangement and style, which render a work acceptable to the general reader …[166]

Dilly answered two days later, saying that he'd have the work appraised and asking how much Pinkerton wanted for the copyright.[167] A family dinner at Dilly's followed in mid-September, at which terms were discussed. Dilly would take four months to make a decision, and only after some stiff bargaining with the author.[168]

It was quite a moment in Pinkerton's life. The history he'd written was in a way the culmination of his first fourteen years as a literary man in London. He of course expected to benefit financially, and by no small figure. How much of his private income remained is unclear. By 1790, when his factor William Buchan had died, his Edinburgh houses were likely more a nuisance than a worthwhile investment.[169] Prior to 1794 the semi-annual support payments to his mother came from those rents, but in that year he bought her an annuity that cost him £277/15/6, a considerable outlay.[170] Since 1782 he'd spent about £40 a year to have books and manuscripts copied in Scotland and elsewhere.[171] The *Iconographia Scotia* had left him "£50 out of pocket."[172] And life had grown more costly. His wife's health was weak and she had expensive tastes in clothing.[173] More children were arriving.[174] Twice before, in 1788 and 1790, Pinkerton had considered seeking a position at the British Museum on hearing of possible vacancies there. None had opened up.[175] Early in 1795 one did, when Richard Southgate died. Pinkerton asked Joseph Banks for help in getting the position, noting that he was "advancing in years" and "his whole life" had been occupied in "that kind of literature, which, while its severity injures the health, contributes nothing to the fortune." Filling Southgate's position as Keeper of Manuscripts and Medals would be "a stimulus to continue my favourite pursuits," he said.[176] Indeed it would have been a good fit for him. Pinkerton asked Walpole to intervene with the Archbishop of Canterbury, John Moore, who had the chief say over who got the job. He didn't get it; Robert Nares, a literary vicar, "principal editor of the *British Critic*," was hired.[177] Walpole had interceded with the Archbishop before for Pinkerton, to no avail. He hated doing it again, but did try, getting not "a syllable of answer." Nor did he expect one, he said in his last letter to Pinkerton, since the Archbishop couldn't be expected "to become the patron of one who had made himself obnox-ious to the clergy."[178]

In January 1796 Dilly agreed to publish *The History of Scotland from the Accession of the House of Stuart to that of Mary* – the Queen of Scots being thus smuggled in[179] – and made an offer of £400 for the copyright. One thousand copies would be printed, and if these were sold within two years Pinkerton would be paid £200 more for additions and corrections to the second edition.[180] Pinkerton was flattered. Dilly, he told Buchan, "has paid liberally."[181] By February the work was in the press, Pinkerton was collecting material to put in appendices, and Dilly made it clear that the *History* was to be an elaborate production. He decided that it would be published not in one volume quarto but two, and it would be printed in elegant typeface on the finest paper. Pinkerton spent almost a year supervising the press. In January 1797, the book was ready for publica-tion, but Dilly delayed the release until Pinkerton could persuade John Gillies to puff the work in the *Monthly Review* and try to have it noticed as soon as possible in the *Critical*.[182] Friends were asked to speak well of it in all companies. On 1 March Pinkerton gave the public "the greatest labour of his life."[183] For the first time his portrait[184] was prefixed to one of his books. His hopes were stated, but not so boldly as to give offence:

> Had the author's abilities been equal to his ambition, it was his object and wish, to have rendered the work so complete a model of modern history, so perspicuous, interesting, various, animated, and elegant, as to merit general approbation, as to appear on the toilette as well as in the library: while at the same time, the events, chronology, references, and appen-dixes, should present an exactness impenetrable by the sharpest spear of the sternest antiquary. Vain hopes![185]

The book sold well for three days. "I am persuaded the demand will continue and be increased," his friend, the MP Andrew Stuart, told him on 11 March.[186] Instead it declined. In May Gillies's review appeared in the *Monthly*, but he had gone so far to avoid charges of puffing that his praise was faint and therefore damning. Much the same could be said of the *Critical's* comments, the first half of which came out also in May.[187] Yes, the reviewers said, there is much information in the work, much to praise. But the words "great historian" were missing. The *Gentleman's Magazine* and *European Magazine* didn't review the book, and other reviews didn't appear until 1798 and 1799. That of the Rev. John Whita-ker in the *British Critic* was a mixture of blame and praise, but rather more of the former. He challenged two of Pinkerton's novelties: giving a character of the monarch at the beginning of his reign rather than at the end, and interspersing the narrative with long accounts of social and cultural conditions. He also found "low," "barbarous, "vicious," "vulgar," and "affected" elements in the style, but conceded it was "vigorous and pointed."[188] This would hardly send a reader out to buy the book. Far from paying Pinkerton for a second edition, in 1798 Dilly had to resort to advertising to sell off the first.[189] No second edition was called for. Pinkerton's *History of Scotland* was a flop.

One possible cause of poor sales was a pamphlet called *Answer to an Attack, made by John Pinkerton, Esq. of Hampstead, lately Published, upon Mr. William Anderson*. The story behind it dates from 1796 when Pinkerton began corresponding with the Edinburgh publisher Archibald Constable with the aim of promoting *Iconographia Scotica*. He also wanted copies of documents from the General Register House for his *History*. Distance from Edinburgh was once more a problem. Constable referred him to Anderson, a writer (in the legal sense) with access to the documents, and in June Pinkerton contacted him with a list of queries. Others followed, Anderson sent a paper on the state of the Scottish records, and all went well until August when he sent a bill for just under £12, including charges for each letter he'd written. He also asked Pinkerton to strike out anything personal he might have said in his paper on the records before publishing it. In his answer Pinkerton pretended he hadn't received the bill:

> I wish you would, in the mean time, give me a previous notion of the expence, as I often requested, for the difference between Scotland and other countries is, that *we* are apt to charge too much, *while I can hardly get my correspondents in England or abroad to charge any thing.* But I need not *remind you* that this is a mere literary matter between *two amateurs* (at least as Mr. Constable represented you to me), and of *my injunctions* not to send the chamberlain's accounts, if the expence exceeded *two guineas in all.* … The trouble of an *amateur* is nothing; and I *daily sacrifice my time to literary acquaintance*; but as you are in business, I must request you to accept a couple of guineas for your trouble. But, when I tell my *English acquaintance* that 30 written pages (price here thirty shillings) have cost me five or six guineas, *how they will stare*!

Anderson replied at once, saying he hoped that Pinkerton had received the account:

By the by, that account will explain the nature of my work, which you seem to mistake; for I am not in use to be told, 'that my charge will make people stare.' And if you are apt to stare at such a trifle, you should in future avoid giving trouble to those who only value your correspondence in proportion as they are paid for it. My account is exceedingly moderate, and had I followed out the investigations you suggested, it would have been moderate though ten times the amount.[190]

Pinkerton protested that he hadn't received the account. Anderson sent him another, adding "a few shillings" for the extra trouble. A month passed without a reply, and Anderson commenced a suit, arresting some of the rents on Pinkerton's houses. On 1 October Pinkerton wrote to express his astonishment at the high charges. If Anderson would not take back some of his "paltry" scraps of paper and deduct the charge for them, he would print the paper on the records "*literatim*" with Anderson's name, would print Anderson's letters on the neglected state of the records, and comment on the "extortion" in the account. Anderson replied, saying he had stopped the rents. Pinkerton paid up.

Anderson said he'd managed to get "the celebrated, the chronological, the romantic author of the History of Scotland bound up like a cat in a bag."[191] How far he acted alone in binding up the cat is a good question. Perhaps Constable had a part in it.[192] Anderson replied to the letter of 1 October, saying that if the threat in it were carried out he would publish an answer that included Pinkerton's correspondence. Pinkerton printed the paper on the "*Present State of the Scotish Records*" in his *History*; it contained a reference to the "enormous" charge that Anderson (he is named) made for "a trifling labour."[193] In his pamphlet Anderson printed all of Pinkerton's letters, with remarks on their contents. Pinkerton had made comments on his mother, his houses, and his friends – he had failed to proceed with his usual caution. One month after the publication of the *History of Scotland* Anderson's pamphlet was in the hands of his reviewers and enemies. It was a "shameful predicament," George Steevens, an enemy,[194] told Percy on 12 April, and then promptly sent the pamphlet to Richard Farmer in Leicester.[195] In May the *European Magazine* said the *Answer to an Attack* presented "the most illiberal and sordid controversy we ever saw, disgraceful alike to both parties."[196] The historian who'd hoped his work "would appear on the toilette" as well as in the library once again found himself in a wrangle, brought on largely by his tightfistedness. Anderson's pamphlet is another piece of evidence showing how deeply felt was Scottish hostility to Pinkerton at this point in his career.[197] He had complained ten years earlier, while attempting to get copies of the makars' poetry from Edinburgh, of being treated "like dogs in the manger."[198] He was in the doghouse now, and a history of the Stuarts, no matter how philosophically composed, would not get him out of it.

The quality of Pinkerton's work as history remains to be noted. The book is not a general history of the period covered, but rather a scholarly history written for an educated readership, which, of course, is not the same thing. Walter Scott's *History of Scotland* belongs in the first category, Pinkerton's in the second. Pinkerton's is scholarly because much of it is based on footnoted original documents, that is to say such documents as were available to someone writing in London in the 1790s. A history of Scotland from 1371 to 1542 constructed mainly from those sources was bound to have limits and weaknesses. But they were not his only sources. He got some help from Scotland, as noted. The early historian/annalists had to be brought in too, weighed carefully so as to separate fable from fact – no easy matter – and used to construct a narrative. Pinkerton's work was an amalgam of these materials. His references were as careful as in any piece of modern history. His judgment of the early historical authorities was mostly sound, though he placed less reliance on Robert Lindsay of Pitscottie's and George Buchanan's sixteenth-century histories of Scotland than, perhaps, he should have. He could be over-scrupulous. There is no apparent misquotation or misrepresentation of sources in the two volumes. His transcripts of letters in his appendices seem to vary only slightly, in matters of punctuation and capitalization, from transcripts in official state papers. He was the first historian of the early Stuarts to make extensive use of the collections in the British Museum Library – his annotations are scattered throughout the Caligula and other MSS. He was the first to print a large selection of documents related to the period. In his *History* the period he covered was established on a foundation of fact, though not cleared of all error. The clearing of error in history is arduous and ongoing. Nineteenth-century historians such as Patrick Tytler and John Hill Burton made use of Pinkerton's work and were best able to appreciate the patient amassing of evidence and the critical approach to sources that made it a pioneering effort.

Pinkerton's interpretations of what happened during the reigns of the seven kings will often be found to differ from those of later scholars and writers. This is not surprising. Those who followed him had his work to build on, with a larger supply of documents at hand in libraries and archives and, at least for academic historians, years of salaried employment to work on amassing and studying data, together with research assistants to do some of the hard slogging. By the twentieth century historiographic progress and fashion had made Pinkerton's work seem merely quaint. That doesn't mean he was always wrong. So little is known for certain about many episodes and personages in early Scottish history that right and wrong hardly enter into it. Pinkerton, doing his own slogging, made what sense he could out of the materials he had. Nor was the way he approached the writing of history without its attractions when compared, say, with the work of later economic and social historians.

His treatment of the first two Stuarts, Robert II and Robert III, may still be read by a student of history without going too far astray. Much the same can be said of the entire work. Readers have to be on their guard in perusing Pinkerton's or any eighteenth-century history; but then they have to be on guard in reading twentieth- and twenty-first-century histories too. Part of his account of Robert III's reign will illustrate the flavour of some of his writing. Robert III (who had taken the name Robert as king though his name was John) succeeded to the throne when he was over fifty. The real power in the kingdom was held by Robert, later Duke of Albany, the king's brother, third in line of succession, who was in effect regent or governor. Ahead of him were the king's sons, the heir-apparent David, who became Duke of Rothsay, and James, the future James I. In 1402 Albany imprisoned Rothsay in Falkland Tower, where he soon died. What had happened to him? Mystery surrounds the whole affair, but to Pinkerton it was simplicity itself. Albany was a murderer and Rothsay an innocent victim. Rothsay would have made an ideal king:

> That warm effervescence of vigorous youth, which tamed by reason, experience, and time, affords mature materials of firm and spirited character, had led him to some excesses, especially of the amorous kind, which afforded pretexts of constraint from his uncle the governor, and of reproof from his royal parents. A fondness for riotous pastime and arch roguery were also laid to the prince's charge; who, to candid eyes, sufficiently compensated these youthful and trivial defects by his good qualities. Endued with a comely person, an honest heart, an able head, a most sweet and affable temper, and even deeply tinctured with learning for that century, his virtues, and not his vices, attracted the regent's enmity.[199]

This is an expansion of a passage in Wyntoun's *Cronykil*. It is certainly not a complete sketch of Rothsay's character; nor is Pinkerton's view of

Albany, who is accused of "insatiate ambition, unrelenting cruelty, and its attendant cowardice,"[200] what historians today would call a nuanced conclusion. What we are seeing in the sketch is a romancer wishing to find a noble prince, a dramatist looking for a hero and villain, and a writer trying to imitate Gibbon. (Not to mention an author with his eye on sales.) His reading of events is not, however, entirely made up. Not enough is known about the incident to exculpate Albany. Scottish history through the years covered in Pinkerton's book is a story of conspiracy, sibling rivalry, Highland uprisings – which Pinkerton made too little of – bloodletting, backstabbing, and treason. Excesses "of the amorous kind" that Pinkerton was always inclined to forgive also figure in it. Much conjecture was and still is needed to find a way through the entanglements. The most satisfactory chapters in the *History* are those on James IV and James V, for both of which Pinkerton had ample documents, but not a complete set. That of James V "is almost wholly composed from the original letters of the chief actors; and is perhaps the first attempt of the kind in any language."[201] Yet there are flaws and omissions in both accounts; sometimes Pinkerton was so bogged down in detail that he failed to see connections and developments.

What may irritate some readers of the *History* – but possibly charm others – was the leaven of "philosophy" he inserted, in hopes of raising his work to a high level, to make it "a model of modern history," as he phrased it. It wasn't philosophy he put in as much as extraneous moral and social comment. At the founding of the University of St Andrews, for instance, he says "It is pleasing to consider the dawn of the light of learning, rising as after a tempestuous night of discord and barbarism, and commencing its salutary influence."[202] On the execution of the Earl of Douglas in 1440 he comments: "The want of wisdom in the government, upon this occasion, exceeds belief; but it is easier to commit a murder, than to perform an action of common prudence, and crime ought never to infer ability."[203] He admits that the chastity of Mary of Gelder, James II's queen, "was dubious," but warns that "nothing can be more unjust than to infer that the loss of female modesty is the loss of every virtue."[204] An incursion into Scotland in 1528 evokes this remark: "Whether we regard prudence, or justice, the conduct of Angus on this occasion was contemptible."[205] This peremptory *judging* goes on through his book, not always consistently. After Flodden the nobles were "generous and valiant"; after the murder of James III they were "rebellious" and their "sanguinary lust of power … cannot be too severely reprobated."[206] Under James V the clergy were "friends of order and peace, and distinguished by talents and learning"; elsewhere they were "voluptuous, illiterate, and remiss in their duties."[207] At times he combines this instinct to judge with ill-advised conjectures about the motives of historical figures. Thus Margaret Tudor assented to Albany's assuming the regency in 1514 because she "was disgusted with the vain exercise of an authority merely nominal," while Albany in 1522 decided not to take action against his enemy Angus, because he "had discovered from experience that violence is but a resource of weakness, while a firm government is only to be founded on conciliation, and views of general interest."[208]

The shade of Gibbon hovers over this *History of Scotland*. Like Gibbon's, Pinkerton's characters "tremble to behold," "behold with despair," "sigh," "smile with contempt," "blush with shame," are "disgusted with vanity." Gibbonian words – rapacious, insinuate, warlike, salutary, numerous (as in a "numerous body of soldiers"), equity, specious, complacent, desultory, opulence – are sprinkled about. The sentences and paragraphs ape Gibbon:

> Great was the general expectation of Albany's presence; and the delay of a year had raised that expectation to eager desire. His arrival was fondly longed for, as the sole remedy of the public disorders, as the sole pledge of a tranquillity, so much the more happy as it was to succeed the darkest tempests of anarchy. The young, and the turbulent, were not averse to the novelty; the old, and the peaceable, sighed for the protection and stability of a regular government. This event at last happened. Albany arrived at Dunbarton on the eighteenth of May with eight ships, apparently part of the fleet of James IV, laden with ammunition, and warlike stores, and with yet stronger implements of government, the gold and luxuries of France. The peers and chiefs crowded to his presence; and his exotic elegance of manners, his condescension, his affability, his courtly deportment, won all hearts.[209]

The polished style is more appropriate for a Roman emperor's visit to Pompey than for a Scottish duke's trip to Dunbarton. But here, as well as in the description of James V's voyage to the northern extremities of his kingdom,[210] in the account of Scotland's reaction to Flodden,[211] and in certain other spots, Pinkerton carries it off rather well. One notices the word "apparently" slipped in: one of the few signs, amidst the flourishes, of scholarly hesitation. Behind the "fine" writing is a small thin man with green spectacles "on 'is nose," pondering a document, worried about the fact he is about to put down on a page of history.

4 Reviewer and Geographer, 1798–1802

Pinkerton attributed the poor reception of his *History of Scotland* to the "general impoverishment and want of commerce" caused by the war with France[1] rather than to any defect in the work itself. The times were "so strange and portentous," he said, "that literature must go to bed and sleep for a few years."[2] There was something in this perception. With habeas corpus suspended, Europe convulsed in war, and all eyes on Napoleon, Joséphine, and Nelson, it did indeed seem an inopportune moment to drop a two-volume quarto history of the first seven Stuart kings into the mix and expect to be noticed. In the aftermath of his book's failure Pinkerton seemed for a while, understandably, to lack purpose. To write a good book, or even a not so good one, and find nobody reading or, worse still, buying it, can be a bitter blow. He became subject to "fits of indolence"; he no doubt missed having no other "great historical work"[3] to occupy his time. Announced plans for more publications in Scottish history were set aside.[4] The *History* had been his finest effort. The scholarship was close to impeccable; the style was what he wanted it to be, though evidently not what readers wanted it to be. He could do no better. It had got him nowhere. He complained later that his efforts in Scottish history had been received "not only with ingratitude, but with calumny."[5]

He turned again to portraits, no doubt partly to make amends for *Iconographia Scotica*, but also to make money.[6] He had "many mouths to feed," he commented in 1800.[7] *The Scotish Gallery: or, Portraits of Eminent Persons of Scotland* appeared in numbers from September 1797, to June 1799. Edward Harding was the publisher and the work proceeded without interruption or quarrels, the result being another contribution to Scottish elite iconography, with more letterpress and generally better engravings than what was in the earlier work. Pinkerton opened with a "Chronological List of the Portraits" and an "Introduction to the rise and progress of painting in Scotland." The introduction starts well but deteriorates, padded out with a reprint from the periodical *The Bee* and a list of Jameson's works sent by Buchan. The printer's devil must have been on the doorstep! Towards the end it turns into a sort of preface where Pinkerton expressed his satisfaction that "amid many disappointments, inconveniences, and obstacles" he'd been able "to form such an assemblage of Scotish Portraits, as is contained in this and a former work," the last words being the only reference in the book to *Iconographia Scotica*. "An authoritative series of the kings has been recovered, from Robert I," he boasted, "after fictitious sets had been current for a century and a half." It is clear that he intended *The Scotish Gallery* to be a supplement to the former work; he specifically referred to fulfilling commitments made in the 1794 prospectus. His "only motives were his own curiosity," he said, "and a warm inborn wish, which has stimulated most of his literary endeavours, that his countrymen should not neglect their national productions of art and literature." He added, with fingers crossed, "To emolument he never looked"; after "the sacrifice of so much time" he expected "but a small sacrifice of gold." He claimed, in private, to have lost about £20 "by the perverseness of Edinburgh artists."[8]

The pictures he found for *The Scotish Gallery* of Robert I (the Bruce) and David II completed his line of kings. He added another of Mary, and one each of James IV, who had four in the previous work, and James VI, who had one. Mary of Guise also reappears. His Bruce came from a painting of Jameson, copied by the young artist Robert Johnson of Newcastle from a collection at Taymouth Castle in Scotland. (Johnson, who had been assigned the task of copying Jameson's works for Pinkerton, died after completing "about fifteen," leaving four undone.)[9] In his commentary, Pinkerton said the prototypes of Jameson's Bruce and certain other royal portraits were reputed to be "old limnings" of Dumferlin Castle. "They bear indeed every mark of authenticity," he said doubtfully. But "No portrait can answer the *idea* of Robert I" more than Jameson's. Thus he strained to get his line of kings settled and in order, starting from the Bruce. Putting the "idea" of a king in a collection was not much different from Buchan's depicting the "soul" of someone and not worrying about the wig. More of Buchan's sketches were used in *The Scotish Gallery*, including that of the patriot Andrew Fletcher of Saltoun whose "soul" he'd striven to catch. Pinkerton acknowledged Buchan's "many services," one of which was a letter from Dalrymple containing a few amusing anecdotes about Fletcher.

Rather than give "a dry abstract" of Fletcher's life, Pinkerton just printed the letter. Among the others on the list of fifty-two engravings in the book were the "exquisite poet" William Drummond of Hawthornden, James Gregory, inventor of the reflecting telescope, George Buchanan the historian, the antiquary Sir Robert Gordon of Straloch, and Anne Cunningham, marchioness of Hamilton. The length and content of Pinkerton's comments varied considerably. As in *Iconographia Scotia*, he disdained repeating supposedly well-known biographical and historical facts. "Who hasn't heard of John Knox?" he asked at the beginning of his piece on that luminary. He was more forthcoming about the warlike Anne Cunningham who, differing from her son in politics, declared on his turning up with an English fleet in the Firth of Forth that "she would be the first" to shoot him "should he presume to land and attack his countrymen and country." Of the painter Sir John Medina he icily remarked that a descendant of his was "not many years ago *judiciously* employed in repairing the forged set of Scotish kings, in the gallery at Holyrood-house." Descriptions of two notables, Mark Ker, the Prior of Newbottle, and the Earl of Ancrum, were sent in by their families.

Splicing the *Scotish Gallery* together likely required little effort from a scholar accustomed to long hours toiling over historical sources, ferreting out their meaning if they had one, and writing down his findings. Pinkerton continued reading at the British Museum, but from force of habit rather than with any fixed intention in mind.[10] Around March 1798 he printed a small volume entitled *Other Juvenile Poems*, containing verse from his earlier volumes that he thought worthy of reissuing and a poem called "Public Happiness."[11] He wrote letters to his main correspondents, Malcolm Laing and Joseph Cooper Walker. Laing, an Edinburgh lawyer who was to become MP for Orkney and Shetland, solicited Pinkerton's correspondence in 1797. He had been "bewildered with Celtic poetry and tradition," he confessed, until the *Enquiry* and *Dissertation on the Scythians or Goths* had "disabused" his mind.[12] He visited London in 1798, met Pinkerton, and became a confidant. Walker, an Irishman, corresponded with Ritson, Percy, and (after 1794) Pinkerton. He managed to maintain contact with all three, and he too was a useful friend. He replaced Percy as Pinkerton's connection in Ireland, just as Laing supplanted Lord Buchan, whose friendship was lost in 1799.[13]

A symptom of Pinkerton's lack of direction in 1798 was his attempt to get his tragedy, "The Heiress of Strathern," on the stage. He was rummaging through some old papers, he told Douce in March, when he came upon it. He found it "very pathetic and interesting," but deficient "in incident, knowledge of dramatic effect, and <u>action</u>." He'd rewritten it and intended to bring it to Richard Sheridan to produce. If he did, the spectacle would "rival the magnificence of the ancient Greek stage."[14] Douce went to Hampstead in April, ate mutton with Pinkerton, and received "much pleasure" from hearing the play read to him.[15] Encouraged, Pinkerton soon afterwards gave the play to the poet Laetitia Barbauld, who lived in Hampstead; she promised to share "the entertainment it will afford"[16] with her brother John Aikin, the physician and prolific writer whose name was among the scorners of *Letters of Literature*. Aikin now edited a new journal, Richard Phillips' *Monthly Magazine*. In July Pinkerton dined at Phillips' home in the company of Steevens, Alexander Geddes, and Thomas Holcroft. Steevens rattled on, Holcroft noted in his diary, "but had read and remembered. Pinkerton said little."[17] But he was not one to let an opportunity slip by. A week later he called on Holcroft and asked him to read his tragedy. Holcroft did so, was impressed by some of its qualities, but noted that it was "occasionally verbose" and had "not enough of soul,"[18] which was not a word of a lie. He was persuaded to correct the whole play and didn't finish making his pencil marks until 1799.[19] "The Heiress of Strathern" then disappears from view for another thirteen years.

On 30 July 1798 Pinkerton learned in a note from Scotland that his mother had died on the 25th; she was buried in the Canongate Churchyard.[20] Only one short letter from the 1790s from her to her son has come to light. In it she alluded to a house, doubtless one of his, where "some smok comes in to the back room[.] theirs no curin it unless that vent could be stopt." She added, "I have pead sevrell pounds for repairs," and signed off with "I am your loving mother Mary Heron."[21] Owing to his long absence from Edinburgh, some of the worry about houses had been shifted onto her in her old age. Yet he had provided for her. Mary's death meant that Pinkerton could now consider disposing of his property in and near Edinburgh. Laing inspected it with Pinkerton's new factor, James Gibson, in April 1799, and advised "converting the whole into money." Some of his houses were in "ruinous" condition.[22] A year later Pinkerton authorized Gibson to "sell & dispose of the whole property belonging to me in Scotland."[23]

Through 1798 Pinkerton was a visitor at the homes of Holcroft, Phillips, and William Godwin, all three, by contemporary standards, political radicals. He met Godwin at Holcroft's in July. Godwin immediately set out to read his *Enquiry* and other works and found in them "veins of candour ... more pure than I recollected in the writings of any other man."[24] Holcroft was also impressed, conceiving that Pinkerton was "pleasant in manner, and apparently not ill-tempered."[25] (But Isaac Disraeli, encountering Pinkerton at Phillips' house, insulted him to his face.)[26] As Godwin's diary[27] reveals, Pinkerton's association with these figures was rather frequent from July 1798, to October 1799. They were not too jacobinic for him; they would have been for many. But he was drawn into their company likely because of his work for Phillips, in whose *Monthly Magazine,* starting in March 1798, he was writing a series of twelve articles on Walpole. That ended in May 1799, and Phillips then published them as *Walpoliana* (1799). Walpole had died in March 1797, two years after his last short note to Pinkerton which marked the end of their friendship.

Walpoliana, published anonymously, was a "little lounging miscellany" containing table talk, anecdotes, and seventeen of Walpole's letters to Pinkerton, with extracts from others. To the material in the *Monthly Magazine* Pinkerton added an introductory "Biographical Sketch, in fugitive crayons" of Walpole, and some miscellaneous matter, mainly citations, to fill the two volumes. Not all the *bons mots* were from Walpole. Some were from *The Treasury of Wit*[28] and other jest-books. As he slyly noted in his preface, in such a work "A tale told fifteen years ago may innocently be ascribed to a wrong person; or an expression misstated."[29] Yet a core of authentic material remains that is an invaluable source of material about Walpole. The fifty-page biographical sketch, among the finest things Pinkerton wrote, was the first "formal biography" of Walpole prior to 1851[30] and the most reliable account before Austin Dobson's biography of 1890.

If Pinkerton's biography was biased, it was biased in Walpole's favour, as in this description:

> His engaging manners, and gentle, endearing affability to his friends, exceed all praise. Not the smallest hauteur, or consciousness of rank or talents, appeared in his familiar conferences; and he was ever eager to dissipate any constraint that might occur, as imposing a constraint upon himself, and knowing that any such chain enfeebles and almost annihilates the mental powers. Endued with exquisite sensibility, his wit never gave the smallest wound even to the grossest ignorance of the world, or the most morbid hypochondriac bashfulness: *experto crede*.[31]

The last phrases were personal, an unusual admission by Pinkerton, even if only in an anonymous text, of his social awkwardness and hypochondria. (Though "bashfulness" didn't keep him away from dinners and teas with the Godwin coterie.) He'd made similar confessions to Lord Buchan earlier, terming himself "a *homo umbratilis*, of a hypochondriac, unsocial disposition" and "a recluse LITERATO."[32] There was a good deal more of Pinkerton displayed in this biography of Walpole, as for instance in the praise bestowed on the sceptic Hume – "that real, mild, unfanatic, unenthusiastic, and universally tolerant, philosopher."[33] Elsewhere he advised that iced water had a salutary effect on the bowels! He admired Walpole and defended him against some of the charges levelled by contemporaries. Walpole was innocent of any cruelty in the Chatterton tragedy, he said. Nor was it Walpole's fault that he and Gray had quarrelled on their Grand Tour, since Gray, though "a man of great genius and erudition," was "haughty, and impatient, and intolerant of the peculiarities of others."[34] The deception practised by Walpole in the first edition of his gothic novel *The Castle of Otranto* (1764), which was then said to be a translation of an Italian work dated 1529, Pinkerton claimed was innocent. He passed over Walpole's religious views in the biography as "moderate and rational, not enthusiastic," but later in the book recorded a statement by him that "Fontenelle's Dialogues on the Plurality of Worlds, first rendered me an infidel." Yet he made it clear that Walpole was no atheist.[35] He said Walpole's poetry "seldom rises above the middling," his play *The Mysterious Mother* "aspires to the praise of real genius," and *The Castle of Otranto*, "proves the power of his art."[36] He dismissed Walpole's antiquarian works ("Fashionable company, and luxurious ease, are not schools of accuracy") but rightly noted the importance of his letters, saying "The amiable ease, and playful elegance, the striking expression, ready sense, and graceful turns of his language, were singularly adapted to epistolary correspondence."[37] He pointed to the paramount "pride of birth and rank" in Walpole, which was so strong that he considered himself degraded when assuming the character of author. He thought himself far above "vulgar people, vulgar painters, vulgar authors," who were "made, God knows how, on the fifth day of the creation." Pinkerton regarded such aristocratic airs as "adventitious and absurd."[38] Overall, the depiction of Walpole's character and personality seems accurate. The writing was far removed from the often stilted mannerisms of the *History of Scotland*. Ketton-Cremer praised the account of "a private rainy day of Horace Walpole" as "especially vivid."[39]

Among contemporary reviews, only the *British Critic* saw that the biography was "sensible, judicious, and generally correct."[40] A more typical reaction was that of his old enemy, the *European Magazine,* which placed Pinkerton among "the parasites of the living, and the defamers of the dead."[41] The book provoked a short burst of obloquy and suspicion. Douce thought Pinkerton had expected a legacy from Walpole and "when he found he had none abused him."[42] Edmond Malone gave his views of "that detestable fellow, Pinkerton" in a letter to Percy.[43] William Cobbett, who according to Malone was "well disposed" to expose Pinkerton, printed a paper against *Walpoliana* in *The Porcupine,* written by one John Blakeway.[44] Some of the abuse was directed, not at the compiler of *Walpoliana,* but at Walpole. The four letters sent to the *Monthly Magazine* during the first publication of Pinkerton's work all attacked Walpole.[45] Charles Burney in the *Monthly Review,* while not failing to note the "want of wit, humour, mechanical order, novelty, or arrangement" in *Walpoliana,* concentrated his attack on Walpole's "frivolity," "vanity," and "pretensions to modesty."[46] Pinkerton inadvertently touched off the anti-Walpole sentiment that was prevalent through much of the nineteenth century.

With *Walpoliana* and the *Scotish Gallery* off his hands, or about to be, Pinkerton looked for other projects. Early in 1799 he'd arranged with James Tassie, the modeller,[47] to publish a series of images of Scottish kings taken from medals.[48] Tassie's death in June ended that scheme. (Thomas Astle had already produced such a work.) Pinkerton then tried to get Laing interested in collaborating on a third volume of portraits. He'd said in the *Scotish Gallery* that, if encouraged, he could produce one more from the "the most curious of remaining portraits," meaning ones left over from the collecting he'd done for that book and *Iconographia Scotica.* He made that announcement with no great enthusiasm, conceding the volume could be done in Scotland "with more ease, and at less expence." Laing was dubious but at length agreed to participate.[49] Many letters were exchanged on its contents in the summer and fall of 1799 as Pinkerton grew busier with unrelated literary tasks and hatched a new scheme. (See below.) Nothing was to come of the book with Laing.[50] In effect, Pinkerton's scholarly work on Scottish literature and history was over. Which is not to say that he lost interest in those fields or stopped commenting on them. Here we must return to David Macpherson, whose strictures on Pinkerton have been noted. In 1799 lies the source of that enmity.

In June 1799, the historical and antiquarian critic in the *Critical Review* apologized to readers for not having reviewed earlier James Andrews'

History of Great Britain (1796) and Macpherson's edition of Wyntoun's *Cronykil* (1795). It would be more "consonant with our own wishes," the critic explained, to review books as near the publication date as possible, but given the "vast number" of British publications and the interruptions in "literary toil" caused by "sickness, business, necessary relaxation, or various accidents," readers should not attribute the delay to neglect. We have to read between the lines. Someone had been taken on as reviewer of historical books that nobody had been qualified to assess before, or nobody had bothered to. That someone was Pinkerton, who reviewed Macpherson's book and very likely Andrews' too.[51] Macpherson's appeared in May; that month may be taken as the date of Pinkerton's return to the *Critical Review*. How to explain that return? He was certainly available, having finished a big book and being on the lookout for new sources of income.[52] As noted earlier, the operations of the *Critical* are obscure, but it is evident that changes in its management were again occurring in 1799, or perhaps at the end of 1798. An upheaval of some sort was brewing. By May 1799 the two brothers Archibald and Samuel Hamilton had "separated,"[53] the word implying they had run the journal in tandem prior to then. Samuel would evidently be sole proprietor from then on. Whether Pinkerton's restoration had anything to do with that change is unclear. But he was back, and would be a presence of note in the *Critical* until he left England in 1802.

Macpherson,[54] a Scot whose early occupation was that of land surveyor, was ten years younger than Pinkerton. He had visited America, had a wife and family, and had suffered financial losses that compelled him to move to Kentish Town, where he lived "in a small way" and took up the profession of author. Having learned that Pinkerton did not plan to proceed with editing Wyntoun, as he'd said he might in *Ancient Scotish Poems*, Macpherson set aside other literary projects and in 1792 turned his attention to that difficult task, working on it for two and a half years. It was published in April 1795, in two volumes. He arranged to have printed 250 copies in octavo, and twenty-five in quarto on expensive paper that he had been "rashly persuaded" to use. Reluctant to proceed by subscription, which he was too proud to try, he accepted a guarantee from his friend George Chalmers to cover him against any loss. Soon after publication the stationers demanded payment for the paper from Thomas Egerton, the publisher – so-called, although the work was in truth self-published – and Chalmers provided it. Macpherson managed to repay Chalmers and in due course cover the other expenses incurred in the work. Twenty-two of the quartos sold; at Macpherson's death in 1816 twenty-six of the 250 octavos remained unsold, and had to be disposed of at a low price by one of his sons. Close to 1,000 copies of Macpherson's other works also remained unsold at his death.

That, in brief, is the story behind Macpherson's edition. It was pains-takingly done. The first volume had a thirty-eight-page preface intro-ducing the work; a series of "General Rules" for reading it, consisting of a discussion of pronunciation and of orthography, which was preserved virtually as written in the original; and a glossary. Four hundred pages of text followed, given "verbatim et literatim,"[55] with very minor tinker-ing. Four hundred more pages of text were in the second volume, which ended with a list of variant readings from manuscripts other than his main source, sixty-five pages of notes, mostly historical in nature, an index, and a list of corrections and additions to the notes. The work of scholars was acknowledged throughout. Dalrymple got more than his fair moiety of praise; the "improvement of historic knowledge" since the time of Hector Boethius should, Macpherson says, "in a great mea-sure be ascribed to the valuable Annals of Lord Hailes."[56] The "excellent author," that "good man"[57] Hailes was *hic et ubique*. Chalmers got a good word too; his *Life of Thomas Ruddiman* (1794) Macpherson termed the *Literary History of Scotland*.[58] Innes, Ruddiman, Ritson, Percy, Callander, Maitland, Astle – all these and more were acknowledged.

Pinkerton was not mentioned. What was more, Macpherson airily gave advice to historians of Scotland as if Pinkerton never existed. "Here I cannot help observing," he says at one point, "that they, who wish to understand the history of Scotland, will employ their time much better in studying the Norwegian and Icelandic authors along with the old English writers and the few authentic monuments of Irish history, and comparing them with the old domestic authors and such original charters and other authentic documents as are accessible, than in bewildering themselves in the fictions of Boyse [Boethius] and his followers."[59] The *Enquiry into the History of Scotland* had disappeared in a puff of smoke. Celtic smoke!

Pinkerton began his review by (unfairly) questioning the importance of Wynton's *Cronykil*, saying that all the "memorable facts" in it had been previously extracted. He then quoted a sizeable portion of the preface in which Macpherson asserted the usefulness of the work to the historian, the compiler of Scottish peerages, and the lawyer. He gave no other extract, saying "few of our readers will peruse the old language of Winton." Retaining the "Saxon letters" was a "ridiculous absur-dity," he said, rendering the text "unintelligible." He went on: "Should

Mr. Macpherson publish any more old works, we would advise him to print all his pages in strict *fac-simile*; and then he may remain in mist like a highland seer, inaccessible to profane eyes." A "more heavy charge" followed: "the editor's preface and notes abound with plagiarisms from the productions of Mr. Pinkerton. Mr. Macpherson affects never to mention that author. Like the ostrich, he hides his head, and thinks that no one sees him."[60]

Macpherson had to be scarred by the charge of plagiarism and the curt dismissal of his fine piece of editing that took so much toil, and cost so much money that it might have pauperized his family. Since he was a friend of Ritson's he must have known or at the very least suspected that Pinkerton was a *Critical* reviewer, and that he reviewed his book; but he didn't make that clear when he wrote the annotations on his copy of *Ancient Scotish Poems* that W.A. Craigie published in a *PMLA* article in 1927. (Craigie included a few of them as well in notes to his edition of the Maitland Folio.) One of Macpherson's annotations was this: "The Gentle Shepherd [by Allan Ramsay] is admired by all readers; and has never been spoken ill by any but one single critical reviewer, whose want of taste can only be equalled by his universal malevolence."[61] The sentence, which seems to show he knew the reviewer's identity, referred to Pinkerton's remarks on "The Gentle Shepherd" in his review of Chalmers' edition of Ramsay's *Poems* (1800) in the *Critical Review* of 1801.[62] The odd reluctance to finger his enemy left Craigie wondering whether Pinkerton "viewed with equanimity" seeing Wynton pass out of his hands into Macpherson's. Macpherson's annotations might not have seemed as damaging if the full story had come out. His motive would have been divulged. It wasn't just distaste for Pinkerton's editorial practice or moral outrage over his private life. It was a form of payback.

One of Pinkerton's reviews in the 1799 *Critical* reflected the decision he'd made about his future as a writer. He would turn geographer, not a geographer of Scotland but one of the whole earth, a "universal" one. He would say later that he'd been "conversant" with geography "from his early youth,"[63] and there is little reason to doubt his word. The *Dissertation on the Scythians or Goths* was grounded on vast reading in both history and geography. But in all likelihood the main reason he switched genres was to seize an opportunity. William Guthrie's *Geographical Grammar*,[64] the standard current world geography, had been published in 1770. Multiple editions had appeared by 1800, but it was reasonable to argue that at the start of a new century, after so much tumult and change in the recent past, a replacement was in order. If well done, a new geography would sell and Pinkerton knew it. Yet it is wrong to say he was motivated just by the prospect of making money. His mind was always active, and in the late 1790s he was turning away from history and antiquities towards science, or what he understood to be science.

Once he made his decision, he felt compelled to make his claims known to the geographical circle in London, centred on the African Association,[65] whose founding president was Joseph Banks,[66] his occasional correspondent. Banks' close friend and neighbour in London was James Rennell, a pre-eminent and pioneering figure in English geography owing to his maps and studies of India and Africa. Pinkerton's arrival in their compound was via the *Critical Review*. One of his friends was the African explorer William Browne. Pinkerton had known him since 1792, and Browne, while making his treks through north and east Africa, had sent him long letters giving details of his discoveries and plans. On Browne's return to England in 1798, Pinkerton helped him find a publisher for his *Travels in Africa, Egypt, and Syria*, corrected his style, and took on the "vexatious task" of supervising the press.[67] It could have been during this work with Browne that he conceived the idea of writing a geography. In June 1799 rumours circulated in London about Browne and his book, and Browne, away in Suffolk, complained to Pinkerton that some individual had been "reporting much more than he knows, and considerably more than the truth."[68] The gossip apparently had to do with the arrangements he'd made with his publishers, including what he was to be paid. Pinkerton himself was suspected of being the guilty party, but as soon as he got Browne's letter he wrote to Banks and asked if he had spread the rumours. Banks denied it in a sharply worded note.[69] Immediately Pinkerton wrote to George Nicol and accused him of doing the dirty deed.[70] Nicol shot back, saying Pinkerton was the "convicted Slanderer" of Browne.[71] Friendships dropping by the day! Pinkerton had become agitated, partly because he suspected the geographical elite in London led by Banks and Rennell were prejudiced in favour of Browne's rival Mungo Park, whose famous excursion to the Niger River in 1795–7, sponsored by the African Association, was narrated in *Travels in the Interior Districts of Africa*.[72] It was published in 1799 and reviewed, in the first of two instalments, in the *Critical* in July.[73] In the August number, Pinkerton wrote what Robert Southey called "the puff superlative" of Browne's *Travels*.[74] Browne's "profound learning and undaunted enterprise," Pinkerton said, "combine to produce a work perhaps *unique* in its kind." His journey to Darfur "has rendered more essential service to the geography of Africa than any attempted since the great discoveries of the Portuguese." He was forced to smile, therefore, "when major Rennel [*sic*] speaks of future discoveries as mere supplements to those of Mr. Park!" He went on to offer his own ideas about the progress of African discovery. He could not congratulate the African Association on its efforts. What was needed was to dispatch "sensible Mohammedans" into the interior of that continent, or else travellers "with every external badge of that religion" and "a complete knowledge of Arabic." Rennell was "the greatest of living geographers" but several of his "lapses" needed correction, as they "impede our knowledge of Africa." Pinkerton corrected them, or thought he did, and advised Rennell "to *read* books instead of merely *consulting* them."[75]

Having thus unwisely trumpeted his challenge, Pinkerton went ahead with his new scheme. In October he wrote to Cadell and Davies, the firm that published Browne's book, and outlined his plan for a work on geography. He'd been thinking of such a plan "for these two years," he said, and he thought he could write on the subject in a way that would be honourable to himself and lucrative to the publishers. His idea was "to publish a System of Universal Geography, in two volumes 4to – with an abridgment, in one volume 8vo, for the use of schools." With "great industry," made easier "by a previous general knowledge of the subject, and of the books to be used," and "by employing several amanuenses to make extracts, &c. &c.," the work could be completed "so as to publish in winter 1801, the first year of a new century."[76] Cadell and Davies agreed that such a work was needed, and Pinkerton drew up an outline of the procedure he would follow.[77] The publishers consulted William Vincent, headmaster of Westminster School, an authority on ancient geography, who said Pinkerton's plan was "a very good one" and that he seemed "perfectly possessed of his subject."[78] A contract was signed in March 1800.[79]

According to John Murray, the upstart Pinkerton was thought of "contemptuously" by Rennell, Banks, Capt. James Burney, mapmaker Aaron Arrowsmith, and the hydrographer Alexander Dalrymple, Lord Hailes' younger brother – in short, by much of the geographical establishment. "They all refused to give him any countenance," Murray wrote, "and told him he knew nothing of the subject he was about to undertake."[80] Murray exaggerated slightly. Arrowsmith could not have been wholly antipathetic, since he agreed to supervise the maps for Pinkerton's work soon after the agreement with the publishers had been reached. Even so, the relationship was not congenial.[81] Pinkerton stirred up opposition and resentment by his supposed effrontery in assuming he could walk in and take over geography. That, anyway, was how it looked, to those who assumed *they* owned the subject. It is hard not to feel some sympathy for Pinkerton's position as this new avenue of ill feeling opened before him. He was a scholar, he had the idea to write a geography, why couldn't he try it? But the geographers had a point too. Had he been a more clubbable man, his entry on the scene might have been easier.

Among the first to show him the entry wouldn't be easy was Banks, whom Pinkerton approached politely in January 1800 for access to his library and permission to consult him on certain geographical issues.[82] Banks agreed to let him use the library, but alluded to his imprudence in beginning a geographical work so soon after making mistakes on "Geographical Chronology" in his "criticisms" of Rennell and the African Association, meaning those in the *Critical Review*. Some "discredit" on Pinkerton's "geographical reputation," he warned, would be forthcoming in the refutation Rennell was about to publish.[83] No doubt fearful that his connection with the *Critical* would be exposed, and unsettled by Rennell's anticipated assault perhaps damning him as a quack, Pinkerton replied squirmingly. He had merely "furnished some materials" for the review of Browne's *Travels*, he said, and "such garblings, additions, and alterations were made by other persons concerned in the Journal, that I hardly recognized my part." He had not written a word in the *Critical* during 1794–9 and had always spoken with respect of the African Association. He'd been all along a warm panegyrist of Rennell.[84] So he claimed.

Rennell's *Geographical System of Herodotus*, published shortly afterwards, contained a pungent note on the *Critical*'s strictures, but Pinkerton's name wasn't mentioned in it. The *Dissertation on the Scythians or Goths* indeed was lauded as "excellent" in the book.[85] Pinkerton could now respond more confidently to Banks, who he thought had insulted him in his earlier letter. The "great error" Rennell had detected in his review was no error at all, he said. And as for mistakes in "Geographical Chronology," he'd seen one in Rennell's new book amounting to 650 years.

> … I suppose you will of course write to the Major, as you did to me, to abandon his researches after so gross a lapse. But as I impute that unjust

and uncandid letter to mere misinformation, and the misrepresentation of literary sycophants on questions which you could not have the smallest opportunity of examining, I beg pardon for this remark, which is wrung from me by the novelty of the charge. I shall only further observe, that I hope you will not again commit your name on such questions; and that the Major and his friends will remember that a <u>monopoly of science</u> can no more be established than a <u>monopoly of light</u>. A monopoly of ignorance may indeed be firmly fixed.[86]

To which Banks replied, ending "the intercourse which formerly subsisted between us."[87] Pinkerton had not finished with Rennell. In May and July 1800 he devoted twenty pages of the *Critical Review* to the *Geographical System of Herodotus*. He warned Rennell that he had access "to few sources of information which are not open to every literary man," and that the *Critical* reviewers "detest despotism and monopoly in any branch of science." Rennell, he noted, should not have attempted to discuss the Scythians and Sarmatians because that subject "has embarrassed even learned writers."[88] And he should have read Gibbon's *Decline and Fall* before beginning to write on his subject.

The *Critical Review* thus became an organ for Pinkerton's opinions of books pertaining to geography, together with something else he had a hand in,[89] the "Review of Maps and Charts" which, after 1800, replaced the "Review of Public Affairs." But science was still just one of his interests, though a commanding one at this stage. In 1799 it appears he was given a wide freedom, and not just to select books for review. The fact that he could insert a bald "Linnaean Table of the Nations and Languages in Europe and Asia"[90] shows what latitude he had. He was quick to take advantage of this position, as the reviews of Macpherson's, Browne's, and Rennell's books demonstrate. As in his earlier tenure with the journal, he stayed on the watch for personal insult or praise, and rarely left insults unanswered. He had a record of literary accomplishment, and those who followed behind would ignore or slight him at their peril. Scots especially would have to be wary.

Robert Heron was a Scot whose writings have sometimes been attributed to Pinkerton owing to the nom de plume used in *Letters of Literature*. The first volume of Heron's *History of Scotland,* covering the early period, was published in 1794 by the Morison firm at Perth. The *Critical* reviewer took a soft line, no doubt because the work "chiefly followed" by the author was "Pinkerton's Inquiry into the History of Scotland." Heron "thinks for himself," the review said, but did not have "sufficient erudition, or exactness, to justify his dissension."[91] This was Pinkerton in a gentle mood. By 1799 four more volumes of the Heron history were in print. In the recast first volume there was now a sixty-page preface where the author lugubriously disclosed, among much else of a personal nature, that he had written the last two volumes while overwhelmed with sorrow over the death of his sister Mary. Imperfections, he said, should be ascribed to that rather than to incapacity or negligence. It was a large, inaccurate, sloppy work. Heron had done little if any original research. In writing of the Stuarts, he said Pinkerton's *History of Scotland* gave too much attention to mere details and not enough to the spirit of the various transactions. "I have been obliged," he noted, "in too many instances, to differ from him. I differ from him, with hesitation, with reluctance, with deference, with respect."[92] No Caledonian venom here. This was friendly. Yet with a sharp eye for the numerous errors, Pinkerton proceeded through Heron's volumes, denouncing as he went. This "literary quack," "mere smatterer," "pretender to science," he wrote, despite his claims to the contrary, didn't see Pinkerton's *History of Scotland* until the part of his own work dealing with the same period had been printed off. That fact "can be evinced by letters from Scotland." No one "who has read Pinkerton's history" could have repeated "the numerous old fables and errors, exploded by that writer, but which Mr. Heron has preserved just as he found them in preceding compilations."[93] That Heron was an object of pity was clear, or should have been, from his preface and from his book generally. Pinkerton showed no mercy.

One consequence of the treatment of Heron was the breach of another friendship. Godwin, dining at Pinkerton's home in October 1799, asked if he had written the review, meaning the first half of it,[94] which had appeared in the August issue of the journal. Pinkerton said no, claiming he had not been connected with the *Critical* for the "last five years," and a day or so later sent Godwin a surly letter, saying that "in scarcely one principle of religion, morals, politics, or literature, is there a shadow of agreement between us." Godwin replied at once: he found too few men of Pinkerton's "extensive information," "industrious research," "power of investigation," "principles of honour," and "general candour of mind," not to "cherish their intimacy when I find them."[95] The relationship petered out. The incident shows how close to the vest Pinkerton kept his link to the *Critical Review*. Yet the self-promotion must have revealed his authorship to any close observer of the literary scene.

In dealing with Pinkerton's work in both the first and second stages of his association with the *Critical*, a distinction has to be made between his treatment of a work like Heron's and the reviews where he was mainly motivated to hit back at an enemy. His severe response to Alexander Campbell's *Introduction to the History of Poetry in Scotland* (1798–9),[96] where he was branded a "foul-mouthed cynic,"[97] illustrates the latter category, but there were various other instances, as has been made plain. In the condemnation of Heron's latter volumes he was chiefly concerned to expose an incompetent historian. His impulse there was scholarly. It is a truth widely acknowledged that though it takes so long, and so much hard work, to master, the field of history attracts charlatans. The sad sack Heron was among them. Others like him were on the scene in eighteenth-century Britain, often well-meaning local figures who might write a book called *Views of Devonshire*, say, and fill it up with talk of Gogmagog.

By the late 1790s[98] Pinkerton's hostility towards Celts had mellowed but still showed now and then in his writing. Ossian he now viewed with utter contempt. He dismissed all claims the Irish made to a pre-Christian alphabet and other, to him, visionary evidence of an early Hibernian civilization. Nowhere was his sceptical approach to Celtic matter more pronounced than in his attitude to the Welsh bardic tradition. In 1788 he'd provoked a dispute in the *Gentleman's Magazine* by arguing that all Welsh poems so far published and ascribed to centuries preceding 1200 were not genuine.[99] In his view the whole question rested on one main point. He said he'd found in the work of historian Giraldus Cambrensis (Gerald of Wales), who wrote mainly in the late twelfth century, evidence that rhyme was unknown in the Welsh poetry of his day. So, to Pinkerton, rhyming poems by bards of a supposedly earlier date could not be genuine. He stated his case forcibly, even though he did not know Welsh. Knowledge of that language, he said, was "not sufficient to decide a subject which requires skill in the literature of the middle ages."[100] He returned to the Welsh claims in his edition of the *Bruce* where he discussed the history of rhyming poetry in Europe. The earliest rhyming poet in any language, he said, was Otfrid, a ninth-century German. Rhyme did not appear in Anglo-Saxon poetry until the eleventh century. There was "a noble specimen of Anglo-Saxon poetry"[101] in the British Museum from the tenth century: it didn't rhyme. He concluded that "they who believe in the riming Welch poetry, ascribed to Taliessin and other bards of the sixth century, may enjoy their own credulity."[102] In 1793, in his review of William Owen Pughe's edition of *Heroic Elegies, and other Pieces, of Llywarch Hen*, he denounced the Welsh Triads, a series of traditional pieces supposedly of great antiquity now appearing in print, as a "fabulous authority,"[103] and brought the same objections to Llywarch's poems as he'd urged against others in the 1780s. In January 1800 he turned to the first volume of Sharon Turner's *History of the Anglo-Saxons* (1799). Turner, an English lawyer, was a researcher of note and his work was a pioneering study. He had evidently mastered the Anglo-Saxon language, as Pinkerton had not,[104] but in his book he made extensive use of such Welsh sources as the poems of Llywarch, Taliesin, and Aneirin, and the Triads, placing them alongside more conventional historical sources. One product of this Welsh influence was the intrusion of the mostly mythical[105] Arthur into a supposedly scholarly history. He alluded to Pinkerton's "great learning" in a footnote.[106] In the *Critical Review* Pinkerton said Turner had too much "imagination" to be an historian. He didn't know how to handle sources: "Authors of great reputation, authors of none, authors of veracity, authors of falsehood, authors of ancient times, authors of recent description, are blended in one confused mass of equal quotation, and equal confidence."[107] He once again denied that the poems of the Welsh bards were genuine sources of history, returning to his argument from Geraldus Cambrensis:

> It follows, therefore, that all these pieces ascribed to the early Welch poets, are posterior to the days of Geraldus; and, as they are equally unknown to Nennius, Geoffrey of Monmouth, and Caradoc of Llancarvon,[108] who, if they had existed in their time, would have been as eager to have rifled them for historical sweets as Mr. Turner himself, we may conclude with mathematical certainty that they are modern fabrications. They may therefore repose under the grey stone of Ossian, in spite of the vociferation of modern enthusiasts, strangers alike to real antiquarian lore, and to historical precision.[109]

On the appearance of two more volumes of Turner's history, Pinkerton again censured him for using these materials. He could not, he wrote, allow Turner and the Welsh antiquaries to "correct the history of England" by reference to such fraudulent sources as the Welsh Triads.[110]

William Owen Pughe thought Pinkerton's review of Turner's first volume "altogether an attempt to destroy the book."[111] Southey said Pinkerton was wrong and Turner "certainly in the right." Pinkerton's assertions, he said, were made "merely for the sake of contradiction," he being "very often a pretender in literature. ... I catch him sometimes in the Critical telling downright lies."[112] Turner wrote a *Vindication of the Genuineness of the Ancient British Poems* (1803) to answer Pinkerton's aspersions, charging, among other arguments, that he had misrepresented what Geraldus had written, even to the point of deliberately mistranslating his Latin.[113] Who was right, and who wrong, in this set-to? The answer is: the poems of Llywarch, Taliesin, and Aneirin were not forgeries; on this, Turner was right and Pinkerton wrong. But if we judge by the scholar R.H. Hodgkin, whose magisterial *History of the Anglo-Saxons* appeared in 1935, Pinkerton was partly right in the general point he was making, or at least right to be worried. Hodgkin said the poems of the early Welsh bards "have something to contribute," but are "disintegrated," "overlaid with later Bardic additions, and often unintelligible." He allowed them two short paragraphs in a work of 700 pages and said they afford "side-lights, however flickering"[114] to his history. This is a step-down from Pinkerton's outright rejection of the sources, but even Hodgkin's brand of scepticism has faded over time into what appears to be a wary acceptance of much Celtic source material, not just Welsh, as useful to historians of early Britain.[115] In English history, for instance, one notes references to Aneirin's poem *Y Gododdin*, which is a description of a sixth-century battle in Yorkshire, albeit preserved in a thirteenth-century manuscript.[116] Pinkerton rejected such sources too peremptorily and somewhat surprisingly in view of his earlier happy discovery and use of the *Albanic Duan*. Yet if by his series of questions and attacks he helped to stir up debate and clarified in a small way the sources of British history, his efforts were far from insignificant. He was wrong to discredit the bards but he wasn't lying, as Southey implied. He had questions and doubts, and expressed them too strongly, as was his wont.

We may take it that Pinkerton did not begin work on his geography until the contract was signed in March 1800. Haste was needed. There was "danger of being frustrated," he'd told Cadell and Davies earlier, "as the idea of a new system being necessary has become general – and of course the more expedition is used the better."[117] Yet he obviously couldn't put pen to paper on it straightaway. Preparation was required, besides which he had other demands confronting him in the spring and early summer of 1800. His correspondence reveals a continuing interest in collecting portraits for an envisioned future publication. Also on his mind, almost at times consuming it, and in his letters was a subject of growing importance in the intellectual life of the day: mineralogy.[118] He had in view creating a new system of classification of minerals; a

preliminary *Sketch* of it, printed for him by Samuel Hamilton, was circulating by September.[119] His work for the *Critical Review* continued.
He had to correct an essay on the Gowrie conspiracy he'd written for
Laing.[120] He had bouts of illness.[121] In May a house he owned in Somers
Town, an area on the northern outskirts of London, was destroyed in
a fire, a severe loss. He carried no insurance.[122] Amidst all this, news
reached him of literary activity in Scotland, some of it upsetting. He was
under attack, and had to decide what to do in response. Comments on
him by Chalmers led him to make this statement to Laing, who at this
point was a friend to whom there was no reason to put on airs or lie:

> As, from a constitutional irritability of nerve, I have in my earlier produc-
> tions shown much controversial asperity, it would be ridiculous in me to
> complain when I am paid in my own coin. Were I revising my books, I
> should dash out all such passages, which I never see without disgust. I can
> only say they are the products of infirmity, and not of malice.[123]

It is another rare instance of self-revelation by Pinkerton. His remorse
didn't mean, however, that Scots and others who challenged him would
go unanswered.

Late 1801 or early 1802 was the date set for completion of the geography. The period from mid-1800 on was full of furious activity. Pinkerton
had a number of helpers and amanuenses. The astronomer Samuel Vince,
a Cambridge professor, agreed to write an introduction to the work.
Arthur Aikin provided botanical information. Arrowsmith took responsibility for the maps; they were engraved by Wilson Lowry. The writer
A.F.M. Willich supplied translations from books in German. Librarian
Samuel Ayscough, John Aikin,[124] Arthur's father, William Ouseley the
orientalist, mineralogist John Mawe, and various others gave assistance.
It was altogether an impressive stable of associates. Vince's introduction was so elaborate that he could almost be termed joint author of
the first volume. In the event, Pinkerton was the main author. He had
many areas of expertise, a "tenacious memory,"[125] and two literary skills
already amply demonstrated: he could write well and write fast. His
method was as follows. When he began to compose about, say, Spain,
he sent Cadell and Davies a list of required books on that country. They
sent the items they could locate and he, using them, his own library, and
whatever other information came his way, wrote the account of it. What
he'd written was then forwarded to the printer, Andrew Strahan, and
the books returned to Cadell and Davies with a list for another country.

Region by region, month after month, he laboured at his geography, feeding copy to Strahan on Printer's Street and, for the *Critical*, to Hamilton in Falcon Court. In the spring of 1801, his health damaged by the strain, he went to Margate, the seaside resort town in Kent, to recover. In June he reported from there to his employers:

> I find the sea bathing has greatly restored my health so that I shall continue it for a week or two longer. It is now 20 years since I was forced to have recourse to it before and I hope I shall again be set up for some time.
>
> When I return I shall resume my task with fresh spirits and have no doubt I shall complete it with some ease by the time mentioned. The printing has gone on pretty well but there has been lately a pause I know not for what reason.
>
> I have taken a house in Eaton Street Pimlico where I shall be nearer the press &c. Be so good as to get Arrowsmith to forward some drawings to the Engraver who says he is at a stand.[126]

He had moved from Hampstead back to central London. In a normal life, such moves occur without signalling crisis within a family. With him they spelled trouble. Six months passed. In January 1802 he told Cadell and Davies that he had "finished the Geography and begun the Abridgement."[127] He – with considerable help – completed a two-volume, 1,450-page description of the world in less than two years, all the while maintaining his position on the *Critical Review*. It was a feat of hard-driving, demanding authorship.

We must now look more closely at the book. The "novelty" that Pinkerton boasted of in the preface to the work and elsewhere referred to his arrangement of the contents rather than any new theory of geography. He had no new theory. He thought "theoretic geography" was "always useless."[128] On meeting geographer Jean Nicolas Buache in Paris, he noted that he (Buache) was "rather too fond of theories in a science which of itself admits of nothing but facts and discoveries."[129] *Modern Geography* presented "the most recent and authentic information concerning the numerous nations and states who divide and diversify the earth,"[130] which was more or less what Guthrie had tried to do in his *Geographical Grammar* and what contemporary readers still expected from such a work.[131] Pinkerton explained in his preface why he thought a new geography was needed. There were "numerous and gross mistakes" in earlier systems, he said, and in them "the chief geographical topics have been sacrificed to long details of history, chronology, and commercial regulations." In addition, "the most recent and important discoveries" were either omitted from them or imperfectly presented. Discoveries in the Pacific Ocean and elsewhere "within these few years" now had to be "admitted and arranged, in a regular and precise distribution of the parts of the habitable world." There had been "surprising changes" even in Europe that made a new work indispensable. "Whole kingdoms have been annihilated; grand provinces transferred; and such a general alteration has taken place in states and boundaries, that a geographical work published five years ago may be pronounced to be already antiquated." He made no bones about where he got most of his new data. He compiled it. In his work "the essence of innumerable books of travels and voyages will be found to be extracted." This again reflected a current notion of the geographer's task.[132] Pinkerton claimed that "Every great literary monument may be said to be erected by compilation, from the time of Herodotus to that of Gibbon, and from the age of Homer to that of Shakespeare." He saw no need for fieldwork of the kind Rennell had done. He read and digested Rennell, and made use of his analyses in discussing Hindostan (India) and Africa. He was a geographer of the closet. He said what he was doing was "science," but by the standards of the twenty-first century he appeared to have a limited understanding of that word as applied to geography. (Not that there is just one definition of geography.)[133] Some might refer to what he did as bookmaking rather than scholarship. But Douce, a stern critic, found "elaborate researches in every page."[134]

To present his information Pinkerton divided up the world, first under continents, then under certain heads, where appropriate, for each "nation and state." Those heads were: 1. historical geography, 2. political geography, 3. civil geography, and 4. natural geography. This was his "arrangement," which he admitted was "in part suggested" by the French geographer Robert de Vaugondy.[135] O.F.G. Sitwell, who has written a notable paper on Pinkerton's work, says that No. 4 on the list corresponds "fairly closely to what is today called physical geography."[136] Pinkerton thought in 1802 that "natural geography" could not "be admitted to a prominence,"[137] and in his second edition (1807) restated the view, saying that "all that relates to man and human history" had to be dealt with first.[138] Though it came fourth in his estimation, as his book proceeded "natural geography" tended to require, or at least be given, more space than the other three, thereby asserting its paramountcy. That it was inserted at the length given to it, and separately described, was a step forward for geography. The *Edinburgh Review* pointed out that the division was a major contribution of the book. Most of the topics covered in it, the journal said, were "entirely omitted, or very imperfectly detailed in former works. What Mr. P. denominates the physiognomy of the country; the hills, vales, and rivers; their size, direction, and length; the nature of the soil and state of agriculture; the component parts of the mountains, their general appearance and height above the level of the sea; Botany, Zoology, and Mineralogy, – form the most original articles in this division."[139]

Sitwell says Pinkerton was "very little inclined to see the physical environment as a causal factor in either human nature or human activity."[140] That comment is valid in the sense that it wasn't a central purpose of Pinkerton to make such analyses. His approach was not as focused as that. But he did bring it up.[141] In his general piece on Europe he said that "freedom from the excessive heats of Asia and Africa has contributed to the vigour of the frame, and the energy of the mind." Again, he noted that the "number and extent" of Europe's inland seas are "chief causes of the extensive industry and civilization, and consequent superiority to the other grand divisions of the globe." In his treatment of England he said the "moist and foggy climate conspires with the great use of gross and animal food, to produce that melancholy, which is esteemed by foreigners a national characteristic."[142] He doubted North America was first settled from the north of Asia, "as the progeny of so cold a latitude is ever found rare, feeble, and unenterprizing."[143] Comments of this nature, mostly guesswork, were sprinkled here and there. (Such observation is a commonplace of classical literature.)[144] If the "essence" of geography is "the nature of the adjustment of man's activities to the physical environment,"[145] Pinkerton will not be thought a pioneer of the discipline. He was well aware of what French thinkers, including Montesquieu, had said about the connection between climate and human character. He'd devoted a sceptical essay to Jean-Baptiste Dubos's climatic theories in *Letters of Literature*,[146] at the end of which he shied away from "airy speculations" on the subject.

Under historical geography Pinkerton dealt with: the name of the country; its extent and boundaries; original population; "progressive geography," meaning the history of geographical knowledge of the country; and historical epochs and antiquities, i.e., history. Under political geography were: religion; ecclesiastical structure; government; laws; population; colonies; army and navy; revenues; and political importance and relations. Civil geography treated: manners and customs; language; literature; the arts; education; universities; cities and towns; edifices; roads; inland navigation; and manufactures and commerce.

Natural geography dealt with those topics listed by the *Edinburgh Review*, along with climate; lakes; forests; mineral waters; and natural curiosities. Most of these subheads worked fairly well for developed countries, but when he left Europe and moved into Asia and the Pacific islands they were scarcely useful at all and he dropped many of them, sometimes all of them, adapted them, or added different ones. "Ecclesiastical structure" and "universities" wouldn't open profitable lines of inquiry in an entry on the Solomon Islands of 1800. The subheads were something of a straitjacket even in describing Europe, since they prevented him from writing integrated accounts that a certain kind of reader naturally looks for.

As for the order of subjects to be written on, he stated in his preface that "The States are arranged according to their comparative importance,"[147] by which he meant political importance. Presumably he had continents as well as countries in mind in that remark. In any event, Europe was in the lead, taking up the entire volume one. He gave reasons why Europe came first:

> As Europe is the seat of letters and arts, and the greatest exertions of human energy in every department; and is besides the native region of the chief modern geographers, and that in which the readers are most intimately and deeply interested, it is always the division first treated; though the order be arbitrary …[148]

Within Europe were three "orders" of countries, that is, groups in which countries were comparable in stature. The first "order" in Europe comprised: Britain, France, Russia, the Austrian dominions, the Prussian states, Spain, and Turkey. It would, he said, "be alike idle and presumptuous to decide the precise rank of a state in each order; for instance, whether France or Russia be the most powerful. This part of the arrangement must therefore be elective; and it is sufficient that the states of the same order be treated with a similar length of description." Britain nonetheless topped the list, with subheadings for England, Scotland, and Ireland. He admitted that foreigners could complain that too much space was given to Britain, but explained:

> … the same objection might extend to every system ancient and modern, as the authors have always enlarged the description of the countries in which they wrote. His native country ought also to be the chief subject of every reader; nor can much useful knowledge … be instituted concerning

foreign regions, till after we have formed an intimate acquaintance with our native land …[149]

Asia came after Europe, followed by the Pacific islands, Australia, and New Zealand; then America (North and South); then Africa. The order of politically important states within continents was not inviolable. When he turned to Asia, he admitted that China should be treated first since it belonged in "the first and chief rank, beyond all comparison" with other countries. But it was "estranged from Europe," and it was "preferable" to treat first countries that were "intimately blended with European policy," namely Turkey and Russia.[150]

The following will give an idea of Pinkerton's style of geographical description:

> From the south of the Tweed to Bamborough, extends a sandy shore; and the most remarkable object is Lindesfarn, or Holy Island, divided from Northumberland by a level, which is dry at low water, but out of which the flowing tide oozes suddenly, to the terror and peril of the unwary travel-ler. From Bamborough Castle, to Flamborough-head, are mostly low cliffs, of lime-stone, and other materials; and at Sunderland of a peculiar stone used in building, and which seems the work of marine insects. Scarbor-ough stands on a vast rock, projecting into the waves; but Flamborough-head is a far more magnificent object, being formed of lime-stone, of a snowy whiteness, and stupendous height, visible far off at sea.[151]

Sitwell says *Modern Geography* "closely resembles the national hand-books prepared by the British Admiralty,"[152] and at times it surely does. Pinkerton knew how a geography differed from chorography ("which illustrates a country or province") and topography ("which describes a particular place, or small district").[153] Yet the passage quoted looks as if taken from a common guide for walkers or tourists – he points to "remarkable," "magnificent," "peculiar," and "stupendous" objects, as in such a guidebook, and indeed warns the traveller about the oozing tide. It is close to topography here, less so when he leaves the familiar settings of Britain. But a tendency to point out "remarkable" things is found throughout the book. When there is "nothing remarkable," he can let us know that too. He uses the adjectives "interesting," "curious," and "singular" often, as well as those already noted. In the first sentence of his preface he alludes to "the exuberant variety of knowledge and amusement" exhibited in geography. The notion that geography could

be amusing would not go over well among twenty-first-century practitioners of the discipline. He was conscious of writing for Dr Johnson's
"common reader," much more so than his successors today. His book
was "a work of science" but also "one of general instruction."[154]

Modern Geography belongs to what has been called the "political–
statistical school"[155] of geographers. "Political–factual" might be better,
since statistics, while present, were not primary because dependable
ones were mostly unavailable. The book piled fact upon fact within the
appointed categories, and what a stupendous pile it was. Dr Johnson
once referred to the "solemnity" of geography, as if relating fact were a
serious responsibility, like conducting a ceremony. It was hard for any
writer to make a discourse of that nature intriguing, but Pinkerton was
such an opinionated and odd man that his personality came through
amidst the mass of information, sometimes in quite a diverting manner.
He might call geography "science" but as his summoning up of Homer
and Shakespeare in the preface shows, he saw it as a literary exercise
too. He didn't think he had to hide behind his data, and was never slow
to offer his opinions. Mayhew has traced Pinkerton's political views
in the book, noting their "brand of Whig reformism" and as well the
loss of "faith in the French experiment."[156] That "faith" had never been
wholehearted, as we have seen. It had now turned to suspicion and
hostility. In *Modern Geography* the French Revolution is said to be an
attempt "to extinguish knowledge and civilization."[157] But the "political
parts," as he said later, "form a very small portion" of the book.[158] Much
of the commentary was at a more homely level, amounting to a kind of
non-ideological editorializing, similar to that in the *History of Scotland*,
only more frequent and direct. The reader who was curious about the
Highlands of Scotland found out why Roman Catholicism was strong
in that region, why Highland gentlemen objected to the propagation
of industry, and why a university was needed there; and, further, why
schoolmasters' salaries throughout Scotland should be raised and where
to find more information about the superstitions and manners of the
"peasantry" – in the "exquisite" poems of Burns. It is good to be sent to
poetry by a geographer! In the American edition of the work (1804) the
account of "America" (both continents) was "corrected and considerably
enlarged" by the botanist and professor Benjamin Smith Barton. Pinkerton's opinions collided amusingly with straight-faced academic judgment. Pinkerton wrote that in Philadelphia "The general use of salted
provision must be injurious to health; and it is inconceivable why this
custom should have continued so long."[159] Barton relegated this (to him)
bizarre dictum to a footnote and commented: "There seems to be little

foundation for these observations. The inhabitants of Philadelphia are not remarkable for the consumption of salted provisions, and it appears pretty certain that the use of a portion of well salted meat, during the hot summer months, is very conducive to health."[160] (Barton was a physician as well as professor.) A remark by Pinkerton on a disease called "the black vomit" received a similar rejoinder.[161] When Pinkerton, encountering the name Ekanfanoko or Ouaquafenoga for a swamp in Georgia, became flabbergasted with American geographical terms derived from Indian languages, he exclaimed against the use of such "long and barbarous appellatives." They came, he said, from "savages who have a word of fourteen syllables to express the number *three*," and recommended they be changed by "some learned society." But Barton, after removing the offending passage from the book and admitting the Indian names were sometimes long, said they were "soft-sounding," "beautiful," and "appropriate."[162] The process we see is an American taming the wild man from Scotland, an oddity for sure. Barton said that he'd been asked by the "editors" of the Philadelphia edition of *Modern Geography* to correct the article on America, which had been found to be "in many respects, extremely defective and erroneous."[163] With such a learned editor, and two named assistants, scrutinizing the text, it is a wonder so little error was detected. Pinkerton's preface stated that "Many blemishes will, no doubt, be found in a work of such a multifarious nature."

Pinkerton's theological views obtruded but little into his geography. Polytheism came up occasionally, as when he pointed to the Mongols' "rational polytheism, not unknown to the Jews, who admitted, as appears from Daniel, great angels or spirits, as protectors of empires."[164] We find him noting, in a reference to the Treaty of Paris (1763), that in it "Canada was acquired at the price of about fifty times its real value," the result being "the loss of America." He added: "so incapable is human prudence of presaging events, and so often does Providence effect objects by the very means which men employ to avert them!"[165] Thus he genuflects – a rarity in this book, though that didn't necessarily mean it was insincere. The topics covered in *Modern Geography* offered many opportunities for heretical discourse on the history of the earth and man, but there appears to be just one overt instance of it. It occurs in the discussion of Persia. The distant ancestors of the Persians, the Scythians, Pinkerton said, attacked Egypt "about 3660 years before the Christian aera." The Egyptians had been "civilized" earlier – "their genuine chronology seems to begin about 4000 years before Christ." Pinkerton here was wary of treading on Bishop Ussher's toes. He added, after an allusion to the "venerable historical records contained in the Scriptures," that there is no evidence that "the planet has been inhabited above six or seven thousand years." (7,000 was getting dangerous!) History can account "for every relic that is found." Then: "For the great antiquity of the earth there are many evidences; but none for the antiquity of man."[166] Believers in Ussher and Genesis would have been pleased by the second half of this sentence; some of the more delicate among them would not have been by the first. It was not as sharp a crack as those he'd made in the *Dissertation on the Scythians of Goths*, but it was one just the same. The alert Christian could detect other subtle affronts to Moses in several of Pinkerton's comments on human populations. In some regions he found what he called "primitive," "indigenous," "original," or "peculiar" races of men. In south-central Africa, for instance, "the population appears to be indigenous and peculiar, these being the native regions of the negroes, whose colour, features, and hair, distinguish them from all other races of mankind." This could be thought to conflict with accepted belief about the role of Adam and Eve, who had been his targets in earlier writing. But overall, *Modern Geography* was inoffensive.

Mayhew calls Rennell an "early modern," saying he "clearly divorced theology from geography to an extent which none of his predecessors had." Yet he finds an allusion to "the wise dispensations of The Creator" in a scientific treatise by him published in 1832.[167] Pinkerton has some claim to be crowned an "early modern" too. While he didn't altogether separate theology from geography, he mostly did so. In a work encompassing such a broad spectrum of knowledge, he could hardly avoid questions touching on the beliefs of Christians. By 1802 he knew the peril of saying outright what he thought. He hedged, he had to, being dependent now more than in the past on the proceeds from writing; but in a cautious and indirect fashion, one he no doubt fancied he'd get away with, he let his views be known.

Modern Geography represented no big breakthrough in the geographic discipline, but it was a description of the earth, to the extent that it was known in London in 1800 by a scholarly man of letters and his assistants. It was a timely, thorough, and orderly, sometimes idiosyncratic assemblage of factual description, citation, and opinion; and it must have been a handy compendium in a library.

The preface to *Modern Geography* declared that Europe "reposes in universal peace." This was an allusion to the Treaty of Amiens, concluded in March 1802 between England and France after nearly a decade of war. The peace had an effect on Pinkerton's life. On 15 May he told Laing, "The week after next I think of a trip to Paris, where it will give me pleasure if I can serve your learned researches."[168] His reasons for going were various. The great French museums were one. Having already formed the idea of writing a large work on mineralogy,[169] he'd sought the correspondence of geologists John Mawe and Robert Townson, collected a cabinet of minerals, and, as noted earlier, printed a pamphlet on the subject to distribute among his acquaintances. It was called *Sketch of a New Arrangement of Mineralogy* (1800). Through the early months of 1802 Mawe wrote letters to him from Paris, telling of the mineral cabinets he'd seen and the scientists he'd met. This attracted Pinkerton: he was now an enthusiastic collector of rock specimens and fossils. He also wanted to consult books of travel in the Paris libraries. Yet another reason for going, and likely the most important, had to do with his personal life. It is hard to see into that life to find the source of his unhappiness. Yet unhappy he was. Later he would blame Rhoda Burgess for disturbing his peace of mind. He explained to Macvey Napier that she was "an old maid who must forsooth live in my house and at my expence to please madam and make a domestic hell."[170] He told Douce that he and his wife would have lived in peace, had not Rhoda "excited repeated quarrels by her malignity."[171] We might ask why "madam" would want to keep her sister with her despite her husband's objections, if indeed he stated such objections. We might ask, but there is no ready answer. Perhaps she needed to have Rhoda near her for support. Pinkerton was not a steadfast or dependable figure as a husband and father. He was an irritable, overworked, penny-pinching hypochondriac, with his head always in his books. He must have been hard to live with. A letter from Henrietta to him in 1797 provides a sad picture of her alone at home in Hampstead, tending to their 19-month-old son who was afflicted with smallpox. "I did not undress or change my dress for seven days and nights," she writes. Pinkerton was seventy miles off, perhaps on a holiday.[172] He had evidently made light of the disease in the letter she is answering. She says, "the small pox may seem a bagatelle to you. It never appeared to me such for I always dreaded it myself."[173] The boy must have been one of their "five or six" offspring who died in childhood.

We find evidence of Pinkerton's domestic misery even in the geography where, in commenting on native women of the Society Isles in the Pacific, he says they "seem entire strangers to those unaccountable caprices, sudden frowns, and violences of temper, which form the chief domestic pestilence of civilized society. [They are] easily pacified, never entertaining the sentiments of long and slow revenge, of which the sex seems, in many countries, far more capable than the men."[174] He didn't actually believe this nonsense about primitive life. (See chapter 6.) It was merely a veiled crack at his sister-in-law or wife. Perhaps both.

The sum of all this was, he wished to go to Paris, and go there alone. He planned to take more than an extended holiday. The evidence suggests he wanted a long break from his wife, his children, and Rhoda. He could well afford a trip. His pen had brought him a considerable return during 1800–2. By February 1802, he had been paid £700 by Cadell and Davies for the geography; he would receive another £600 during the six months following its publication, and the prospect of still more for the second and third editions.[175] He had never been paid so much for a literary work. The *Critical Review* yielded further income. In Paris he would have the opportunity to sell to the French booksellers, and to act as researcher and book finder for scholars in England. Yet going to a foreign country for a long period was risky for a man in his circumstances. A writer can't afford to stay long out of sight and out of mind.

By the end of May he was in Paris. Had he stayed a few months longer in London he would have witnessed the publication, good reception, and quick sale of his *Modern Geography*. He not only denied himself those pleasures but also neglected to complete his abridgment of the geography, thereby breaking his contract with Cadell and Davies and risking the loss of £250 (see chapter 6). The remainder of that task was passed to Arthur Aikin,[176] and the book did not appear until 1803. Pinkerton must have been itching to get out of England. Before he left, he said goodbye to certain authors in the *Critical Review*. (He was promised more work for the journal while in Paris.)[177] Three of a number of books passed to him for review dealt with Scottish matter, namely Dalyell's *Scotish Poems of the Sixteenth Century*, Dugald Stewart's *Account of the Life and Writings of William Robertson* (1801), and John Leyden's edition of *The Complaynt of Scotland* (1801).[178] He evidently didn't have time to review the first, which did, however, please him greatly.[179] The other two he condemned, and Leyden's *Complaynt* prompted him to write a piece of autobiography. The man who suggested that the *Complaynt* be published, he wrote, was "the editor of the Poems from the Maitland Manuscript." That editor had claimed that Robert Wedderburn was the author of the *Complaynt*. Leyden tried to confute his opinion, "but certainly without success." One of Leyden's doubts about the authorship "might have been done away by looking at Mr. Pinkerton's History of Scotland." The reviewer had read through the "two hundred and ninety-two deadly pages" of Leyden's introduction, which formed a "prolix, digressive, and retrogressive dissertation." Then he came upon the text of the *Complaynt*, almost an appendix to the introduction, and what did he discover? It was printed "in *fac simile*, with all the confusion of the original edition." We suspect "the editor of the Maitland Poems would have followed a different method," but he is, "we believe," disgusted with the barren field of Scottish poetry and history:

> 'It is an ancient saying, that neither the wealthy, nor the valiant, nor even the wise, can long flourish in Scotland; for envy obtaineth the mastery over them all;' says sir David Dalrymple, in his Annals … translating the words of old Fordun.

By 1802 a new generation of scholars had sprung up in Scotland, the gifted young editor Leyden among them. In July 1802 he exposed Pinkerton in the *Scots Magazine* as "the antiquarian and historical Reviewer in the Critical Review."[180] Too late! Pinkerton had fled the scene. The future of the *Critical Review*[181] was to be uncertain enough after his departure. So was that of his wife and their two daughters Mary Margery and Harriet Elizabeth, left behind in Pimlico.

5 Paris Interlude, 1802–1805

Pinkerton thought his *Modern Geography* would remain authoritative for a hundred years.[1] It wouldn't, but it did bring him temporary fame. It was instantly popular. A print run of 1,500 went on sale in July 1802; by November, 800 were sold and a rumour started that a second edition would be called for. By 1804 the complete impression was exhausted.[2] Second and third editions would be required. A large print run of the abridgment was issued in 1803, found its way into the schools, and sold out by 1805; it too was reprinted.[3] Two adaptations of the abridgment, made independently of Pinkerton and aimed at pupils in primary schools, had appeared by 1820. The American edition published in 1804 was followed by an American abridgment the year after. A French translation appeared in 1804, and by 1827 there had been four editions of the French abridgment. *Modern Geography* was a hit, even with critics. The English reviewers were virtually unanimous in praising the work. William Taylor wrote in the *Annual Review* that Pinkerton was a greater geographer than Strabo.[4] The *Critical Review* thought the work formed "an aera in geography" and fixed "a standard of excellence in animated description."[5] The *Anti-Jacobin Review* recommended it to the public "as a capital production, with which there is nothing in the English language that deserves at all to be compared. It is a monument undoubtedly of singular industry, of extensive knowledge, and of discriminating judgment."[6] The *British Critic* and *Edinburgh Review* approved the work, the latter with reservations.[7] The finest compliment of all came from Robert Southey, a careful book buyer and no friend of Pinkerton's; he thought fit to invest in the "useful" compilation.[8]

Well might Englishmen extol the book! The title said it was "*A Description* of … *all Parts of the World*," as indeed it was, but in outlook it was an English geography and, as noted, England led all other nations in Volume 1. He would later say he did not write "systematically, and in favour of one country," that "he really wrote with general views, and as a well wisher to humanity."[9] It is true that he wasn't a harping anglophile or chauvinist.[10] He had England's interests more in the back than in the front of his mind as he wrote. Yet they were present. She should make colonies in this or that region, prevent Russia from encroaching on "our opulent possessions" in the Orient, do well to trade with China, and so on. A high regard for England did not preclude an admiration for other countries, and, as we've seen, Pinkerton had been so deeply moved by the Revolution across the Channel that he'd written a book on it in 1790. By 1802 such political attraction had vanished, yet curiosity remained. France in the decade following the Revolution had become a marvel among nations. "When expected to fall an easy prey," he wrote in his geography, "she suddenly arose the aggressor, and has astonished Europe by the rapidity and extent of her victories."[11] It was a phenomenon that a universal geographer needed to investigate.

Pinkerton's letters written prior to his departure from London reflected fatigue and fretfulness. Those written from Paris show a far different mood. On 1 June 1802, shortly after his arrival, he wrote to Cadell and Davies, dating his letter by the republican calendar, and asking for copies of his books to give to new acquaintances. "The literati here are so kind and attentive to me," he said, "that I am anxious to make some little return."[12] When he wrote to Douce in October he was still excited by what he was seeing:

> I saw lately a curious collection of engraved gems from Egypt. A M. Torsan has a curious private collection of antiquities with which you would be highly pleased. I need scarcely inform you that the Natural History in the Jardin des Plantes, the Bibliotheque Nationale enriched with so many new acquisitions, and the Gallery of Paintings and Statues [the Louvre] are three objects which cannot be rivalled.[13]

He had leisure to explore not just the big institutions but the nooks and crannies, beauties and blemishes, of the great city. We might examine a few excursions and reflections from the memoir of his visit, *Recollections of Paris* (1806), passing over for the moment the "interweaving Essays" and other "patch-work"[14] that filled out the two thick volumes. The book was no major literary event, but its importance to a biographer is obvious since it provides a revelation of character and personality at a key moment in Pinkerton's life.

One charm of Paris to him was its variety. A visitor passed abruptly, he said, from "a trifling or disagreeable object, to one of greatest utility and magnificence." Yet care had to be taken while out walking. The streets were narrow and filled with "ordure." A passing cabriolet could easily cause an indignity. This ordure was an "abominable" nuisance. No matter where you went, "while your eyes are delighted, your nose is offended." Again, if you strolled in the early evening through the gardens of Paris you could not help but notice the numerous prostitutes, but they "are never rude or indecent, the utmost familiarity being to seize your arm, and one word of dismission suffices." In the Turkish Garden the secret bowers were often "seats of promiscuous and vulgar love," so it was wise to tread more frequented paths.[15]

The public buildings of Paris, he said, were superior in both general magnificence and "infinite variety of elegant architecture" to those in London, the English preferring utility and comfort to frippery and ostentation. Yet to return from Paris to London was like moving to a Dutch town "with little neat uniform houses of brick." Napoleon, though his ambition threatened "to devour all Europe with military barbarism," had to be credited with embellishing Paris with elegant buildings and monuments. He accomplished more in three years than the Bourbons, with their "pimps and mistresses," did in the whole eighteenth century. England, Pinkerton stated, had nothing to equal the Bibliothèque nationale, Jardin des Plantes, and the Louvre. The "three" treasures of the Louvre were the Venus de Milo, the Lycian Apollo, and the Laocoön.[16] The second provoked "tears of admiration"; the third, though the subject was unpleasing, reminded him that "pain is so necessary a lot of humanity, that we must not join those who blame the artist for his choice." He spent many days at the Jardin des Plantes, talked with scientists there, and examined the exhibitions. He was "shocked" to see Louis Jean-Marie Daubenton's brain preserved in a glass jar. That seemed "indecent."[17]

Pinkerton visited the Tuileries, the palaces of Versailles and Luxembourg, the hall of the National Assembly, the Institute of France, and the Paris Exchange. Who but him would have attached special significance to the lack of carpets in the Assembly hall? If a floor is uncarpeted, he said, the feet get cold and blood rushes to the head, making it hot; thus great events arise "from little causes." He investigated the Paris graveyards – doubtless little suspecting he'd end up in one. At the

Montmartre churchyard coffins were covered with "a few shovels-full of sand" and, though he was refused permission to examine them at close quarters, "the smell is offended at the distance of forty or fifty yards, if the wind blow from the cemetery." The dead, he chided, should be treated with greater respect. He gave special attention in his wanderings to characters and events from the Revolution, noting where the "heroic" Charlotte Corday stabbed Marat, where Robespierre had been shot, and where, "if we credit the accusations published by authority, Moreau met Pichegru, and almost consented to the restoration of the Bourbons."[18]

On the outskirts of Paris were taverns where the poorer classes went to drink, eat, and dance. Pinkerton visited one such tavern and reflected on the waltz. "The sexes know each other better, and rejoice in the perpetual contact," he said, but the dance "approaches so nearly to the lascivious, that one would not wish to see one's wife or one's mistress thus in the arms of another." The subterranean passages of the city were a curiosity: he explored them for "three or four hours," holding a candle to light his way. At Grignon northwest of Versailles there was a spot where varieties of sea shells, underneath "a solid limestone rock," could be dug out of the earth. He went twice, collected shells, and speculated about how they got there. Voltaire had said they were thrown away by pilgrims on their way back from the Holy Land. That was absurd, Pinkerton thought; they were clearly an ocean deposit. The limestone rock above the sand showed that "it was the work of ages, and not of years." But why was this "pure sea sand," with beautifully preserved shells, laid down only in certain spots? It was a mystery. "All theories of the earth are here equally deficient," he wrote, slipping again into worm theology, "and the Author of nature has spread a veil over his operations, which that little insect called man, whose existence is but an hour, and whose mental light is darkness, in vain endeavours to remove."[19]

Life in Paris, he fancied, was healthier and less inhibited than life in England. The "iatrical, or medical arrangement" of dishes at mealtime, for instance, was better: "A person who leaves England with so weak a stomach [as he did?] that it has long refused the luxury of two dishes, may, without inconvenience, taste of twenty at a French repast." Or take divorce. In England, divorce was permitted only on the grounds of infidelity. In France, although a woman could sue for divorce if her husband kept "a servant maid or other concubine in the same house," infidelity was "a matter of laughter." In the new civil code one of the grounds for divorce was "a complete dissonance of temper." And "certainly," he wrote, likely with his eye on his own situation, "as marriage could never be intended either by God or man, to produce mutual misery, it would seem that no juster cause can be assigned." Again, though French wines were various and delicious, it was rare to see a drunken man in the street, and gentlemen rarely drank more than six or eight glasses in an evening. The practice of medicine in France differed from that in England. Nervous disorders were rare because Frenchmen drank little tea or spirits, but when they did occur they were treated by "the products of the orange-tree." French doctors emphasized the blood:

> In case of [a] cold, even the warm bath is not permitted, as it is supposed to direct the disorder towards the breast. For a head-ach the immersion of the feet in salt and water is recommended, in order to divert the progress of the blood towards the head. In general the French medical language often refers to the blood, which seems almost forgotten with our physicians.

Still more evidence of hypochondriasis intruded, as it did in so many of his books. French public baths were healthful, he said. In England he had experimented with all forms of bathing and found none satisfactory; in Paris he often lay for "an hour or more" in the tub and never emerged "without feeling a marked increase of appetite and health."[20]

Pinkerton still preferred the English to the French, or let on that he did in the book. The English were solid, honest, and responsible citizens; the French were not. For one thing, they were Roman Catholics and so were guided "by sensations and not by arguments." Their "lively passions" made them prefer "the enjoyment of the moment to any future advantage." They seemed to him "of all possible nations, the least adapted to a republican form of government," owing to their "vanity" and "love of petty and frivolous distinctions." Pinkerton knew of what he spoke, since he had "perpetually mingled in respectable society in Paris."[21] (A boast, no doubt, but likely with an element of truth.)[22] There was a charm in that society that was "nowhere else to be found."[23] This, he thought, was owing to the urbanity that sprang from the company of women, who surpassed men in brilliance of wit and elegance of manners. The French woman was "warm, humble, and alluring," never, like the English female, proud, cold, and avaricious. And although some ninety out of a hundred French women had lovers, "Let it not, however, be supposed, as not infrequently happens to the inexperienced traveller, that the French fair grant their favours without previous selection, difficulty, and devotion":

> Innumerable are the young and beautiful females who preserve the sanctity of the marriage-bed, and amidst a charming freedom of manners, and even a great friendship for another man, are models of maternal tenderness, and conjugal fidelity. "No, my good sir, it would infallibly be the death of my husband, the father of my children, and I should never survive the consciousness of having caused such a disaster," was the answer of an enchanting Parisian lady, after long solicitation, to a youthful admirer.[24]

When Pinkerton published these and similar observations in 1806 it seemed to some English critics that he'd undergone a transformation. The "plodding, pedantic, pugnacious archaeologist," said the *Edinburgh Review*," had become "a Parisian *petit-maître*" who prattled about wines, dishes, ornaments, and "the freedom and gallantry of the French ladies."[25] The *Critical* said that "very amusing" though the book was, "many symptoms of bad taste are every where apparent," together with "an idle parade of learning, and an unreasonable admiration of the French." "What are we to think of this?" the reviewer asked of the enchanting Parisian lady's response to the "youthful admirer." "Must we consent to admit feelings of prudence in place of those of honour?" French women might be good mistresses and friends but were inferior to English women in "the more sacred and important duties of life."[26] The *Monthly Review* seemed more forgiving, but said "some individuals" would find Pinkerton "too *frenchified*."[27] The *Universal Magazine* said the "grave antiquary" had become imbued with "gallic mania" and carried away by "sensual pleasures and erotic delights."[28] The *Scots Magazine* delivered a fierce broadside too.[29] The *Quarterly* had only contempt for a man who, at a time of life "when the passions are generally supposed to subside," threw off his "plodding and laborious habits" to set up as "a Parisian coxcomb."[30]

Pinkerton's references to sexual scenes and issues in *Recollections of Paris* – he even refers to the effects of gonorrhea on young men, and knew more than he should about pudendal disorders of women[31] – thus sent a *frisson* of horror through the British reviewers and distracted them from paying enough attention to the book's other contents. He confirmed the common view of French women's "supposed licentiousness,"[32] though he was not in the least offended by it. The thought could have crossed the minds of some reviewers and readers that *he* was the "admirer" who'd proposed adultery to the enchanting lady, putting in the word "youthful" to throw off the scent, or that he was the blundering "inexperienced traveller" who didn't yet know how the courting game was played à la mode Parisienne. He could have been either; he was not yet fifty, the blood, it hardly needs saying, can still be hot at that age. His *was*.[33] That he needed, shall we say, female companionship of a certain kind, has already been made apparent in this biography, and his attempt at walking out of his marriage showed how far he was prepared to go to get it. In *Recollections of Paris* we come close to the real Pinkerton, and even some of what appears to be kowtowing to convention, as in his reference to the "Author of nature" already cited, could represent what he now really thought. Not that he didn't sometimes try to placate his English readers and critics. He was adventurous, but not as fearless as he once was. Yet Pinkerton was at bottom still unconventional: sexually liberated, eccentric in religion, egotistic, and wide-eyed, almost infantile, in his curiosity. He wanted more out of life than what a sterile marriage offered. He was drawn to France in 1802 possibly because he sensed that someone of his inclinations could live more happily there. That thought would stay with him, as we will see.

Two general points must be briefly made about Pinkerton's thinking as reflected in *Recollections of Paris* (leaving one more for later). The book confirms the turnaround, noticeable in *Modern Geography*, in his view of the French Revolution. He was now drawn near the scenes and characters of the Revolution; he could see what constituted it and what its lasting effects were. "The violent and sanguinary revolution," he wrote, "has rent the ties of friendship, and even the confidential intercourse of relations. The reign of terror has introduced such mutual distrust, that almost every man looks upon another with a doubtful and jealous eye."[34] Of the Republican triumph at the Palace of the Tuileries, he wrote: "the fatal tenth of August, which dissolved the monarchy of France, led to ... useless horrors, and crimes of no avail."[35] The revolutionaries were "barbarians."[36] His disenchantment was complete. The second point has to do with religion. Again, a change of view is noticeable, and that change is a movement – a further movement, since signs of it too are in *Modern Geography* – towards a quirky Christianity. He used the Bible to argue down those who claimed the English and French were natural enemies, citing Luke and Matthew: "The founder of our religion has said, 'Ye know not of what spirit ye are. He that uses the sword shall perish by the sword'"; and he praised the Quakers for their "literal observance of the precepts of christianity."[37] He also looked in scripture for a definition of a just man, and found Solomon, who "had a harem of three hundred wives and seven hundred concubines." (Going on memory; Solomon had 700 wives and 300 concubines.) He expanded on this, noting the absurdity of treating bigamy so seriously, though not specifically mentioning England. The "real christian code," he said, "as presented in the Scriptures, breathes the most mild and tolerant morality."[38] He is here a sort of heretic Protestant. At the beginning of his attack on Rousseau (of which more in chapter 6), we find him, astonishingly, citing Milton from memory, saying

> That to the height of this great argument
> I may assert th' Eternal Providence,
> And justify the ways of God to man.[39]

References to providence are found elsewhere in the book.

He was not long in Paris before he had dealings with booksellers, who apparently tried to "skin"[40] him, whether successfully or not is unclear. He had reason to think that his *Sketch of a New Arrangement of Mineralogy* would be well received. In March 1802 Mawe gave a copy to Eugene Michel Patrin, a scientist, and he had "kissed the book and was delighted: he was in ecstasies."[41] A translation, *Esquisse d'une Nouvelle Classification de Mineralogie,* appeared around June 1803. *Recollections of Paris,* by the commentary it contains on rock structures, rock composition, shells, and cabinets, reaffirms what has been already apparent in earlier works, his fascination with mineralogy and his overall conversion to science after the failure of the *History of Scotland.* There can be no doubting the genuineness of that conversion. Not that he forgot altogether his earlier enthusiasms. His *Dissertation on the Scythians or Goths,* translated as *Recherches sur l'Origine et les divers Établissemens des Scythes ou Goths,* was published early in 1804. He'd made revisions. At about the same time the bookseller Jean-Gabriel Dentu brought out *Géographie Moderne* in six volumes.[42] This book, in the preparation of which Pinkerton, it appears,[43] took no part, and the *Recherches sur les Goths,* now made life acutely uncomfortable for him.

By March 1803 the Treaty of Amiens was ruptured, and in May France and England were again at war. From the middle of May onwards,

English travellers in France were refused permission to return home; many were detained at Verdun and other cities. Pinkerton had been in Paris a year, and so far had given no signs of wanting to leave. Now he couldn't go, and the possibility existed that he would be sent out of Paris and incarcerated. Among his papers is a rough draft of a letter to a French minister, asking to be excused from such treatment:

Citoyen Ministre

On vient de m'assurer que le gouvernement avoit déclaré tous les anglois résidents en France, prisonniers de guerre, et qu'ils recevroient incessament l'ordre de se rendre à Fontainebleu. Je ne réclame point contre la première de ces dispositions, mais j'ai l'honneur de vous observer que vous m'avez autorisé à faire imprimer, à l'imprimerie de la République, des *Recherches sur les anciens peuples de l'Europe*. Cet ouvrage qui est fort avancé, exige ma présence pour être terminer; et je viens vous prier instament Citoyen Ministre, de vouloir bien m'autoriser à rester à Paris, pour achever ce travail.[44]

He was not thinking of asking to be allowed to go home, just not to be sent from Paris to a place of internment. He lay low, and through 1803 was apparently left to his own devices. Then, early in 1804 the translations of his *Dissertation* and *Modern Geography* appeared and his position became less secure.

His geography may have impressed the scientific fraternity in Paris, or some parts of it, but it did little to bolster French national pride. Instead, the author was a sort of amateur politician pointing out how his country could gain an advantage over her rival in future, or how British forces, especially the navy, had triumphed in the past over the French. Of the recent war, for instance, Pinkerton wrote that "by the protection of all-ruling Providence, the British empire rose superior ... and remained free from those scenes of carnage and devastation which attended the French progress into other countries ... Great Britain has less to apprehend from France, than at any former period." France faced "the fixed destiny of inevitable defeat."[45] This was diluted in translation – e.g., Providence's support for the enemy was omitted[46] – but the still apparent air of British smugness must have been intolerable to many delicate nationalist readers. Nicolas Louis François de Neufchâteau, president of the Senate, thought *Géographie Moderne* contained the official policy of the British government, the blueprint of its plans for world domination. Soon after its appearance he published *Tableau des vues que se propose la Politique Anglaise dans toutes les Parties du Monde*, a pamphlet denouncing Pinkerton's book. There could be no doubt about the pernicious policies of the British, he wrote. The political sections of Pinkerton's work made them clear:

> L'auteur nous assure lui-même qu'il n'a traité cette partie que d'après les meilleurs publicistes de son pays. Or, dans cette division de sa géographie, on doit être, je ne dis pas simplement étonné, mais très-scandalisé de l'esprit de discorde que cette politique s'efforce de répandre, des boulever-semens qu'elle suggère ou qu'elle annonce, et du projet trop clair de ramener à elle seule la spoliation de l'univers entier. Ce projet est l'effet d'un plan que se mûrit depuis long-temps. On en soupçonnait l'existence; mais c'est un ecrivain anglais qui en donne la certitude.[47]

He proceeded through Pinkerton's book, noting the (to him) brazen assumptions, twisted logic, and evil intentions that lay behind the project to impose "la domination anglaise" upon the world. France, he wrote, should protect herself against such works, the author being a warmonger who believed that governments "ne sont occupés qu'à se détruire et se manger les uns les autres." (Not that Napoleon refrained from being un mangeur de gouvernements!) He ended with a plea to European nations to awaken to "au chant du coq gaulois."[48]

In normal circumstances an absurd attack of this nature would have caused little resentment against a writer, but François de Neufchâteau was no ordinary citizen and in 1804 "the very air" was "contagious."[49] Pinkerton was threatened not merely with detention but with imprisonment.[50] Out of fear of repercussion he didn't dare publish an answer to the pamphlet – he would later[51] – or apply for a passport to leave the country, something that was now crossing his mind.[52] The Senate president was not his only enemy. Edme Mentelle, a leading geographer, was an opponent, as was Mentelle's associate, Danish-born geographer Conrad Malte-Brun, who spread the word that Pinkerton was a corrupting influence on youth, and whose plagiarisms from Pinkerton's work were later to be exposed in court.[53] The *Recherches sur les Goths* also bred enmity since many Frenchmen thought themselves descendants of the ancient Celts.[54] The effect of the hostility his two books stirred up was that he was thought by many to be one of the enemy's hirelings. Or, as Pinkerton put it, he, "a mere literary man solely occupied in scientific pursuits," was considered "a spy of the English government."[55] It was unnerving, perhaps frightening.

In the rough draft of a letter to Cadell and Davies at the end of November 1804 Pinkerton wrote that he had been compelled to remain in Paris "against my will a [canc.: twelvemonth] year & a half longer than I intended." It was in his interest to exaggerate to the publishers the length of his enforced stay, so the cancellation, twelvemonth, is probably the more dependable statement.[56] If so, he intended to leave Paris late in 1803, about a year and a half after his arrival. Late 1804 was in any case well beyond his projected time away from England, and he hadn't taken enough money with him for such a contingency. By December 1804 his resources were "nearly exhausted," and he was forced "to sell out stock at a great disadvantage" to pay expenses. He worried that he might be "left perhaps to perish in a foreign country."[57] Cadell and Davies, with whom he was involved in a dispute over *Modern Geography* (see chapter 6), refused to send him money. Pinkerton approached the London lawyer John Spottiswoode to try and squeeze money out of them:

> If a lawsuit be necessary, so be it. But as you have known me ever since the year 1780 I wish you would act rather as a friend to both parties. Mr. Strahan who has printed the book is also I believe your brother in law and your son is in his house so that your amicable interference would have the greatest weight imaginable. If an arbitration be prepared pray act as arbiter for me. Pray do not suffer your old acquaintance to perish for want in a foreign country. Have the goodness to desire Mr. Strahan's interference. I hope to get away in the spring and all I want is that they should let me have £100 or £200 in the meantime.[58]

His groans about perishing were probably faked; but that he could be a groveller when facing the possibility of being hard-up would be amply demonstrated later. Spottiswoode died soon after receiving the letter. In January 1805 Pinkerton screwed up his courage and asked the police for a passport to go to England.

The "abominable police"[59] were not easily convinced that he was an impartial literary man deserving of special treatment. Why, they asked, had he been so long in Paris without reporting his whereabouts to the authorities? He answered sternly, or so he reported in 1806, safely back in England: "What can an honest man have to do with the police?" They quickly yielded, saying "Remain in security; there is not the smallest accusation against you; you have behaved perfectly well." The incident was "singularly disagreeable to a man accustomed to reputation and respectable society."[60] Thus Pinkerton's public account of the proceedings. In truth, the police "pestered" him (he explained privately in 1809)[61], evidently searched through his papers,[62] and were reluctant to give him a passport. Pinkerton wrote to a French official, promising, on his arrival in England, to write a book "demonstrative of the advantages to be derived by both nations from a permanent peace, accompanied by a commercial treaty." He added that English booksellers owed him about £1,000 and refused to pay a shilling, that he was living in Paris at an advanced age, "almost destitute of the necessaries of life," and that in England he could afford to live comfortably on £700 a year.[63]

One way to get out of France was through the auspices of a neutral power such as the United States. Around the beginning of 1805, Pinkerton met Gen. John Armstrong Jr, the American Minister to France, his secretary David Bailie Warden, and through them the diplomat now engaged in dealings with Britain, James Monroe. If he could travel with Monroe, he told Warden, there would be no difficulty in getting to London.[64] But in May he was still without a passport. He then tried to have Armstrong appoint him to the post of Secretary Interpreter, which again would enable him to leave the country. It would have another advantage too, "which is, that if I continued to be disgusted, as I have been, with the despotic regimen and heavy taxation[65] in England, I should either wish to revisit France, or to pass to America, I could not be hindered." He was "sad and solitary," he wrote to Warden, who had scientific interests and had become a friend, "as all my acquaintance suppose me in Holland." He implored him to "call as often as you can."[66] It was through Armstrong's influence that Pinkerton at length got a passport. He was deeply grateful. On his return to London he placed a portrait of Armstrong in his "best apartment."[67]

He left Paris in July 1805. There being no direct means of transport between England and France, he was obliged to go to Holland and embark from Rotterdam. The trip northeast through parts of Flanders and Belgium restored his spirits. At Cambrai his passport was demanded but no unexpected difficulties arose. Hal, in Belgium, had a small fourteenth-century Gothic church; Pinkerton visited it and made inquiries. Why, he asked the "extremely ignorant" priest who served as guide, had the decorations been preserved intact for so long, and from which quarry in Flanders did the builders obtain their marble? The priest couldn't answer, and Pinkerton "gave him half-a-crown for no information whatever." One of the treasures in Brussels, a painting by Rubens, provoked a reflection on "the ignorant and fanatic order of

Franciscans" and their "lunatic founder." At Mechlin he nearly lost his life when the four horses drawing his carriage took fright and bolted. A "journeymen baker," however, rushed from his shop and at the risk of his life seized the reins of the foremost horse. An author's life was saved. Pinkerton was forced to stay twelve days at Rotterdam awaiting favourable winds. He could find few literary men to mix with and would have been "devoured with *ennui*" had not a respectable merchant, "descended from a Scotish family," invited him to spend the time at his home. Money and Calvinism, Pinkerton commented, form "the sole meditations of a Dutchman." Then he sailed:

> The passage to Harwich occupied five days … the westerly winds never ceased to blow with some violence; and though the Orion was a new and beautiful packet, and the captain of acknowledged skill, instead of gaining the mouth of the Thames, the first land perceived was Orfordness. The ship having three times touched the ground at three o'clock in the morning, the chief passengers insisted on being landed at Harwich; and the delight with which I sprung upon English ground may easily be conceived. Delivered from the police of Paris, from passports, garrisons, commissaries, and consuls, I felt a flush of satisfaction which doubled the sensation of existence, and hailed a country which retains so large a portion of practical liberty. *Esto perpetua!*[68]

In mid-August[69] this "chief passenger" was back in London, negotiating with booksellers. It cost him nearly £75 to get his books, maps, and other material through customs, and he owed over £90 to a Parisian banker.[70] It would take more labour with his pen to get him back on his feet.

6 The Dishonoured Veteran, 1806–1814

Back in London after his holiday abroad, Pinkerton might have thought he had a lot to be thankful for. In his absence his houses in Scotland had been sold, not without difficulty, to the city of Edinburgh for an annuity of £75[1] – after tax deduction, a small but reliable income for the rest of his life. His reputation as a man of learning had not yet been effaced. One competitor, perhaps misreading the signs, thought Scottish interest in his books had risen.[2] He had been the subject of a complimentary biographical sketch in 1801, a second had appeared in 1804, and a third would come out in 1807, showing he was, at that juncture, still notable in some circles.[3] One enemy, Ritson, was dead, and another, David Macpherson, had decided to hold his tongue. Two works written against him while he was off in France sightseeing, Turner's *Vindication of the Genuineness of the Ancient British Poems* and Coxe's *Vindication of the Celts*, must have injured his reputation to some extent, but their full impact was not immediately apparent. The former was a convincing, if arcane, rebuttal of what he'd written about the Welsh bards, but the *Critical Review* stood by him, saying, after an analysis of Turner's arguments, "We have surely stated grounds of hesitation sufficient to justify the expressions employed by us on a former occasion."[4] The second *Vindication*, another stinging refutation directed mainly at the *Dissertation on the Scythians or Goths*, got a long critique in the *Edinburgh Review*. The reviewer, while by no means flattering to Pinkerton throughout, far from it, came down on his side at the end. The vindicator – Coxe's authorship was still a secret – "completely fails," the critic said, "to overturn Mr. Pinkerton's hypothesis."[5] This judgment must have been a great relief, for Coxe had carried out a close scrutiny of the handling of sources in the *Dissertation*, a punishment few scholarly books have had to withstand and fewer still survive unscathed. These two opponents had fought back, and hard; but few in the literary establishment seemed to realize how well they'd argued. The fact that their vindications came out amidst the accolades Pinkerton was receiving for the geography might explain how little notice they received. Despite all this attention, he would never again play as active a role in English letters as he had in the years 1781–1802. More attacks on his writing *would* damage him. From 1805 on, financial worries plagued him. He had health setbacks and woman trouble. His life was unsettled. Something else: he'd been away too long. His three years in Paris had made him a stranger to recent literary developments in England, and to young writers in the new century he came to be seen, if by 1806 they thought of him at all, as a curiosity. It is surprising how little he figures in the correspondence or publications of the emerging literati.

He was in any case out of sympathy with some of the changes in letters, with the heightened emphasis on novels, for instance. He regretted that no high tax was imposed on "that species of publication, which corrupts and debases the female mind, and renders it wholly unfit for the real business and duties of life." (Another clue, maybe, to his difficulties with Henrietta.) "What novel," he asked feelingly, has ever represented the heroine as seized with a fit of the tooth-ach in the midst of her felicity and golden dreams?"[6] He was not alone in this perception of the novel's supposed baneful influence on women.[7] And as for poetry,

If Thought, if Science, if the sentient Soul,
Arrange not their high powers in rich array;
If Taste and Learning spare their keen controul;
If nought of varied life the page display;
In sweetest melody the sounds may roll
But want the sense that crowns the classic lay.[8]

This, from a sonnet written around 1813, was directed "To some young poets of the present day" who dwelt on groves, streams, and flowers – just as he had done in his long-forgotten *Rimes*. In *Recollections of Paris* Pinkerton gave pride of place to the men of learning, the scholars and scientists of France, above those in the belles lettres. He noted too that "many literary men now hold important offices in the state" there.[9] Part of his dismay with the literary scene in England was a concern that men of learning like himself were ignored, while "sentiment-mongers"[10] were rising in fame and fortune. This feeling deepened as time passed. He was also, in time, repelled by some of the periodicals that were appearing and gaining influence. These new journals, he said, no doubt having in mind mainly the *Edinburgh* and *Quarterly* reviews, displayed "political phrenzy," "envenomed acrimony," and "concentrated malignity."[11]

The six chapter-length attacks on Rousseau in the first volume of *Recollections of Paris* hint at a still deeper estrangement. That Rousseau's radical ideas appalled him was made clear in *Modern Geography* and indeed earlier.[12] In the greatly expanded treatment he got in *Recollections of Paris* he was depicted as "an angel of darkness," "the very leader of the new giants" of philosophy "who attempt to disturb the creation, the beauteous order, the golden chain of Jove." Rousseau "assailed the arts, the sciences, society itself, its progressive improvement, and of course a superintending providence." (It is an oddity to see the once highly sceptical Pinkerton stand up for providence in this fashion!) His point by point rebuttal of Rousseau's positions showed no small skill in argument. He quoted the paeans to the "state of nature," mostly from the *Discours sur l'origine de l'inégalité*,[13] and cited as one proof against them, as he had in his geography, the condition of the aboriginals of Australia who "appear evidently to be in the first rudiments of society," i.e., the first condition lauded by Rousseau, the one prior to that of the savage:

> Their constant conflicts, their sanguinary rites, their revengeful spirit, their gross ignorance, their vanity and affectation, their uncomfortable life, chiefly occupied in procuring scanty and noxious food, their idle fears and superstitions, their violent and depraved passions; in a word, their comparative misery and inferiority even to other savages, all evince that Rousseau's idea of a state of nature is a mere phantom.[14]

He similarly lambasted Rousseau's ideal savage state, also said to be free of many of the ills of civilization, including that created by property. The word property had no meaning for the savage, Rousseau said; "public esteem is the only good to which each aspires."[15] Pinkerton countered that "where nothing exists worth the name, there can be no temptations to theft" and that "the very bent and instinct of a savage is to steal." Rousseau claimed savages lacked the effeminacy created by civilized life. Pinkerton said that among native peoples of the southern Pacific islands and America "abominable vices, and what are called unnatural crimes … are found to prevail." Rousseau argued that the original equality of men was altered by competition and the varying capacities of individuals, so that "inequality insensibly developed." Pinkerton replied that "equality and inequality" were not "accidental to human nature" but instead "necessary and unavoidable, interwoven in our very texture, flowing in our veins"; he added, "we no where find, even in the rudest and most primitive state of human nature, any trace of equality." Another writer who got a severe tongue-lashing in *Recollections of Paris* was the Marquis de Sade, whose novel *Justine* (1791) was termed a "detestable rhapsody" and a vehicle of "such selfish and cruel gratification, as madness alone would conceive."[16] Pinkerton evidently thought de Sade one of the disturbers of the golden chain of Jove, let loose by Rousseau and the revolution he helped to engender.

Rousseau's attack on contemporary European civilization and his preference for an imagined ideal primitive human condition called into question what were now some of Pinkerton's bedrock beliefs. He sensed too that Rousseau's ideas were current and influential, as indeed they were – they lay behind much of the Romantic movement in literature.[17] That Pinkerton was the child of another age was reflected in the range of citation from long-dead British writers through *Recollections of Paris.* Shakespeare, Samuel Butler, Milton, Addison, Pope, Dryden, Johnson, Fielding, Goldsmith, Gray, Robertson, Adam Smith: these were the props he raised to support or ornament his views. The iconoclast we met in *Letters of Literature* has walked offstage.

The two issues immediately facing him on his return from France related to his money and his family. He tried first to settle a dispute with Cadell and Davies over *Modern Geography*, one that had begun before its publication date and was carried on through his time in Paris. He had signed two contracts with them. The first, dated 10 March 1800, specified that he was to receive £1,200 for the first edition, and that the publishers would issue 1,500 copies of the work. In the second, dated 2 February 1802, it was agreed that Pinkerton would be paid £400 for the first edition of the abridgment of the geography and that Cadell and Davies could publish as many copies as they wished of that abridgment.[18] Both contracts were broken. Cadell and Davies printed 1,750 copies of the main text, adding 250 on large paper. In a letter in February 1802, Pinkerton had asked what was to be his "allowance" for the extra copies.[19] He was told that the work had been overprinted by mistake, and the only allowance would be to pay him at once the £200 that would be due him on the publication of a second edition.[20] He went for advice to lawyer Charles Butler.[21] At length it was agreed that Butler and Thomas Cadell (Sr), the founding publisher, who had been succeeded in the business by his son Thomas and William Davies, would act as arbitrators in the matter,[22] but old Cadell's death in 1802 prevented that from happening. Pinkerton then, as noted earlier, broke the second agreement by leaving England before finishing work on the abridgment, having received £150 of the expected £400. Cadell and Davies thought that Arthur Aikin's part in the abridgment was to be only as corrector of the press; it was more than that,[23] but the publishers were, by agreement between Aikin and Pinkerton, to be "kept in the dark" about his bigger role. When Cadell and Davies learned what that role was, they told Pinkerton he had forfeited any further claim upon them for the work.[24] Early in 1803 in Paris, he'd drawn a bill for £200 on them; they refused to honour it.[25] A further attempt at arbitration in March 1804 came to nothing.[26] By 1804 Pinkerton had finished with Aikin. Longman & Co., which had a share in the geography, said in November that they wanted to proceed "almost immediately" with a new edition of the main text, putting amendments in Aikin's hands; he replied that asking Aikin to revise his book was like asking "a sign painter" to retouch portraits by Van Dyck.[27]

On 19 August 1805 a statement of accounts on *Modern Geography* was displayed at a meeting in the Longman premises. Before seeing the facts and figures, Pinkerton, Thomas Longman, Cadell, and Davies signed an agreement. Part of it was this: if it appeared from the accounts that the book (the large work, not the abridgment – Pinkerton had now withdrawn his demand on that) had "hitherto done no more than repay its expences with interest on the money advanced," then Pinkerton would relinquish all claims to the first edition; if it appeared that more than 12 per cent profit had fallen to the publishers, he would receive one-half of the remainder. The accounts were displayed. Proceeds amounted to £5,563/10; total expenses incurred were £5,692/10. The amount "unrepaid by the sale" was £129.[28]

Pinkerton was normally astute, even stingy, in dealing with money.[29] He "had ample worldly shrewdness (in pecuniary matters at least)," R.P. Gillies wrote.[30] He couldn't exhibit much of that quality in dealing with Cadell and Davies over the *Geography*, since they had the upper hand. They were in dispute with a writer approaching fifty who held no office, and who they knew was concerned about his livelihood and prospects. He had broken his contract on the abridgment; they refused to pay for work he hadn't completed. Fair enough. But they too broke a contract by bringing out the extra copies, the returns from which amounted to over £1,100. Surely Pinkerton had been right to expect an "allowance" from that sale. The clue to his decision not to pursue that claim could lie in a clause in the new agreement offered him, which stated that in view of the "very extensive additions" he would make in the revised edition of the geography, he would be paid £600 "at such periods as shall be hereafter agreed upon" and £100 for every future edition in his lifetime.[31] This seemed a big concession. Yet by the original agreement on the geography (voided by the new one) Pinkerton was to receive £200 for the second edition and the same for each succeeding one. Cadell and Davies could make up the extra payment in future editions, especially if the "very extensive additions" Pinkerton made could justify an increased price. (It jumped from £3 to 6 gns. in the 1807 edition.) The stratagem worked. Pinkerton signed. But in December 1805, he wrote again to Cadell and Davies, saying he still felt entitled to "a <u>douceur</u> of £100" from the first edition. In both England and France he had praised the firm's "liberal confidence" in advancing large sums "solely on my literary reputation," he said, and his demands while in Paris were made from misinformation. When he came to London "and saw the facts I immediately desisted as you know." He abjured disputes "as not only hurtful to my nerves but injurious to all parties." If they paid him the £100 "I shall thank you – if not I shall neither speak nor write more on the subject."[32] Cadell and Davies waited a year and paid up. Pinkerton signed a receipt renouncing all further "Claims and Demands whatsoever" on the first edition of the geography.[33]

Cadell and Davies have been charged with "hard bargaining, rudeness, and untrustworthiness" in their dealings with Percy.[34] In 1799 they'd tried to encroach on the Liverpool book trade and made themselves hated by the booksellers there.[35] They showed some tightfistedness in their treatment of William Browne over his *Travels*.[36] Boswell, rumour had it, once stood over their press, counted sheets, and saw type distributed, before being satisfied he wouldn't be wronged.[37] Canniness, toughness, and rudeness on occasion characterized their treatment of Pinkerton. They were surely hard bargainers; but hard bargaining is a practice businessmen often have to adopt. Thomas Cadell Sr had recruited Davies specifically for his "business acumen."[38] The firm was a leading publisher in London, and it is doubtful that it could have risen to that position solely through "untrustworthiness." As for rudeness, Pinkerton could be rude too; in a letter to him in 1803 Cadell and Davies objected to his "offensiveness of language" in both conversation and correspondence.[39]

The other problem confronting Pinkerton in 1805 had to do with his wife and their two children. Henrietta Pinkerton had stayed on in Pimlico for ten months after her husband left for Paris, which suggests, first, that she expected him to return to his family, and second, that he hadn't told her how long he planned to live away. In July and again in September 1802 she came to Samuel Hamilton to ask for money, having been told by her husband to apply to him. Pinkerton had made some arrangement with him to dole out a sum to her, and she needed to pay rent and taxes.[40] It must have been humiliating for her to have to ask for money from a third party. In April 1803 she moved with her children to the Burgess family home in Odiham[41] and was still there when Pinkerton got back. She was, not unexpectedly, worried about money. Pinkerton wrote to her on 20 August 1805, the day after he'd made the new deal with Cadell and Davies. He had "a thousand things to attend to" in London, he said, and could do little at present to relieve her anxiety:

> I give my blessing to the little babes and kiss little doll a thousand times. ...
> If I die possessed of any property to whom can I leave it but to the children? Have they given me any offence? But do drop this anxiety about money, so nauseous in the mouth of a woman, and learn to trust in Providence and to enjoy life. My only property is a little annuity from the city of Edinburgh which you instigated Ingram to arrest,[42] so as to load me with law expences to the amount of £75. How I can settle an annuity I cannot divine. I wish to God you had more heart and more happiness and fewer of those brooding cares with which you are always embittering life and rendering existence a misery. But to drop this be assured I shall always do every thing in my power for the children. Only be chearful yourself and do not send me letters that recall melancholy.
> ... I am sadly out of linnen and know not when I shall know any of my old comforts. God knows I have need of consolation and not of lectures and reproaches. Adieu. Adieu.[43]

Here he was, supposedly leading a bachelor's life at 7, Clement's Inn, fretting over linen after deserting his family. Henrietta stayed at Odiham. He had many projects in mind and underway and, it seems clear, wished to be left to himself to complete them. He sent some money to her, as well as short but not unfriendly notes. He wanted books from his library, much or most of which she had taken to Odiham, manuscripts, keys for his mineral cabinets, and household articles. He was too busy to come to Odiham, he wrote on 7 October:

> I am very glad that you are so good a mother and so fond of our dear children whom I beg you to kiss for me again and again. At present every day is precious till the printers be set a going and I have two new editions[44] and a new work[45] all in the press at once. The pay to my amanuensis ruins me and I know not how to escape to Odiham. Be assured that I shall take the first opportunity.[46]

He was indeed "loaded with literary toil," as he told Walker.[47] Henrietta got such letters until 1809. In no surviving letter[48] did he mention that he wished to have her and the children come to London.

The contract with Longman and Co. on his *Recollections of Paris* was signed in September 1805.[49] The book appeared in June 1806; he got £300 for it. Some of it was clearly flung together in a hurry, but the attack on Rousseau was close to being a finished piece, as were the two chapters in the second volume written (as he'd promised) to recommend a treaty between England and France. There is something very appealing about this latter item, especially his insistence on the futility of warmongering and the need for friendly relations between the two natural neighbours – this lengthily argued while war still raged between them. Many scrappy pieces mingle with these and the edgy memories of Paris to swell the work to two volumes. A slipshod but at times charming hodgepodge, it was just the thing to bring out the worst in reviewers, and we have noted what they made of it. It was a serious error on Pinkerton's part to produce such a book just prior to the appearance of the second edition of *Modern Geography*. It very likely quickened the decline of that work's reputation. Such a decline was inevitable anyway owing to the progress of discovery and competition from rivals, not to mention changes within the discipline; but in this as in so many other instances he brought trouble on himself. In September 1806 the *Critical Review* turned against him and pointed out the "ignorance" in his geography.[50] In November the *Anti-Jacobin*, the periodical that in 1804 had offered some of the most fulsome praise of *Modern Geography*, pounced on him in a long-delayed review of the *Vindication of the Celts*. It was an unexpected and brutal assault, and an omen of what lay ahead.

As noted earlier, Coxe, the vindicator of the Celts, had taken issue mainly with the *Dissertation on the Scythians or Goths*; almost all of the attack was centred on that work. Yet *Modern Geography* was given space in Section 1 of the *Vindication*, as well as towards the end; and the title put the geography before the dissertation, as if it were the primary target: *A Vindication of the Celts, from Ancient Authorities; with Observations on Mr. Pinkerton's Hypothesis Concerning the Origin of the European Nations, in his Modern Geography, and Dissertation on the Scythians, or Goths.* The book had a double purpose: to vindicate the Celts and to expose the heterodox views, expressed forcefully in the *Dissertation*, but obliquely in the *Geography*, about the Book of Genesis. There can be no doubt that Coxe intended to discredit the *Geography* as much as the *Dissertation.* He could not accomplish that by using apostate passages from the *Geography* alone; they were few and innocuous. Longer and sharper quotes from the *Dissertation* were summoned alongside those of the *Geography* to colour the later work and expose its allegedly dangerous character. In effect, *Modern Geography* was blamed for what was in the *Dissertation.* In *Modern Geography*, Coxe said, Pinkerton "rejects the chronology of the scriptures, considers many nations of the earth as aboriginal, and establishes a great Scythian empire in the heart of Asia 3660 before the Christian aera." Since "this opinion … is advanced as HISTORIC TRUTH in an elementary book of geography which, as it has merit in other respects, and is the first of the kind in English, may get into general use; we deem it necessary to combat an opinion which tends to overturn all history, both sacred and profane."[51] This was stretching the nuances of the *Geography* into heresies, and concealing malice by fake praise. It was unctuous and deceitful – an underhanded scheme to destroy the book. Little wonder Coxe kept his name hidden. The *Anti-Jacobin* swallowed the package whole. The reviewer, likely John Whitaker or Richard Polwhele, both men of the cloth, took a quick look back through Pinkerton's record as a writer:

Our motives to action are known only to Omniscience, for even the individual is often blind to his own motives. We will not, therefore, determine whether an overweening conceit, or a passion for notoriety, or both, have influenced Mr. Pinkerton's career; but it is certain that in many of his works he has shewn himself more solicitous to advance what was new and strange, than what was true. – In taste, religion, and antiquities, under his plastic hands, old things are done away, and lo! all things are become new. Virgil and Horace,[52] hitherto the delight of ages, this professor of legerdemain has converted into a pair of blockheads, worshipped by as great blockheads as themselves. In an elementary work on Geography, professedly intended for the use of schools, he endeavours to set aside the Scripture Chronology, and all that is said of the creation, deluge, &c. in the first chapters of Genesis, and thus leads the minds of our youth to reject the religion of their fathers.

The anger mounted as the reviewer went on. Pinkerton's anti-Celtic "dreams" originated in "a diseased and perverted intellect." He had taken pains "to conceal, pervert, or falsify the testimony of antiquity." His account of the Scythian advance through Europe was a "wild, and unsupported hypothesis." The entire "*Gothic* fabric" had been "raised upon the sand."[53]

Worse was to come. In February 1807 the second edition of *Modern Geography* appeared. The already hefty book had been expanded to three volumes. We may safely guess that one reason Pinkerton padded it out was to justify the extra payment he got from Cadell and Davies; yet he argued in the Advertisement that new information had come his way, making the extension necessary. Even poets, he said, revised and enlarged their work, giving Pope's *Rape of the Lock* as an instance. He said that in his new system he had repaired the "brevity and deficiency" in parts of the old, and his revised geography had that harmony "indispensable in solemn and classical compositions." As for style, learned foreigners had assured him that the "purity of the grammar and expression" made the work of superior quality. There were other similar vaunts, including a reference to his "long, sedulous and painful researches."[54] Why he still felt it necessary to brag in this way is hard to explain. In April the *Edinburgh Review* delivered its verdict. It began by questioning the justification for such a huge enlargement of the 1802 edition, which had been presented to readers as "a finished work." The "unlucky purchasers of that complete system have now the satisfaction of hearing its manifold imperfections proclaimed by the author himself." The review then exposed the "unpardonable carelessness and ignorance" in the additions to the text, focusing for illustration on the treatment of Prussia, where "the number of inaccuracies and defects is altogether unaccountable," and on that of the Spanish colonies in America, the newness of which Pinkerton had bragged about. The reviewer pointed to the excessive dependence on excerpts from sources to fill up the volumes, stating that fifty-two of the eighty pages on Australasia, and eighty of 105 on Polynesia were thus compiled. The overall assessment struck at the very heart of the work:

… with all its pretensions, the new portion of the book is a most hasty and slovenly performance; eked out, by more than the excess of the ordinary book-makers' arts; and compiled with so little care or knowledge (where it is not mere transcript of noted works), as to render it at once a most unsafe and most cumbrous guide.[55]

This was followed with a discussion of the style of *Modern Geography*, touched off by Pinkerton's boasting. The reviewer was persuaded that "the discovery of a worse style than Mr Pinkerton's is reserved for some distant age." That line of attack too was copiously illustrated. "Something more than a journey to Paris, and an unshaken faith in his own perfections," the reviewer snarled, "is requisite to make Mr Pinkerton worthy of half the praises he lavishes upon his book, and its style."[56]

Pinkerton responded to the review by sending an ill-tempered letter to John Allen, the Scottish physician and expert on Spain, who he wrongly thought had written it.[57] (Geography, he said meekly later, was "rather approximate than exact.")[58] In the third edition of his book in 1817 he struck back at the "anonymous … nests of hornets" assailing him in new magazines, without naming specific ones.[59] What was needed after getting such a hard knock was to have a distinguished friend stand up for the victim; but friendship is not always available, even if it is the "peculiar gift of heaven."[60] Did the *Edinburgh Review*'s harsh reassessment spread, like some slow contagion? Possibly. It is not surprising to find in, say, the *Quarterly Review* in the years ahead a series of disparaging comments on the book. They culminated in a grossly unfair dismissal in a footnote, where Pinkerton was said to be "a mere dabbler in geography."[61] That was in 1821. By then he had long been absent from the literary scene. The *Quarterly* could sneer without fear of repercussion.

Two projects he undertook after 1807 were extensions of his geographical labours, the first a world atlas, the second a collection of voyages and travels. Plans for the atlas were discussed with Cadell and Davies early in 1807 and an announcement that it was "Preparing for publication" was printed in the new edition of the geography. Each map would be "drawn under Mr. Pinkerton's own eye"; the work would be "engraved with the utmost beauty that the state of the arts can admit; so as to be a national and perpetual monument."[62] He was always a big promiser! The cost to subscribers would be "about twenty or twenty-five guineas."[63] The proposal was to issue the atlas in monthly numbers, each containing three or four maps. It would take eight years to complete. The contract for the travel collection, also to be published in numbers, was signed in April 1807. A prospectus was issued soon after, and the first monthly number appeared in March 1808. It wasn't completed until 1814. Though originally intended to occupy only ten or twelve volumes quarto, it expanded to seventeen, the total cost of the set amounting to 36 gns. The third major effort in this period was *Petralogy. A Treatise on Rocks* (1811), likely the only one of the trio in which making money was not a big consideration.

At the end of April 1807, Pinkerton again experienced "the agonies of removal."[64] His Clement's Inn lodgings being too confined to carry out the work for the atlas and travel collection, he moved to Queen Anne Street near Cavendish Square, renting "a large house with windows of a northern aspect, solely for the accommodation of draughtsmen."[65] Meaning drawers of maps, ones he presumed would work in the house. He stayed at that address for over two years. There, he said, he had the pleasure of repeated visits from "some of the highest characters in this country for talents, probity, and science."[66] Who were they? Occasional assistants, translators, amanuenses – but were these among the "highest characters" he referred to? Pinkerton did have connections with a number of earls, viscounts, and M.P.s at different periods of his life, but their calls on him at this time must have been few, if they came at all. The traveller Browne and chemist Smithson Tennant visited him, perhaps during these years: "Often have Mr. Browne's gravity, and Mr. Tennant's absence of mind, relaxed with the author over a glass of Shiraz wine, the precious gift of a Persian ambassador to our distinguished chemist."[67] One regular correspondent was Warden; their letters were taken up with geographical and geological matters. Warden was useful, living still in Paris whence he could send recent books and maps. There was also his friend from his youth in Edinburgh, John Young, with whom Pinkerton reconnected in late 1806.[68] Otherwise, some notes to Henrietta, letters to publishers, a few odd scraps, and of course his books tell the story of his life in 1807–12. On his return from Paris he'd resumed corresponding with Laing but, it appears, only briefly. Something Laing said in a letter about the *History of Scotland* displeased him, though Pinkerton had asked him to write "very freely" of its faults since he was contemplating a second edition in octavo.[69] Walker too had given offence by saying he'd seen Pinkerton's name listed among prisoners of war returned from France. He was answered sharply: Pinkerton said he had not been treated as a prisoner, he'd been "respected as a man of letters."[70] He followed this with a paragraph of destructive criticism of *An Historical and Critical Essay on the Revival of the Drama in Italy* (1805), a work he'd urged Walker to write but in which his name wasn't mentioned.[71] Walker's mild response was apparently left unanswered. In June 1807 Pinkerton wrote to Banks, asking him for help in getting an acquaintance in Paris back to England. He got an honest but curt reply.[72] It was possibly to curry favour in that quarter that he included an account of Banks' expedition to Iceland in 1772 in the first volume of his *Collection of Voyages and Travels* (1808).[73] Pinkerton at this stage of his life had few friends or intellectual companions. This has to be taken into account in gauging his state of mind.

A partial explanation for his position as a semi-outcast lies in his private life. Soon after returning from France he took into his home, as a sexual partner, a girl named Hester Brown, sixteen or seventeen years old.[74] She lived with him as his wife from 1805 or 1806 to 1809. Two children came from the union: a "beautiful" daughter, who was about four years old in 1813, and a son, with whom Hester was pregnant in 1809.[75] Pinkerton was in his fifties, with a child "wife" in his house, and another wife in Hampshire. This became known, of course; the lawyer George Tennant recorded[76] that he saw Hester Brown in Pinkerton's home. Others must have as well.

Rent and taxes on the Queen Anne Street residence amounted to £100 per year, a sum Pinkerton could ill afford.[77] Cadell and Davies, having permitted him to issue a prospectus for the atlas, waited for more than six months before mentioning the project again, and the contract wasn't signed until the end of December 1807.[78] More delays followed. By June 1808 only one number, comprising maps of France, the West Indies, and Japan, had been published, and Pinkerton, who was paid as the numbers appeared, was worried. He wrote to the publishers about the way matters were proceeding. The engraver, Samuel John Neele, could "easily use more expedition," he said; there should be three or four draughtsmen working on the maps rather than one man (Lewis Hebert, Sr)[79] and his apprentices; and he had thought certain draughtsmen would be employed at his house, solely under his control.[80] That arrangement didn't suit Cadell and Davies. In the three maps in the first number the background to names and other insertions on them was uncoloured. Likely after receiving unflattering responses from early subscribers,[81] Cadell and Davies changed course. When the next six maps were published eight months later on 1 March 1809, all, including reissues of the three earlier ones, were in colour. By then, over two years after the work had been conceived, Pinkerton had received only £200 from an effort he thought would yield a far higher income.[82] Furthermore, once he examined Neele's engravings he concluded that his high hopes for the atlas rested on shaky foundations. He judged them to be inferior work.

Growing frustrations over the atlas might explain another damaged friendship. In November 1808 he sent a scrawled note to Douce, who was now on the staff of the British Museum, asking a favour for an engraver working on the travel collection. Douce was busy and didn't have time to help.

> Mr. Pinkerton's comps. to Mr. Douce. Mr. Cooke the engraver who carried a line from Mr. P. to Mr. D. has just been here with bitter complaints and says that Mr. D. did not even treat him like a gentleman.
>
> Mr. P. thinks his acquaintance with Mr. D. dates from 1784 or 5 – and 24 years form a great part of the life of man. He may therefore mention with the freedom of an old friend that he is truly surprized. If he had applied to Mr. Planta or Dr. Shaw[83] they would have ordered the book to their chambers and the artist would have finished his tracing in one hour. No officer of the Museum ever refused Mr. P. such a favour before, though only known to him by literary reputation.
>
> The monopoly of the B.M. is a disgrace to the country as Mr. D. has often stated in conversation with Mr. P. Every householder pays taxes to support it and has as good a right to its contents as any of the officers, who are paid by the public for its service and not for their own caprices or literary avocations. If Mr. P. were an officer there he would attend to nothing but his public duty – and of this the chief part is to assist literary men in pursuits of public benefit.
>
> Mr. P. has heard before that Mr. D. has already made enemies by his haughtiness in his new situation – and a real friend would wish him to avoid it above all things. If Mr. P. were in his situation he would act towards him on a very different plan; and has never spared any trouble to serve Mr. D. when in his power.

To Douce who, as Pinkerton must have known, was sensitive about his duties at the Museum, this was a grave insult. He responded to parts of the letter, saying the rest was "too savage & impertinent to merit any thing but contempt" – which was a trifle extreme. He wrote this explanatory private memorandum:

> Mr. Planta, to whom P. sent an abusive anonymous letter, would, as he told me have shut his door again to him, & Dr. Shawe would have turned his back on him.

The man, with great talents & learning, is low and vulgar in his manners, treacherous in his conduct & ungrateful for every sort of kindness bestowed on him. Every body hates him & I have often been blamed for keeping with him so long. I have now an opportunity of shaking him off & I am glad of it.[84]

"Every body hates him"! If that were so, it was a time to show a little charity rather than seize upon the letter, likely written in exasperation or in "the fluster of evening wine,"[85] as an "opportunity" to indulge in the secret cruelty of dumping him. "Nothing is so rare as a true friend."[86]

The year 1809 was a hard one in Pinkerton's life. He was plagued with illness throughout it. In March the third edition of his *Essay on Medals* received a lengthy and incisive exposé in the first number of the *Quarterly Review*.[87] It was written by a witty student at Oxford University named Barré Charles Roberts.[88] That same month Cadell and Davies let him know that the atlas was not meeting their expectations: it was costing too much, and they'd heard complaints about the illegibility of names on the maps. There did seem to be too many names squeezed into some of the maps. Just how many names to insert on maps in a world atlas, what kinds of names – counties? roads? promontories? – exactly where to put them (always to the right of the location?), and in what form to put them (e.g., whether or not to use contractions; whether they should be all horizontal), are of course questions of delicacy and strategy, requiring the greatest care. Pinkerton was not fastidious on such points. He argued that people didn't *read* maps; they consulted them to find "one or two positions at a time."[89] How that view justified clutters of names is far from clear.

"I am surprized at your mention of <u>enormous expence</u>," he replied to Cadell and Davies' letter. "I have been occupied about a year, and have received one or two hundred pounds." He added that he was "ever pinching myself, and tremble to draw for £20 lest my funds should be out. I am loaded and harassed with care, and, God knows, often pray for death." The final phrase, in a business letter, seems extraordinary. Playing for sympathy, no doubt. One half of the drawings were now in Neele's hands, he said, and though the draughtsmen were "ignorant and careless," his own efforts were so demanding that "my health suffers from my conscientious discharge of my duty."[90]

In the summer of 1809 he moved again, to Hart Street, Bloomsbury Square.[91] The rent was £65 per year, a substantial saving. He was ill through the whole of September with "a violent and low fever, accompanied with great depression of spirits."[92] In that month he and Hester Brown parted company. The St Pancras parish official John Tims wrote to him on 11 September, demanding that he make "some proper allowance" for Hester and the children. Hester had applied to him "on the subject of your continual bad usage & ill treatment of her, in consequence of which she has left your house with the determination to have no further connection with you."[93] On the basis of this letter, Dawson Turner said Hester "was driven from" Pinkerton "by his cruel usage of her,"[94] which is an interpretation, perhaps justifiable, of what Tims said. But Pinkerton didn't actually kick or force her out of his house; she left him.

After a few requests for an interview with Cadell and Davies, one was arranged for 13 October. The publishers informed him that they didn't believe great labour was required to prepare an atlas – which may suggest they were out of their league in attempting one – that large profits couldn't be expected from such a work, and that scandals associated with his name were injuring their reputation. They also intimated they would engage another author to supervise the third edition of his geography. He wrote to them at once, answering all points. There was indeed much toil in making an atlas, he said, although that might not be apparent to the inexperienced eye. And as for the low profit expected from the project, he had very different expectations:

> Were I to publish an Atlas myself I made a calculation of about £6000 gain. After 50 years of hard labour I have perhaps a right (had it pleased God) to some repose & even some pleasure – but as Gibbon has observed (in an eulogium very different from the paltry voices of the day) I have made enemies by my invincible law of truth – and I shall be persecuted by slander till I am in my grave.

As if these tycoons cared about Gibbon! The claim to his geography he would not suffer to be disputed, he said. He would allow no collaborator to meddle with it, and would publish it under a different title if the publishers persisted with their threat. At his stage in life he expected a few compliments rather than "lessons of mortification":

> All the artists live better than myself; and all take journies of a month or two except me who sacrifice health and pleasure to a constant attendance on my literary duties. Private eccentricities have always attended what are

called genius and talents; and I have perhaps had my share – but I have wronged none, and I am the only sufferer by them.[95]

A door to another source of complicated misery now reopened. In late 1809 or early 1810, whether motivated by Henrietta's pleading, his own genuine change of heart – which seems unlikely – or a need to counter scandal, a reconciliation with his wife and children was effected. Henrietta came to London with her children, stayed for a while at the Oxford Coffee House in the Strand, then moved in with her husband.[96] Pinkerton wrote to Douce, saying "our coldness has lasted long enough. Shall I call on you or will you call on me at No. 40 Hart Street Bloomsbury where I live with my wife and eldest daughter."[97] (The other daughter presumably remaining with relatives.) Douce made no reply. Had he called at Hart Street, it is unlikely he would have found a happy family. Towards the end of March 1810[98] Pinkerton and his wife again separated, this time permanently. Henrietta didn't return to Odiham at once. She remained at the Coffee House, trying to make amends for a mistake she'd made in the process of reconciling. In moving to Odiham in 1803, she took with her Pinkerton's library, or a large portion of it. She sent some of it back to him during the period 1805–9, but retained part of it.[99] Her mistake was that when she moved to Hart Street she brought with her much of what remained in her charge.[100] She now made an effort, through her brother John Burgess, the London merchant – the "oilman," as Douce would later term him – to recapture those possessions and sell them for the benefit of her children. Her effort, it appears,[101] did not succeed. John Burgess was left with an ugly hatred for his brother-in-law. Pinkerton's connection with Thomas Burgess, now Bishop of St David's and one of his correspondents following the marriage to Henrietta, had ended long before this grim episode. He was *persona non grata* with the entire Burgess family.

In February 1810 another interview took place between Pinkerton and his publishers to discuss the atlas. He wrote to them prior to the meeting, to state his case. He was in a combative mood. It was Neele's fault, he said, that disputes had arisen over the work. Neele, therefore, could not be allowed to play an active role in the discussion.[102] The meeting was unpleasant. Following it, Pinkerton addressed his complaints and charges to Longman and Co., which had a share in the work. He had to be given more money; Neele must be dismissed; payments to draughtsmen and engravers had to be made through his hands.[103] Cadell and Davies, embarrassed by this backhanded move, wrote to Longman's in March, claiming that most of his accusations were baseless.[104] In

April Pinkerton told Longman's that in all the original conversations with Cadell and Davies over the atlas, he was to get £6,000 in payment. Davies had surprised him into signing for £2,000 during "half an hour's hurried conversation."[105] He would never rest content with the present arrangement and if further provoked would take his case to the public.

The charge that he had been cheated of £4,000 is hardly believable. No doubt Cadell and Davies were stern taskmasters and might have interfered in matters relating to the atlas that could have been left to him. The final product could have been superior if they'd taken his advice to replace Neele with Wilson Lowry, as he'd suggested. They were indeed slow to pay Pinkerton for his services. Yet they too had reason to complain. His attention to the atlas may at times have been less than he led one to believe. He had been ill, as we've noted; and Cadell and Davies alleged he had asked Hebert to undertake research that was part of his own responsibility.[106] (How often this happened is unclear.) His troubled private affairs may not, in certain periods, have left much room for the seclusion extended literary work normally requires. Still, it is remarkable how much he did produce in 1806–14. It is another illustration, if one were needed, of books emerging from messy households and tormented brains.

The standoff with Cadell and Davies evidently continued through 1810. The eight maps that were published that year may have been engraved from drawings done earlier. Early in 1811, however, the presses were rolling again. By March Pinkerton had completed his work on all but five of the sixty-one maps. He wrote to Cadell and Davies, asking for advance payment of the £500 due him in November. He had just bought a house, he explained. They refused, telling him to deal with Longman's in future on all questions concerning the atlas and geography.[107] The house, on Marchmont Street, cost £1,000.[108] He stayed there for over a year, turning out volume after volume of the travel collection. In December 1811 he published *Petralogy*. "It is quite a new system," he told Warden,[109] echoing the claims made in many other books.

We must pause to look more closely at the three works from this late period of his productive life. The folio *Atlas* of 1808–15, described on the title page as "Directed and superintended" by Pinkerton, is perhaps the most impressive. The first four maps in it are those of the Western, Eastern, Northern, and Southern hemispheres.[110] The maps then "chiefly" follow the order of continents and countries treated in *Modern*

Geography, meaning they are arranged according to what Pinkerton said was the "*order of political Importance.*"[111] Europe is paramount, occupying twenty-five maps, with Asia and other continents and areas getting thirty.[112] The seven maps after the general one of Europe are of Great Britain and Ireland – England gets two, as does Scotland, and Ireland one. France is #13, getting just one map, and so is implied, if one is eager to read an inference in it, to be half as important politically as England or Scotland. Asia follows Europe. China is the thirty-second map, again, to strain once more for inference, just half as important politically as Scotland. Australasia and Polynesia follow as tag-ons to Asia. North America is given eight maps, with the United States getting two and "Spanish Dominions" three. South America and Africa (six each) take their places near the end. Such is what could be regarded as the Eurocentric and anglophile/scotophile world view re-embodied in the *Atlas*. Pinkerton said in *Modern Geography* that the arrangement of continents, despite his placing Europe first, was "arbitrary," and that of countries "elective" within defined groups. Presumably the same applies to the *Atlas*. The lineup of the world's countries and peoples into (roughly) the haves, the have-less, and the have-nots is nevertheless apparent to the twenty-first-century reader who may, from searching out the behind-the-scene purposes of the mapmaker, construe ordinary writerly practices of an earlier time as chauvinism or something worse. It is wise to be wary of "retrospective impositions of contemporary definitions,"[113] and to try "to understand the scientific endeavours of the past on their own terms."[114] That said, it is probable that the ordering of continents and countries reflects a set of assumptions and preferences Pinkerton shared, if not with all geographers, with most of the educated elite of Britain, who were, after all, his primary target as buyers.

All but two maps are marked as drawn under Pinkerton's direction by L. Hebert. In the two not so marked, termed on the maps Northern Hemisphere and British Isles (#s 3 and 6) the omission could be accidental. On all maps Neale is named as "sculpt.," i.e., engraver. Each map, normally rectangular in shape and averaging 20 inches in width by 25 in length, occupies (with margins) the verso and recto pages on the inner forme of a sheet, with the pages at the back being blank. Thus, each map takes up four folio pages, for a total of 244; there are also blank sheets between the used sheets,[115] plus an introduction of seven more pages. And the sturdy cover, of course. The *Atlas* is big, thick, and heavy.

Pinkerton said in his introduction that "The chief object of a Map is to be clear and easily consulted."[116] His maps are not easily consulted, given, first, the sheer bulk and weight of the work they are in. Also, though a list of the maps is provided at the beginning of the book, neither the maps themselves nor the pages they are on are numbered, and sometimes the names on the maps are not the same as those on the list. This can cause delay and irritation. A third difficulty is the scarcity of direction for users. While scales are provided on each map and Pinkerton wrote a brief introduction, there should have been much more letterpress giving an explanation of symbols and other detailed information. He was ordinarily a garrulous commentator on geographical matters; there is little sign of this predilection in the *Atlas*, though he listed and discussed maps and mapping elsewhere.[117] Not before map #43 (Spanish Dominions in North America) did he provide a key explaining nine "signs" he used, intended to apply to those specific "Dominions." (Some are used elsewhere.) In a number of European maps he inserted symbols for forested and swampy areas, leaving the user to make his own interpretations. It isn't until map #42 (United States of America, Southern Part) that the meaning of the swamp sign becomes clear: it is on a big area between West Florida and Florida. (He might have assumed users would be familiar with the symbols.)[118] He also used symbols irregularly, e.g., by inserting crosses to mark rocks along some coastal areas but not on others where there must also have been rocks. There is not a consistency of symbols, or for that matter of performance, through the sixty-one maps. This is no doubt due to the way the *Atlas* was created – over a seven-year period that was fraught with trouble, with a rush at the end to get it finished. One further point on the issue of ease of consultation. Pinkerton chose not to insert roads on his maps, giving as reasons "the constant and daily alterations in the line of modern roads" and the confusion they would cause by their similarity to rivers and canals.[119] Yet roads could be helpful in finding locations on a map, especially in a crowded country like England (as in Southern Part, #7); but they would add to the clutter too. He said his maps were not road maps, which he thought were in a lower category and easily available to travellers.

Nor was the stated "object" that a map should "be clear" always realized. Pinkerton was proud of the way mountain ranges were presented on his maps – not as a series of "mole-hills" but instead "as seen from above," as "continuous ridges." He claimed to have introduced this practice in his *Enquiry into the History of Scotland*, whence it spread to cartographers.[120] Yet the hachures stretching on either side of the implied escarpments to convey the notion of height, and often the accompanying shadow (as if the sun shone from the northwest), frequently obscure place names and almost blacken certain areas (as in Portugal, #24, Spain and Portugal, #17). This is not an uncommon phenomenon in map history.[121] The same darkening effect is created at the seashores, where lines parallel to lines of latitude are drawn out to sea to give an impression of coastal declivity; there too place names suffer. It is questionable whether such lines were needed. The colours throughout the *Atlas* – shades of pink, blue, green, yellow, purple, orange – also contribute on occasion to blurring names. The colouring is the most striking feature of the work. Pinkerton in his introduction said it was "indispensable in an Atlas of this size,"[122] without explaining further. There seems to be no overall consistency in the way colours are used. If they are "metaphorical,"[123] the metaphor's meaning isn't clear. They are applied with considerable skill in a number of maps, and quite clumsily in others. At shorelines, where a stronger shade of colour sets off the coast from the ocean, and at boundaries between countries and areas, particular delicacy was required. In too many cases, e.g., Europe (#5), Ireland (#11), Remote British Isles (#12), and Turkey in Europe (#18), to mention just a few, the edges almost seem to have been applied rather crudely with crayons – too heavily, with much bleeding of one colour over another. The colouring is most successful when only one tint dominates a map, as in the pink Arabia (#36), where the pink, together with touches of yellow and green outside the country, is applied only faintly. Perhaps the best maps produced for this *Atlas* were the three uncoloured ones alluded to earlier. These are to be found only in the American edition of the work.

Colouring was mainly Neele's and the publisher's responsibility. Despite what is said on the title-page of the *Atlas* about Pinkerton's superintendence and direction of the project, it is doubtful if he had much influence, if any, on the actual engraving of the maps. Cadell and Davies said that "not one of the maps has been considered as finished, till he gave it his full and unqualified approbation," but that his role was "to watch over the accuracy of the engraving." The "style and manner" of it was to be left in their hands.[124] Pinkerton could have had no hand whatever in "watching over" the engraving of the fifteen maps published between November 1812 and June 1814, since he was then in Scotland. (Though he continued his work on both the *Atlas* and *Voyages and Travels* from that distance.) He was right to claim, in private, that Neele's work left much to be desired. In his introduction he noted that Hebert's drawings were "finished with great care and accuracy, joined with a neatness and beauty,"[125] but made no specific mention of Neele. Yet at the end, when he indulged the hope that the work, despite its "errors and defects," might still be considered "the most magnificent and complete Atlas that has ever been published in any age or country," he added that the praise belongs to "the skill of the Artists and liberality of the Publishers."[126] Neele's work was thus indirectly acknowledged, along with the contribution of Cadell and Davies and Longman's.

The *Atlas* presents the world as it was known in the early nineteenth century – as known, that is, to one veteran London-based scholar widely read in geographical literature, including maps. His images were as complete and accurate as they could be, given the extent of his knowledge and the rapid progress of discovery. Maps give off an impression of authority and finality, much more than books; but they are equally artefacts, subject to the normal human limitations, one of which is ignorance. Pinkerton did not try to conceal ignorance. Much of the world shown on twenty-first century maps is missing or blank in the *Atlas*. There is no Antarctic continent.[127] Northern Canada is sliced off, receding into white emptiness (as in #46, British Possessions in North America); the northern extremities of North America are a muddle.[128] "Unknown Parts" are in Brazil, "Unexplored Parts" in Peru (#48, South America). Virtually all Australia (termed New Holland,[129] in #38, Australasia) beyond the coastline is a blank. The term "Unknown Parts" occurs twice in Africa (#54), encompassing vast regions of the interior. Not just ignorance, but error, is on display in Africa, as in the mythical "Mountains of the Moon" and the "equally fabulous" Kong Range of Mountains stretching across the continent, the latter put in place by Rennell in 1798.[130]

It has been said that in "nineteenth-century British maps of African territories, decoration plays a part in attaching a series of racial stereotypes and prejudices to the areas being represented."[131] This is not so in Pinkerton's maps of Africa, where there is no decoration except for a small cartouche on each bearing the name. No decoration hides the blank spaces, in these or other maps. He was trying to get things right rather than decorate or impose an "interpretation"[132] or "theory."[133] What is most clearly observable in the *Atlas* is this concern to be accurate and up to date. It can be seen in many of its features, including nomenclature, where we find an awareness of correct regional names for locations. He gives the sources of some African names: thus Kawar (of Edrisi) and Kuku (of Edrisi), citing Muhammad al-Idrisi, the twelfth-century geographer (Africa, #54). Maori names are used for New Zealand's North

Island and South Island[134] (in Australasia). A similar knowledge of aboriginal names is found in his depiction of Labrador and elsewhere. He was not above giving his own preferences for names. He liked Notasia for Australia (but used New Holland); he tended to call the Pacific Ocean the "Grand Ocean," but not always.[135]

The concern for accuracy is reflected in glosses he made directly on maps. In the emptiness of northern Canada, for example, he pointed to "The Sea seen by Mr. Mackenzie in 1789" and "The Sea seen by Mr. Hearne in 1771," naming this sea "Arctic Icy Sea" (#1, Western Hemisphere). In the map of Turkey in Asia (#31) a dotted line has under it "Dry R. Tirtar of Danville," alluding to the work of the great French cartographer[136] he admired and imitated. On the map of Japan (#33) two islands in the Sea of Japan are glossed "Jootsi Sima according to Perouse" while a nearby island has "Jootsi Sima according to Roberts," referring to conflicting reports from Lapérouse's voyage round the world in 1785–8. He didn't know which was right so he inserted both. Along the coastline of New Holland are mentioned the travellers who reported on or named what they saw: thus "Lewin's Ld.," "Dunning's Land," "P. Nuyts Land of 1627." Comment is especially frequent in the African maps.

Off the Orkneys (#10, Scotland, Northern Part) a North Shoal is traced, and an untypical comment added, "upon it are Caught Cod and Ling." He occasionally inserted – sometimes warning of danger – not just rocks but shoals, banks, reefs, sand bars, and even what seem to be continental shelves. His work thus shows an interest in hydrography, though Pinkerton didn't consider marine charts "as essential in the study of geography."[137] On the east of North America is a sketch of the "Great Bank of Newfoundland" with smaller banks near it (as in #3, Northern Hemisphere).[138] These banks had been shown in earlier British maps such as Herman Moll's; Pinkerton of course made use of these, as well as maps from the continent of Europe.

Despite its defects, the *Atlas* was an ambitious new depiction and naming of the land masses and oceans of the earth. Some of the maps are quite stunning. The effort to be current and accurate is admirable. As in the case of *Modern Geography*, the achievement was not an advance in disciplinary theory and practice as much as progress within accepted parameters. *A Modern Atlas* has got hardly a mention from those who've studied the history of British maps and atlases. Arrowsmith and Rennell have received the notice they rightly deserve,[139] and their work has overshadowed the contribution of their near contemporary. After his initial skirmishes with Rennell, Pinkerton was ready to acknowledge the great geographer's qualities, saying he "introduced the science of geography into England, in a form at once inviting, exact, and scientific, by his memoir and map of Hindostan."[140] He gave Arrowsmith only a passing grade. In his "Memoir on the Recent Progress, and Present State, of Geography," in the 1807 edition of *Modern Geography*, he pointed to Arrowsmith's "narrow jealousy," lack of education, and the "many gross errors of projection and even of latitude" on his maps, while conceding that his delineations of mountain chains were "on the real geographical plan, as describing the nature and appearance of the earth." In Arrowsmith's maps for the geography, he wrote, "scarcely a drawing could pass without many corrections and improvements by the author."[141]

We turn to *A General Collection of the Best and most Interesting Voyages and Travels in all Parts of the World* (1808–14). This massive work in quarto, with each of the seventeen volumes averaging about 800 pages, is of a size that no twenty-first-century commercial publisher would attempt, and some might wonder why Longman's, with Cadell and Davies, whose name also appears on all volumes, would devote so much of their capital to it over so long a period despite dealing with an irascible editor, one, moreover, supposedly of ill repute. Not only did they allow the work to grow to such an inordinate length, they also put in 202 plates, many of them exquisite. The reasons must be: (a) they made money from it, not just by the sale of the final product but continuously through the serial publication – by a clause in the contract for the work, they could stop production at any point to cut losses; and (b) Pinkerton was one of the few scholars in London, perhaps the only one, equipped and available to conduct such a project. They needed him as much as he needed them.

The 1807 prospectus explained how Pinkerton intended to organize the collection and what his basis would be for selecting the contents. The material would be arranged geographically rather than chronologically, he said; "sometimes one single volume, sometimes two, may be allotted to particular countries and portions of the world, and form as it were complete edifices in themselves."[142] What he proceeded to do, as in the *Atlas*, was roughly to follow yet again the sequence of continents and countries he'd established in his *Geography*: Europe first, comprising six volumes as one of the "complete edifices," then Asia (four), Asiatic Islands (one), North America (two), South America (one), and Africa (two), with the final volume reserved for scholarly apparatus. As for what to include in the book, he proposed "to pass uninteresting details, and those which are superseded by further discovery and the progress of knowledge," and "to form a selection only of such works and parts as will ever interest the general reader, from the intrinsic value of the narrative, as forming an epoch in the progress of discovery, as being remarkable for fidelity or animation; and, in some few and short instances, from their great variety."[143] His quasi-scientific commitment to pass over the voyages "superseded by further discovery" was sometimes pointedly honoured, as in Volume 11 where he omitted "the more ancient accounts of the Asiatic Islands," choosing to begin his selection with William Dampier on the Philippines.[144] In general he gave a much higher priority, hence more space, to recent travels than to early accounts. A reader looking in Volume 2 for the twelfth-century *Itinerarium Cambriae* of Geraldus Cambrensis will not find it; instead we get H.D. Skrine's *Two Successive Tours through the whole of Wales in 1795*. But Pinkerton was attentive too to his second stated objective, supplying what he knew would "interest the general reader." The work was intended "for popular amusement, as well as general information," he wrote in his preface.[145] He knew the "general reader" would be interested in a classic item such as Marco Polo's journey to China, so he included it. That same reader would want to see stories of early exploration in America, Africa, and elsewhere; so he inserted accounts of Columbus, Frobisher, Cartier, Magellan, and others in that category, drawing often on earlier collections of travels for texts – starting Volume 1, for instance, with eighty pages derived from Richard Hakluyt's *Principall Navigations* (1589)[146] on the English voyages of Sir Hugh Willoughby and others into Russia and Siberia. Hakluyt was a major source for him, as were John Harris's *Compleat Collection of Voyages and Travels* (1705) and John Churchill's *Collection of Voyages and Travels* (1728). (Volume 12 is straight out of those three works.) The general reader would also expect to find what had become standard works in more recent English travel literature: Mungo Park's explorations in Africa, Cook's adventures in the Pacific, Thomas Pennant's tours in Scotland, Arthur Young's tours in Ireland and France, William Coxe's accounts of Russia, Denmark, Switzerland, and Norway,[147] and such like. He supplied those, sometimes reprinting whole books of great length. Young's *Travels in France* takes up 600 pages of Volume 4. Items of 100 to 200 pages or more are common throughout. Yet there are numerous extracted and abstracted items, some quite short. The table of contents for the whole work at the beginning of Volume 1 has 145 titles, but some of these feature more than a single voyage or item. In Volume 2, for instance, attached to Stebbing Shaw's *Tour to the West of England* (1789) are unannounced extracts from W.G. Maton's *Observations on the Western Counties* (1797), while, similarly, the second edition of William Bray's *Sketch of a Tour into Derbyshire* (1783) has appended to it an unexpected short description of a "Subterraneous Cascade" from Richard Warner's *Tour through the Northern Counties of England* (1802). He also prints dissertations, letters, petitions, letters patent, documents of various kinds that are in the sources but have only marginal relevance to travel. A number of items might be thought not to belong in a collection called *Voyages and Travels*. Johann Jacob Ferber's work *Essay on the Oryctography of Derbyshire*, in Volume 2, is a treatise on geology,[148] one of several pieces – too many, one suspects, for most "general" readers – indicating Pinkerton's fascination with that subject. Capt. John Smith's 250-page *General History of Virginia*, in Volume 13, though it has journeys in it, is more or less what it says it is, a history; and the *Report of an Engagement* (with the Armada) in Volume 1 is military history rather than travel. These are not the only instances of genre expropriation in the work.

The reader who tackles the collection volume by volume is overwhelmed by the sheer number of voyages and pieces in it. It is a library of (mostly) travel literature, and not just of English travel literature. Quite a few items come from continental Europe, especially France. A number of these had never been translated into English before, for example an extract in Volume 15 from the *Relation* about Egypt of the medieval Arabian physician Abd Allatif that had been translated into French by Baron Silvestre de Sacy and published in Paris in 1810. English translators of this and various other items throughout the work are not named.[149] The de Sacy text is one of a number of entries that had been translated from another language into French prior to re-translation into English (another being Dr Johnson's mangled[150] version of the Portuguese Jesuit Jerónimo Lobo's *Voyage to Abyssinia* in Volume 15). The emphasis on French travel literature, reflecting Pinkerton's reading during his sojourn in Paris, is an outstanding feature of the collection, starting with Volume 1, which has a half dozen texts from French sources, among them what appears to be the first English translation of the comic poet Jean-François Regnard's *Journey through Flanders, Holland, &c.* The three voyages to the north by the "Dutch and Zealanders" in the 1590s, led by Willem Barentsz, had appeared in English well before Pinkerton's day, but his text, also in Volume 1, was "newly translated" from a French collection. Elsewhere, we find him overlooking an available English text of a voyage in favour of a more recent French edition, as in the translation of the Chevalier de Bourgoanne's *Travels in Spain* in Volume 5, where he prefers the Paris third edition of 1803 as a source over ones published locally. Germans, Italians, Portuguese, and other travelling foreigners play their part in the work too, their accounts sometimes being newly translated and other times taken from available English texts.

Pinkerton's editorial practice in the work can best be described as minimal. The contents, he said, were presented "with their original authenticity; so that this collection may be used with the same confidence as the original works themselves."[151] As a general statement that is accurate, though with qualifications. This was not the Pinkerton of *Ancient Scotish Poems*. The language of Young's *Tour in Ireland* in *Voyages and Travels* is Young's, not that of a tinkering editor. And so it is elsewhere: the words of the main text on the printed page are the words of the author or translator. This text is authentic, as he said. Not that there is no interference whatever with the items he chose. From some texts he cut a good deal of what he thought was "prolix," tedious, inferior, "of a local nature," done better elsewhere in his collection, extraneous, or (in at least one case) obscene,[152] and usually signalled that he was doing so. But sometimes he lopped off material without letting on he did. In Volume 16,[153] for instance, both the second half of Liérain B. Proyart's *Histoire de Loango, Kakongo, et autres Royaumes d'Afrique* and the fourteen opening chapters of Park's *Travels in Africa* are omitted without acknowledgment. In these cases the cutting was done to save space.[154] Park's chapter 15 became chapter 1 in Pinkerton's version, which illustrates another of his practices, i.e., reconnecting the pieces remaining in the chopped voyages so that they appear in consecutively arranged texts in his volumes.

He tended to reprint – but not always – the source's prefaces, introductions, and some or all of the footnotes. He explained his own "plan of annotation" in his preface, saying a "difference of opinion" between an editor and the "original author" could "scarcely be of such moment as to divert the reader from the course and interest of the narrative." It would have been "remarkably easy for the Editor to have given many corrective notes from his own Geography," but the result would have been "enlarging a work, by information, already in the hands of the Public."[155] (There appears to be just one reference to *Modern Geography* in the notes.) The reader therefore could expect few footnotes, other than those identifying the source of the text being offered. If we compare

Voyages and Travels with his histories and geography, it becomes evident that he did indeed refrain from indulging in the practice. In some texts, even extensive ones, there is hardly a sign of his hand. Yet he *did* insert quite a number of his own (normally unidentified) footnotes – it wasn't possible in such an immense work to refrain completely – and it is often difficult to distinguish his from those of the author being presented, or those of the original editor or translator (or translators!) of that author, typically included as well. This mixing of footnote writers, and the occasional careless practice of leaving unchanged the notes by original authors that require adaptation,[156] are sources of confusion.[157] It is not the only area where a lack of direction from Pinkerton is felt. In only two instances did he provide introductory comment for texts, and those were a few lines long and insignificant. His intention was to leave extensive commentary for the final volume, where he provided a "Retrospect of the Origin and Progress of Discovery" that one reviewer thought "a masterly performance."[158] Here too were a critical "Catalogue" (i.e., bibliography) of travel literature – "copious, highly systematic, and rich in inaccuracies," according to G.R. Crone and R.A. Skelton[159] – compiled largely by the German-born Henry Weber,[160] the antiquary and associate of Walter Scott, and an Index to the complete collection, prepared in-house by Longman's. The "Retrospect" was no substitute for regular, cogent introductions to the selected texts, even if they consisted of only just a paragraph. Baldly dropping a 100-page travel book onto the reader without comment apart from identification of the source was no way to conduct the editing of a work of this nature. Nor was any commentary provided at the beginning of any of Volumes 2 to 16, where it was needed as well. The reader opened each volume to find not even a list of contents,[161] merely the general title-page, and on page one the beginning of the first text, under the headings "Travels in France," "Africa," and so on.

   Pinkerton's collection must be considered useful simply because it incorporated so much literature of travel, a lot of it consisting of obscure texts that his "general reader" – supposing he or she could get access to such an expensive work – would not find elsewhere. It also had a convenient assemblage of standard works and remains a handy source today. The texts are uncluttered and on the whole can be trusted, though some, as noted, have been excerpted or abbreviated. Most sources are clearly indicated (some are not). The giant index in the final volume makes the contents accessible. Certain texts have no doubt lost their appeal after the passage of two centuries, but others, especially the African volumes, portray scenes that still chill the blood. *Voyages and Travels* helped to keep Pinkerton's name alive through the nineteenth century. In the 1880s Henry Morley, who edited the widely distributed cheap Cassell's National Library, selected *Early Australian Voyages* and *Voyages and Travels of Marco Polo* from his work.[162]

We now turn to *Petralogy. A Treatise on Rocks* (1811). A word is needed at once about the vocabulary Pinkerton used in this field. He divided what today is known as geology or earth science – what he called mineralogy – into three "provinces": petralogy ("the knowledge of rocks or stones which occur in large masses"); lithology ("the knowledge of gems and small stones"), and "metallogy," meaning metallurgy ("the knowledge of metals").[163] Thus, to him, petralogy (the more normal spelling is petrology) was a division of mineralogy. The term "rocks" in his book referred to virtually all[164] solid natural structures of the earth; at one point he even alluded to "rocks of ice," but didn't pursue the topic. He reserved the word geology (or geognosy, a word imported into English usage from the German scientist Abraham Gottlob Werner) for a theoretical discussion of the formation of the earth.[165] His book was an effort to sidestep such "general geognosy," which in his day, as he well knew, was embroiled in controversy, and to provide the basic knowledge and vocabulary of his discipline. It was "necessary to begin," he said, "with an elementary work."[166] *Petralogy* presented a classification, definition, and naming (often renaming) of what he termed rocks. It comprised two volumes and 1,250 pages, about half of which[167] consisted of quotations from authorities in the field. He had spent "the intervals of ten years" writing the book, he said, the "three which he passed at Paris" being especially productive. He had there conversed with "men eminent in the science" and seen "the most opulent cabinets."[168] Among the authors he quoted in his book, Frenchmen are prominent. He justified his excessive use of translated extracts from these and other foreigners by saying that it was "scarcely to be expected" that they would otherwise be known to the English reader.[169] It seems evident that he translated many of the extracts.

Porter in his book *The Making of Geology* notes the efforts of eighteenth-century European scientists to identify and classify minerals and rocks through the use of blow-pipe analysis and chemical tests. In this endeavour the Scandinavian and French "schools," he says, tended "to pursue chemical analysis," while the German Werner favoured "the use of external characteristics as equally reliable but much simpler." Of "supreme importance," he adds, "were attempts to establish comprehensive classification systems," that being "a matter of urgency, with the continual discovery of new minerals, and the painful aware- ness that the mineral kingdom ... lacked sharp natural boundaries."[170] It was into this supposed breach that Pinkerton strode in his *Petralogy*. We must briefly – for we needn't dwell on it – look at the classification he made, first noting that the distinction between rocks and minerals did not exist for him. The two "substances" were jumbled together in his book, minerals were called rocks and vice versa, and that of course made his classifications and definitions of no ultimate value. He also tried to work out a system of classifying what he termed rocks without relation to their formation over aeons in the earth's crust, their history. That was a critical error. Maybe Bishop Ussher was looking over his shoulder again. He didn't completely avoid questions about the ori- gin of rocks but relegated them to minor status. What he did was this. He created twelve "rock" categories called Domains. The first six he termed "substantial," meaning that they were based on the nature of the substances themselves, i.e., their chemistry; the second six he called "circumstantial or accidental [*sic*]," meaning they were "distinguished by circumstances or accidences of various kinds," a somewhat opaque category adopted, it appears, mostly for convenience. He devoted the first volume to the first six Domains, according to the chemical ingre- dient that was paramount in it. Thus in the "siderous" Domain iron was dominant, in the "siliceous" silex (silica) was, in the carbonaceous carbon, and so on. These domains divided into "modes," meaning the "mode of combination" of the chemical constituents; modes were fur- ther often divided into "structures," and the structures into "aspects." The siderous mode occupies over 140 pages and incorporates sixteen modes, including jasper, which is indeed a rock and has iron in it. In his account of jasper Pinkerton says that it is not "inconceivable that the surface may even attract more iron from the atmosphere, where atoms of that substance constantly float."[171] He claimed that "the whole sci- ence of mineralogy rests on chemistry alone,"[172] and blithely assumed that he understood chemical processes well enough to proceed with his classification.

The second volume dealt with the six domains of the "Accidental Rocks," each of which he divided into, not modes, but Nomes, further subdividing these into Hyponomes and Micronomes. (His experiment- ing with nomenclature persists throughout the book.) In these domains he lumped "rocks" that didn't seem susceptible to, or capable of, use- ful chemical definition, but were peculiar in their own way, and had somehow to be accounted for. Here were Composite or Aggregated, Diamictonic, Anomalous, Transilient, Decomposed, and Volcanic "rocks." The word diamictonic, derived from Greek, was one of his numerous neologisms. It "implies that two or more substances are so thoroughly mingled, or, in the language of chemistry, so intimately combined, that the rocks cannot be arranged under either Domain, either from predominance or preponderance."[173] In this domain he named some rocks after eminences in chemistry. Thus we have, among others, Baconite, Kunkelite, Pottalite, and Klaprothite. These names did not take hold. In the volcanic section he assembled "substances of quite distinct natures, but all altered by fire."[174] He here hesitantly admitted basaltin (a form of basalt) into this category, saying "it is at least sometimes volcanic," having said earlier that "it would seem an infallible inference that [basalt] cannot be of volcanic origin."[175] (It is an igneous rock.) The contradiction and hesitation hint at the fundamental weaknesses of his book.

We need go no further with his classification, except to say that it was an honest and laborious effort of a scholar who tried to make sense of a hard subject. Pinkerton was essentially an amateur approaching an area of knowledge that was too far beyond his reach, an area requiring much more field work than booklearning. He did little geological field work, though he did do some. He remained a bookworm. He fancied that science progressed as the study of history did – by the slow accretion of knowledge, with one generation adding to what previous writers and thinkers had said and done. And so it did and does. But science also advances explosively and, sometimes, accidentally – through discoveries or theories about the world made by men and women of genius that transform the discipline they work in. The late eighteenth and early nineteenth centuries witnessed just such a development in the field of geology. It was no longer possible in 1811 to write an introductory book on the subject that would have some standing without taking note of what that new approach was, fully absorbing it, accepting it, and recognizing that previous notions not based on it had to be dismissed. The theorist in question was the Scot James Hutton, the founder of modern geology,[176] whose name was not mentioned in *Petralogy*, though the Huttonian theory was, and it is clear that Pinkerton knew what that theory meant. It was available to him, not just in Hutton's own writings but in John Playfair's *Illustration of the Huttonian Theory of the Earth* (1802) and John Murray's *Comparative View of the Huttonian and Neptunian Systems of Geology* (1802). (The Neptunian, Neptunist, or Wernerian system being the one to be discarded.) Two major elements of Hutton's theory, often called Plutonist, simply stated are: that the earth is affected by erosion and deposition which are at work now as they have been for aeons of time, denuding land forms and forming rock strata in seabeds – the principle of uniformitarianism; and that subterranean heat has been the main instrument in the fusion, uplift, and positioning of the rock strata. Among the Neptunist British opponents of Hutton were the influential professor Robert Jameson, another Scot, and the Irishman Richard Kirwan;[177] they derived their ideas from Werner, who posited a vast primeval ocean which, prior to receding, deposited the rocks of the earth, including even basalt. The idea of the vast ocean as the instrument of geological change tied in somewhat with the Noachian flood, making it more acceptable to the orthodox. Pinkerton has been placed with the Wernerians by one recent historian of geology,[178] and it is true that he gave much more room in his book to Werner's disciples than to Hutton's. He'd met Werner in Paris and been impressed by the great man. But on occasion he criticized Werner's system; and he stated in his book that he didn't wish to take sides in this battle between the two sets of theorists. He wished to operate outside both.

Petralogy is mainly of interest for what it discloses about Pinkerton himself. It reveals, more clearly than even *Recollections of Paris*, the conversion to science that he'd undergone in the years following the publication of his *History of Scotland* – a conversion that had taken shape prior to his going to Paris, but that had been made more profound by his experiences there. The book is loosely put together, sometimes contradictory, often more a set of notes than an orderly text; but Pinkerton was entranced by his subject and loved writing about it. Ill temper, a feature of his earlier writing, almost vanished. There were traces of humour. The book contains more remarkable evidence of his latter-day religious views, as in this comment on iron:

> This subject cannot be quitted without the observation, that there seems a most manifest indication of MIND and DESIGN, or in other words a great Creator, in the peculiar distribution of this metal in the northern parts of Europe; where He knew, to whom all times are present, that it would be necessary for the industry of the inhabitants. In like manner the increased thickness of the fur, or of the feathery down of animals, can scarcely be attributed to climate or chance: not to add another simple observation, but which does not seem to have been made, namely, the superior size and strength of the female, when compared with the male, solely among the

birds of prey; as it was necessary that she should both protect and feed her voracious offering.[179]

Nor is this the only such statement in the book.

By early 1812 Pinkerton's personal life and finances were in crisis. In April half his library was sold for him by auction at Leigh and Sotheby's. (The other half, featuring his geographical collection, was auctioned nine months later.)[180] The treasures in the first sale, rare editions of the early Scottish poets, were bought by Chalmers and Richard Heber. Most of his books were not items to catch the eye of a bibliophile – they were eighteenth-century erudite works, the tools of the working scholar. The London booksellers bought these at low prices. Early in 1812 he put his house on Marchmont Street up for sale and moved to nearby Tavistock Place. On 30 April, risking a charge of bigamy since Henrietta Pinkerton was still living and they hadn't divorced, he married Elizabeth Brown Becket,[181] who was known to her family as Betsey. Pinkerton later commented on the relationship, breaking his prior "indispensable secrecy" about it:

> We were married according to her church, that of England though she, her father, mother, and uncle (who is rich) knew well and long that the divorce begun by my first wife on account of my infidelity was not completed. I regarded this divorce as an insult after a separation of ten years on account of her bad temper and a still worse sister … so in wrath married another bitch of the modern hot-bed race quite unfit for the society of men of sense or spirit.[182]

What a shambles! No document relating to a projected divorce from Henrietta has come to light. What the Burgess family contemplated instead was a legal separation, on terms first proposed by Pinkerton. A draft separation agreement was in the hands of lawyers for the two parties in 1813–15.[183]

In 1812 scandal about Pinkerton's sexual transgressions, alluded to earlier by Cadell and Davies, was still circulating in London.

> Pinkerton was a singular and degraded man. I (Britton) was made too well acquainted with him for my own reputation and for my own domestic comforts. He rented, and occupied, for a short time, a house, No. 9,

Tavistock Place, next to my own. His home was frequently a place of popular disturbance, by females whom he had married, or lived with, and
deserted. When in want of money, or over-excited by drink, they knocked
at his door, broke the windows, or otherwise behave[d] riotously. He was
a disreputable character.[184]

Thus John Britton, possibly sensationalizing for effect and hitting back:
his pretensions to antiquarian knowledge were ridiculed in the review
of his *Beauties of Wiltshire* (1801) by the *Critical*,[185] at a period when
Pinkerton's connection with the journal was being mooted about. Yet
some of what Britton said has a ring of truth. Pinkerton had defended
himself against such charges in a letter to Longman's in 1810, saying he
might "at some distant period" write his memoirs which would "over-
whelm all my enemies with utter confusion."[186] But he did not publish
any memoirs apart from *Recollections of Paris*; and it is apparent that
in 1809–12 the sins of his past were rebounding on his head. If he was
indeed chased and harried by women, we need look no further for reasons for his leaving London, or for Henrietta's retreat to Odiham. In 1810
Pinkerton had thrown out a suggestion to John Young that he wished to
return home. In 1812 he had it in mind to try and find paid work in Edinburgh, an idea Young discouraged. Situations Pinkerton would want, he
said, went to "the Clergy and Professors who have the eyes of hawks for
them."[187] Undeterred, Pinkerton decided to go. Prior to leaving he had
another personal matter to attend to: his son by Hester Brown, named
John Pinkerton. He tried to have the child accepted into the Foundling
Hospital. When that failed[188] he placed him under the care of one Mrs
Parsons at Bexley, a village southeast of London.[189] Then, Betsey on one
arm, the manuscript of "The Heiress of Strathern" under the other, he
fled to his native city. He arrived there in November 1812.[190]

"He must have been above 70," a young Scottish author recalled,
"a little withered old yellow man much marked w[ith] smallpox, with
the ugliest of scratch wigs and green spectacles."[191] Old! He was fifty-
four; but the years had taken their toll on him. Pinkerton was now
a veteran wine drinker, and the word "yellow" could be an indication of incipient liver disease. At any rate, he could hardly expect a
warm welcome in Edinburgh – the sins of Pinkerton the Goth hadn't
been forgotten.[192] But some of his countrymen had come to realize that
the impetus he'd given to Scottish historical and literary studies was
deserving of respect. Among them was Walter Scott, who had sent him
a copy of the *Minstrelsy of the Scottish Border* (1802) on its appearance.[193]
Scott saw the value of the *Enquiry* and *History of Scotland*. The latter

he thought "a work of meritorious labour, but not delivered in a very pleasing style or manner."[194] He used it to prepare his essay on Border history in the *Minstrelsy*, and later in 1822 a passage in it provided the inspiration for his "dramatic sketch" *Halidon Hill*. We need hardly add that he drew on it for his own *History of Scotland*, a flimsy though elegantly written work that one is surprised to see given credence in some current historical writing, while Pinkerton's *History* of 1797 is forgotten. When Pinkerton sought Scott's help and encouragement he was received with courtesy.

He set out to make his presence felt. Soon after arriving he wrote to Henry Siddons, manager of the Theatre Royal, asking him to read "The Heiress of Strathern" in hopes of having it staged. Siddons rejected it, noting that it required a splendid setting and doubting if actresses could be found in Edinburgh to play the leading parts.[195] There was another drawback. The tragedy, set in thirteenth-century Scotland, had an incest theme – not incest averted or suspected, but incest commit-ted and exploited. Pinkerton retrieved his manuscript and sent it to Scott, who read it twice "with great pleasure," praised the "accurate and interesting description of the manners" of its period, and said he, "being intimately acquainted with your early poetical efforts," found what he expected, "much beautiful & appropriate poetical diction." As for the "deeply & painfully interesting" developments in the latter part of the play, i.e., the revelations about incest, he was "not quite sure that these 'giddy paced times' would not be as much terrified as pleased with it."[196] That was as near a warning as Scott chose to come. Early in 1813 John Philip Kemble came to Edinburgh to act in Siddons' com-pany. Any worries Siddons had about drawing crowds were set aside; the great tragedian packed the house again and again.[197] Pinkerton had known Kemble in London,[198] and it was through his intercession that Siddons was led to reconsider "The Heiress of Strathern" and decide to stage it.[199] Scott persuaded R.P. Gillies to write the epilogue.[200] Pinkerton wrote the prologue. The playbill appeared, advertising "The Heiress of Strathern, Or the Rash Marriage. Written by a Gentleman of Edin-burgh." The play opened 23 March.

The house was full, the prologue was spoken, the curtain rose. Strath-ern, an aged earl, delivered the opening speech. His army has gathered at his beckoning and he, because

… the snow of years
Covers my head, and shews that winter reigns,
That winter never visited by spring

resigns the command in favour of young Methven, his kinsman, who will lead the men into combat against the English. Strathern's son had been killed many years before:

> ... ye know that cruel chance
> Bereaved me of a son, who might ere this
> Have trod the paths of glory: in one daughter
> Ends the high lineage of our ancient race.

His daughter is Finella, whose hand is sought by Comyn, a young, haughty, and "warlike" earl; he is loved by Aslinda, Finella's hand-maid. Finella detests Comyn, and at the end of the first act it is revealed that she is secretly married to Consalvo, a foreign knight who had recently defeated Comyn in a tourney. In the second act Finella, having resisted Friar Anselm's entreaties that she join a nunnery, openly rejects Comyn's suit; Strathern tells Comyn the full romantic tale of his son's seizure by Moorish traders; and Consalvo arrives at Strathern's castle. Strathern welcomes him:

> Meanwhile you shall abide with me a season;
> My lands abound with game, the deer, the roe,
> The badger, fox, and wolf. My dogs are fleet,
> Of every kind, of chosen scent and strength.
> Steeds of all countries fill my ample stalls:
> Even if you choose a Turk or Barb they're ready.
> If falconry you love, my birds are chosen
> From Norway and the Orkneys, of all wings;
> The lanner, the goshawk, the peregrine,
> The jerfalcon, the sacre, and the kestril;
> For all prey from the ptarmigan to the heron,
> From grouse to the cock o' the forest. We shall sport
> All day; &, when the evening comes, of war
> And feats of arms discourse.

Something too much here of zoology and venery. The main plot had not been greatly advanced by the end of the second act. The acting was poor at times, but Siddons as Consalvo, his wife Harriet as Finella, and Daniel Terry as Strathern played their parts with spirit. Hissing had already begun. In the first scenes of the third act a new plot was introduced involving the character Voda, who, on the day appointed for her to take final vows at the convent adjoining the castle, reveals to Aslinda that she had been ruined by Comyn. To warn Aslinda of Comyn's wiles, she has arranged to interview him while Aslinda, hiding in an alcove, listens. Comyn arrives.

> VODA: Where were thine oaths? Didst thou not still refuse
>   To ratify our loves by holy rites?
> COMYN: Thou wert impatient: my stern sire in life,
>   I dared not on the pain of his displeasure:
>   I vow my love is ardent as at first.
>   And now I would – thou art a nun forsooth!
>   'Tis wild impatience ruins half thy sex:
>   The man is blamed, because his mate is frantic.
> VODA: What do I hear? Wouldst thou then wed me now?[201]

It had been obvious from the beginning of the play that part of the audience had come simply to jeer and hiss. So far their efforts had been insufficient to raise the house. But something in the Scottish character could not suffer seeing a nun propose marriage to a lecher. From the moment she spoke her line "an uninterrupted series of hisses, catcalls, and laughter, drowning all sounds of plaudit, continued to shake the house for nearly a quarter of an hour. The doom of the piece appeared to be sealed."[202] But the din fell off, and the remaining scenes of the third act restored order. A love scene between Consalvo and Finella gave Harriet Siddons a chance to show her powers. Pinkerton was not in the theatre, but two spies sent him notes at the end of the third act. J.M. (probably Dr James Millar, at whose home he was staying) reported that "3 Acts passed off well – one or two expressions excited some noise." H.W. wrote "All goes on extremely well. Mrs S. acts delightfully. A few drunken fellows hiss but are soon overpowered."[203]

In the fourth act came the fatal discovery that Consalvo was Strathern's lost son. Consalvo's agony, Strathern's bewilderment, the revelation to Finella that her love was incestuous – all this created "a considerable tragic effect" and the hissers were silenced. Had the play ended there, Pinkerton might have been the man of the hour. But a fifth act followed: Finella was made to suffer (she poisoned herself), Comyn was persuaded to marry Voda, and Consalvo, though he tried to stab himself, was left to carry on the Strathern tradition of chivalry. Fainting spells, ejaculations of "Oh sin, Oh sin," "O Villainy," "To horse," "Charge, Charge!," "O mercy, save him!," "O poison, poison! Dear Finella! Poison!" – together with the objectionable turns in the plot – brought on a theatrical riot:

But, in the fifth act, all was changed, hiss succeeded to hiss; and, before its termination, the house was quite in a state of tumult. At length the curtain fell, and Mr. Siddons appeared, to announce the repetition of the play. The confusion of sounds which followed can be compared only to that which was heard in chaos before the reign of order commenced. Hands clapping, feet beating, hisses, catcalls, hear! hear! off! off! resounded from a thousand quarters at once. The manager stood for some time in silent amaze; but he soon became sensible, that a great majority was on the side of condemnation. He retired. An attempt was afterwards made, by the party friendly to the play, to call for it, but was drowned by a new tempest of wrath. Another drama was then announced.

Scott had foreseen what would happen. Much of what he'd written to Pinkerton earlier was humbug. The characters come and go and speak notable things "in good blank verse," he wrote to Joanna Baillie a couple of days before the play went on stage, but there is "no very strong interest excited" and the plot is "disagreeable."[204] Some "strong interest" *was* excited; but he was right about the plot. The *Scots Magazine* said the play's theme was "shocking and disgustful."[205]

The hostile reaction to the play was not caused solely by the piece itself. Pinkerton's general behaviour appeared to offend some of those he met. John Gibson Lockhart's memory of him remained vivid as late as 1840:

I when a very young fellow met that old captious scamp pretty often & retain a lively disgust of his affectations, conceit, dogmatic insolence, & audacious nastiness & blasphemy. He came to pass a winter in Edinr bringing w[ith] him a blazing damsel whom he had picked up here [London] & introduced there as his wife – after some months it turned out that his real wife was an old woman living neglected on the continent & he was expelled from Edinr w[ith] fierce indignation.[206]

Some of this is mere gossip, but not all of it. R.P. Gillies recorded an instance of Pinkerton's "sneers against religion" at a dinner, greatly offending another guest and provoking an argument.[207] Standing up for providence in print, as was now his practice, didn't mean he couldn't lip back at the pious in a dining room.

Meanwhile, in London, Longman & Co. proceeded with the *Atlas* and *Voyages and Travels*, on both of which, though nearing completion, some work remained undone when Pinkerton left London. The publishers gave the task of maintaining liaison with him to Thomas Rees, the younger brother of Owen Rees, who was a partner in the firm. Thomas Rees was also involved in the sale of the house on Marchmont Street[208] and other affairs of Pinkerton's. The Index for Volume 17 of the *Voyages and Travels* was well advanced; otherwise nothing else for it was at hand. Rees had to put the other two sizeable segments of the volume, once he'd got them, through the press himself without the benefit of an author close by. Concluding travels in Volume 16, which dealt with Africa, he also had to cope with, following instructions Pinkerton had left for him, though making adjustments.[209] He also made a couple of delicate decisions about the *Atlas*.[210] Various hitches arose. The Scottish runaway had to be prodded to answer his mail and send material south, or send back what had been forwarded to him. Amidst all the goings on in the north, Pinkerton found time to write not only the introduction to the *Atlas*, the lamentable brevity of which could reflect haste, but also the impressive introductory essay in Volume 17 of the *Voyages and Travels*. He also had a share in preparing the catalogue for Volume 17, although, as noted, Henry Weber did most of the work on it.

Following the disaster of the "Heiress," Pinkerton quit Edinburgh for a tour. (He had sent Betsey back, pregnant, to her family.)[211] In April we find him in Raith, a village in Fife,[212] where he stayed for a period with Robert Ferguson, the wealthy mineralogist and M.P. He went northwards from there to visit lawyer George Dempster in Dunnichen, Angus. Pinkerton had known him in London in 1789. He interested himself in Dempster's agricultural schemes[213] and left an unpleasant impression on the Rev. James Rogers:

> At Dunnichen House my father met some distinguished persons, among others John Pinkerton the antiquary. Mr. Pinkerton was, in my father's estimate, self-opinionated and unamiable. He talked much about himself and his writings, was inclined to oppose the general opinion, and would not tolerate that any statement or theory of his own should be impugned or questioned.[214]

At Dunnichen he ran out of ready money. He wrote to William Blackwood in Edinburgh, one of the publishers of the second edition of the *Enquiry into the History of Scotland*, and asked him to remit £20. In his reply Blackwood must have said Pinkerton was a bookmaker, or words to that effect. He got a tart reply. Pinkerton wrote that he would stop the publication of the book if he didn't get the sum demanded. And as for his being a bookmaker, he had enriched his booksellers "to the tune of £60,000, of which he has at least a tenth, but never was employed by them." Blackwood's "impertinence and ignorance are of a piece."[215] Blackwood must have relented: the book came out in 1814. Pinkerton got £50 for it. A "chief object" of the edition, he explained, was "to remove some asperities in the language, which ... always excited his regret after the first warmth of composition." Yet little holding back was evident in the new Advertisement, where he took shots again at Ossian and Mary Queen of Scots, fiercely damned his opponents as plagiarists, ridiculed "the imaginary Celtic of the fourth century," and declared that Scotland was "that country of Europe, in which the least erudition has appeared."[216]

To return to 1813. In August Pinkerton left Dunnichen to visit R.P. Gillies at his estate near Arbroath. For three months the antiquary and the young poet were bosom companions. "Perhaps the less that I say of John Pinkerton the better," Gillies wrote in his *Memoirs*, "inasmuch as my notice of him cannot be very favourable." He then proceeded to give an account of the pleasant hours he spent with Pinkerton among the ruins and quarries in the vicinity. The two would never meet in the morning, but around noon the carriage would be called and off they'd go to explore Roman camps or search for geological specimens:

> But of all our haunts, that which best pleased my eccentric guest was the agate quarry, near the town of Montrose, where he would work with his hammer for hours together among the green earth[217] and plumb-pudding[218] blocks; indeed, would not have come away till dusk, if I had not reminded him that dinner would be spoiled.

They argued often but didn't quarrel; and Pinkerton's conversation, Gillies wrote, to someone not disposed to question his "dictatorial manner and bizarre notions," afforded "an inexhaustible stock of amusement."[219] They would often talk of poetry. To merit immortal fame, Pinkerton said, a poet had only to produce one sonnet that was perfect of its kind. His volume *Rimes* he now viewed with pity and contempt, but he thought some of Gillies' sonnets tolerable. He also related anecdotes of figures in Scottish history that the scenes and ruins brought to mind. The pleasant afternoons, "cheered by autumnal sunshine" and conversation, passed by. Pinkerton also visited the farmer George Robertson, author of *General View of the Agriculture of Midlothian* (1793), at Benholm in Kincardine, and saw the thirteenth-century church at Arbuthnott.[220] It does not appear that he made it as far north as Aberdeen. In November he returned to Edinburgh.

He would have liked to stay there. He explored it again in 1813, perhaps retracing some paths of his youth. In the dungeons of Craigmillar Castle, the ruin that inspired one of his earliest efforts in literature, he discovered a human skeleton, enclosed within a wall, standing upright.[221] He visited Holyrood House, the Advocates' Library, and the Register House, and sought out the literary men – chemist Thomas Thomson, Scott, Weber, John Jamieson, Playfair. Late in 1813 he applied for a position at the Register House. He could not get it.[222] He stayed in Edinburgh for the first few months of 1814. In April the *Edinburgh Review* blasted his *Petralogy* with sixteen pages of ridicule and rebuttal.[223] (Though it missed the book's main weaknesses.) It must have been hard to show his face after such a savage attack. He quit the city in May, no doubt having heard of Napoleon's abdication, which left France once again open to visitors. In Paris, he said bitterly in August, there was much cooperation and friendship among literary men; in London there was less; but Edinburgh "is about a century behind either."[224]

On his journey south he paused at Liverpool and met the historian and banker William Roscoe;[225] he stopped at Stratford-on-Avon too. In late July he was back in London, living at Sloane Street. Where could he turn now? Clutching at one remaining straw, he tried to revive the plan for publishing the monkish historians. He wrote to Lord William Wyndham Grenville and asked for his patronage. Grenville seemed interested in the project,[226] but when his brother, Thomas Grenville, reported that Pinkerton was "a mere book-maker, and one of the most irritable of the *genus irritabile*,"[227] he backed off. Longman's declined to get involved.[228] Around October Pinkerton wrote to George, Prince Regent of England, on the subject, but was not favoured with an answer.[229] In November he told Roscoe of his latest plan:

> I trouble you with these few lines to inform you that I mean in January to visit France. If you know any wealthy man who would wish for an intelligent companion I flatter myself I could be of much use; and I need not be ashamed to confess to you that my finances are very low from various unhappy accidents to which literary men are peculiarly subject owing to their seclusion from the world and its ways. I believe that like Goldsmith I shall die of vexation. Excuse my low spirits and trembling hand.[230]

Late that month Pinkerton was told that his son John was in good health.[231] He appears to have considered taking the boy with him to Paris, but decided not to. In February 1815, shortly before Napoleon's escape from Elba, he again crossed the Channel into France.[232]

7 A Banished Man,[1] 1815–1826

On arriving in France in 1815 Pinkerton settled into a Paris hotel and, once spring came, took a country lodging as well at Clamart on the southern outskirts of the city. The scenery there was beautiful, he had likely salvaged a few hundred pounds[2] from his last years in London, and he had his annuity. He had the prospect of paid literary work; Macvey Napier was his good and influential friend in Edinburgh, and he likely thought Longman's would keep him in mind as well.[3] His peace of mind was restored. "Immersed in misfortune as I have been I have found change of place a main relief," he told Warden, "I can <u>change my mind</u> and reject all afflicting thoughts. 'Never think of it,' is a good plain maxim."[4] He moved around Paris, visiting acquaintances, reading at the Bibliothèque nationale, and examining curiosities here and there. In early April he was in the lower galleries of the Jardin des Plantes, talking to a professor of biology and watching experiments. Suddenly a shout of "Vive l' Empereur" came from the garden outside, and Napoleon, back for his fatal 100 days, entered the room with his bodyguard. For two hours the professor and Pinkerton accompanied the Emperor through the rooms of the Jardin. Pinkerton was not called on to describe any of the exhibits, but he noted the particulars of Napoleon's dress, manner, and mood and left a memorable account of the occasion.[5] And so the first few months of his stay in France were agreeable. A letter to Warden in May from Clamart illustrates his tranquil spirit and serves to show he was capable of charity:

> I shall be happy to see you any day to eat a bit of dinner, but I have no bed. Come at 1 or 2 that we may talk and perambulate these enchanted grounds where a thousand nightingales sing night and day. It is paradise

itself. … You will be glad to do a good action. Stop then to my Hotel No. 41 Place St. Germain L'Auxerrois where you will find a worthy gentleman of Champagne on the Third floor, door on the right, called M. de la Rue. He is threatened with the loss of an eye. Use my name, tell him your own case, recommend your oculist, console &c &c.[6]

The Abbé de la Rue had been Pinkerton's companion at the British Museum in 1790. They had not lost touch in the interim.

Pinkerton fades from view for over a year, and when he returns a different note is sounded in his letters. Through 1815 and 1816 his attempts to earn money evidently received little encouragement. Longman's, though willing to consider him as an agent for buying books[7] in Paris, no longer wished to make use of him as an author. White and Cochrane, the firm that published *Petralogy*, went bankrupt early in 1815.[8] In November 1816, feeling the need for a regular correspondent in England, Pinkerton reopened the correspondence with Douce. He did not believe, he wrote, that "a few hasty words should obliterate an acquaintance of 30 years." He was unwell and unhappy, and was living now on the Rue des Moulins in the centre of Paris, not far from the Bibliothèque nationale. On his rambles he often bought many curious early printed books "for a mere trifle." Recently he'd found an octavo printed in 1527 and an early quarto.[9] He gave the impression that he could supply Douce with scarce works to add to his meticulously assembled library. Douce took the bait, answered, and a correspondence ensued that lasted until 1824.

Pinkerton made his intentions clear in his next letter to Douce, in February 1817. He had become almost a complete recluse, he said. "I am so sick of this world, that if there were a monastery I would retire into it." He had "no resource" for his old age, only "the sad prospect of chilling poverty and the jail." He gave a list of the financial losses he'd sustained, and asked for help.

> During two years I have picked up about 200 volumes all rare or singular, wooden prints (very rich and numerous) magic, MSS. classics. Will you have them for I should be pleased if they passed into your hands and any duplicates you could send to Legh and Sotheby.

His "few resources" wouldn't last beyond a year, he said. He asked about the Royal Literary Fund, a source of help for indigent writers since its foundation in 1790. "Does the literary fund supply loans or temporary assistance for I should hope to repay, if not it should not be from want of industry."[10] The letter was not wholly above board. He didn't mention his Edinburgh annuity to Douce until 1824. He didn't in fact have many rare books, prints, and manuscripts to dispose of. Once Douce received some of these "singular" items he saw that Pinkerton had overrated their scarcity and value. Yet he had some small success in this role as book finder. The number of requests for help soon increased. Would Douce arrange for the sale of the piano he had left in London? Would he recover and sell copies of the *Atlas* and other volumes scattered about? Would he collect small debts owing to him? Such commissions were complex and troublesome and Douce executed them unwillingly. But he could hardly ignore appeals that they were "dictated by necessity."

Pinkerton kept on working. In 1817 he proposed to Longman's that he translate for them Pliny's *Natural History*, "the most valuable work of antiquity and one of the most difficult."[11] They demurred. (But they did buy 300 books he'd collected since coming to Paris.) Also in 1817, through the good offices of Napier, he began writing a series entitled "Anecdotes, Historical, Literary, and Miscellaneous" in the *Scots Magazine*. These appeared at irregular intervals from November 1817, to December 1819.[12] It is possible that he wrote similar pieces in other outlets. He was paid £100 for additions to the third edition of *Modern Geography* (1817), where he complained of the "impudence" and "audacity" of his enemies and admitted he was living in "poverty."[13] In 1818 he published anonymously a Latin version of his *Spirit of all Religions* under the title *Religio Universalis et Naturalis: Disquisitio Philosophica, ex exemplis omnium nationum et rationibus variis consolidata*. It contained the most sublime system of religion ever offered to the world and formed "an epoch in the progress of the human mind." So he told Napier.[14] His authorship he not only kept hidden but, in an introductory note, distanced himself from it by saying it was "transcribed from a manuscript of the early 1800s" that was "difficult to read in certain places." The implication was that it was not his own work. He ended his publishing career as he'd begun, with a (sort of) forgery!

This expanded edition of the pamphlet repeated the polytheistic tenets of the earlier work, but there was less worm theology in it and much more meditating on topics such as why the soul is distinct from the body, the power of God, how matter was formed, what light is, and the nature of æther[15] – all the while his head being full of what the scientists of his time had learned, and what they hadn't. In effect, he tried to marry teleology and science, and from the linkage find his way to an ethic by which mankind could live. One target of attack was the atheist, as we see in a passage from his chapter on "Materia et Creatio" (matter and creation):

> … it is far easier to believe that the making of matter was accomplished by superior beings, than by some mysterious property intrinsic to itself, as the atheists would have it; for, from the very beginning of time through to the insights of tradition or history, nature's powers and phenomena have continued indefatigably; yet it is doubtful whether they were actually able to produce microscopic insects. And it is plausible that those insects, observed by many in animal sperm, are only particles inflated by a kind of life-giving vapour … And such is the case with the vapour of life in birds' eggs, which are brought to life out of a small empty space, next to the side of an egg yolk … yet an invisible source of life is stamped onto it by the male. We are convinced by this example that our knowledge in regard to science is modest … natural phenomena are much easier to explain, if we recognize an invisible architect, rather than certain powers of nature, equally invisible and equally unknown to our reason.[16]

This kind of thinking led him towards Christian orthodoxy, though not wholly into it owing to the polytheistic underpinning which, however, receded as he worked through his arguments. He brought into the discussion the classics, Locke, and even Aquinas; Jesus and the Apostles were present too. He quoted the opening words of the "sacred book of Genesis" and argued that the word "created" in Genesis should properly have been translated as "made" or "formed," so that God made the earth not out of nothing but from an already existing chaos. This "is readily accepted by pious men who wish to reconcile natural phenomena, and who consider mountains that are very high and pristine, made up of seashells; those mountains sufficiently reveal a state of chaos over a very long age: beyond six thousand years, the human race has handed down no records, and no traces."[17] Thus he tried to reconcile the record implanted in the earth's rock strata with the dicta of Bishop Ussher. It was ironically to religion, not science, that he reached for solace as his life was ending. Behind the speculations, digressions, and peculiarities, there is in the pamphlet a search for meaning, a willingness to believe that God is at work behind all the mysteries confronting humankind, that He "is present in all things, as religion teaches."[18] How far is He present in the life of individuals? We "should not inquire into divine laws," Pinkerton warned at one point, "but often, when we see that good and just men are the most wretched in this life, we are permitted to hope for justice."[19]

Justice for him, at least in his own eyes, remained elusive. As time passed his need for money became more acute and his pleading to Douce more ardent. Until 1818 his health remained fairly sound despite periodic illnesses and complaints. That now changed:

> Your last stares me in the face with the date 23d. April and I begin this the 15th. August. Ever since the 2d. June I have been confined by a violent attack of the gout and my left leg is still swelled and painful. This is my first regular martyrdom though Dr. Willan[20] <u>alone</u> suspected many years ago that my flying pains in the breast stomach bowels were of a gouty nature.[21]

He had his own ideas on how gout should be treated. Temperance, Barclay's Pills, and "fish pudding and potatoes," he'd told Warden in 1806, comprised the only effective palliative for the disease.[22] No doubt he now tested the validity of the theory. He eliminated almost all meat from his diet, kept his leg wrapped in a stocking of waxed linen, and likely tried to reduce his consumption of wine, though he regarded it as "one of the necessaries of life."[23] Perhaps because he failed to stick to a regimen of temperance – at his death he owed a big debt to his wine merchant – the gout, or whatever it was, soon assumed alarming proportions. In October 1818 he was struck by an attack in the stomach "which was thought mortal,"[24] and was provoked by it into making disclosures to Napier about his personal life. By mid-November he could no longer walk about without a cane. Early 1819 brought on a second seizure.

> I was slowly mending of the first attack in my left leg when behold a far more violent in my right which confined me to bed for six weeks and since to my sofa. The pain especially in the hip the true hip-gout emphatically called <u>passio sciatica</u> was so extreme that every motion was agony and I could only compare it to being torn in quarters by horses as in cases of high treason. At present the whole has descended to the leg and foot which remain swelled and feeble.[25]

Yet another infliction ("for five days it was a struggle between life and death")[26] came in July 1819, and from then on, to judge from his letters, little could be done to relieve his pain. By May 1820 the gout had become "habitual"; swollen legs, "great weakness," and pain in his side obliged him to remain in his chambers. Any form of movement was agony. He could travel in carriages, but not over cobblestones. In December 1820 he complained to Douce that "I have lost the use of my legs."[27] He was an invalid. He was, he told a friend in 1823, "toujours prisonnier de la goutte dans ma chambre."[28]

Douce kept up the correspondence during this period and managed to sell Pinkerton's leftovers in London for £20 to £30. He was also growing old and becoming sick of the world. He'd had an attack of gout too, and while trying to rise after "stooping to replace a folio" on a low shelf, had fallen to the floor, after which he remained in "agonies" for a fortnight.[29] No doubt trying to comfort Pinkerton, he painted a black picture of contemporary England. The political situation, he noted in November 1819, "is horrible to the philosophic & quiet spectator"[30] – no moderate and rational sentiment was left in British politics. London was noisier than ever, intolerable to all peaceful men. The British Museum reading room was "cold" and "vile." London publishers were paying too much for trifling books. Murray, for instance, offered Southey £4,000 for any half-dozen lives he chose to write, and Scott was getting thousands for his tales. Pinkerton once read him "a tragedy that is worth all those heroes have written; but you are not a laureat nor a baronet, & I doubt whether Murray will therefore deal with you for <u>hundreds</u>." The people of England, he conceived, were "wonderfully changed & not to my humour, from what they were 40 years ago, aye, even 10." The one thing he contemplated with pleasure was the idea of visiting Pinkerton in Paris and enjoying, as of old, his "chearful & interesting conversation."[31] Such grumbling was very much to Pinkerton's taste and he answered in kind, holding forth on triennial parliaments, the scandal involving George IV and Caroline, and the afflicting state of Europe. He had lost little of his old inquisitiveness. Regrettably, the correspondence did not stay focused on literature and politics.

Through 1821 and 1822 Pinkerton grew increasingly anxious. After some rarities he'd collected in Paris were auctioned by Sotheby late in 1821,[32] there was evidently little else left to sell. What was he to do? He was "always in a lingering malady," he complained in March 1822, "my legs discoloured and inflamed, and so feeble that I can scarcely creep about … I feel weaker and weaker every day and have a constant heart-ake that devours my life blood."[33] In April he moved into cheap rooms on the Rue de Ponthieu. Late in the year he decided to turn to his brother-in-law, Bishop Burgess, of whom, he told Douce, he had always spoken with respect and who had no reason to be hostile towards him. He asked Douce to intercede for him and so he did, approaching the bishop early in 1823 and getting £50. Pinkerton asked Douce to express his gratitude. "I pray God to bless him and grant him long life. I never dreamed that I had any claim upon him but merely his benevolence to a kinsman and a literary invalid."[34] Douce wrote to the bishop once more, thanked him for the money, and conveyed another request from his invalid friend, which was to ask for the addresses of his daughters. He received no answer; and that, he thought, was an insult. "I shall take no further notice of him," he said.[35] But he continued to seek relief for Pinkerton, and in March 1823, with the help of the auctioneer James Christie, got £25 from the Royal Literary Fund.[36]

A month later Pinkerton asked Douce to contact John Burgess, the London businessman we met earlier, to get his daughters' addresses. Surely, he said, they would not suffer him to die in misery for want of £100 a year. In the same letter he said he'd compiled a book of anecdotes entitled "Literary Singularities," which Longman's declined to publish. He wanted Douce to claim the manuscript and find another publisher.[37] But Douce, having already meshed himself in a few delicate negotiations on Pinkerton's behalf, was reluctant to act or even to reply. In July 1823, Pinkerton sent him a frantic note, asking the reasons for the delay. Douce offered various excuses, but he was clearly tired of the various jobs being unloaded on him. He would not, he said, help in getting the "Literary Singularities" into print (but the manuscript did find its way into his hands) and he had no time to advert to the many "urgent matters" in the latest letters.[38] Pinkerton pressed on with more appeals. Early in 1824, acting on Douce's advice, he composed a letter to Thomas Burgess, part of which is as follows:

> I shall be sixty six years of age next month and have been ever since June 1818 that is five years and six months confined to my chamber and after to my bed by a lingering disease (by some supposed the gout) which has deprived me of the use of my legs which are always swelled or discoloured. Repeated attacks in the bowels and loins have twice reduced me to the gates of death and I only may live for a year or two more by daily medicines and strict regimen.
>
> [I su]pplicate Your Lordship whose beneficence in your [diocese] has been mentioned even in the French newspapers [to ex]ert it in my behalf by stating my miserable situation [to] my daughters. They are or will be

rich heiresses and are surely bound by the eternal laws of God and nature and still more as Christians educated under a pious eye in the precepts of the holy gospell not to let their old father perish from want in a foreign country to which he is confined by disease. If each will only grant £50 a year for the short time I have to live I should be able to supply all my little wants or rather necessities. This sum of £100 to be paid half yearly in January and July and to complete the favour in advance as I may be dead before the first half year falls due. Surely they would feel much compunction and remorse to learn that their father died from mere want nor would such an event fail to darken their reflections or affect their reputation when too late to be remedied. In all events I pray God the father of us all to bless them and to grant them all the happiness which has been wanting in my portion of existence.

I have lived for many years on the most moderate plan and amidst many privations. The rent of my little apartment is £12 a year and the rest in proportion but my disease often violent constrains me to pay a physician and a poor old sick nurse as fainting fits and other infirmities render me very helpless … But all I have said is, alas, too true at the same time that it is most mortifying to me after having passed all my life in my little independence.[39]

This was sent to the Bishop on 4 March 1824. Three months passed; no answer was forthcoming, nor had Douce replied to a letter of that date. Pinkerton now asked Douce once again to approach John Burgess for his daughters' whereabouts.[40] Douce wrote a short memorandum, describing his interview with Burgess:

I applied to the oilman who almost insulted me at first, but very soon found whom he had to deal with. It was to no purpose & he refused to tell me where the children lived. His abuse of P. was outrageous & whether well founded or otherwise I know not, but it sickened me completely from interference in cases of this kind.

On 5 July he wrote his last letter to Pinkerton:

You had in a former instance involved me almost in Ecclesiastl. censure from an appeal energetically & humanly made in yr. favour, & you have now a second time, & I fear insidiously, almost exposed me to insult on the part of the Oilman. I shall not describe the disgusting interview I had with this person – it is sufficient to state that all access to your children is prohibited, that you have nothing to hope for, so far as I coud. collect, from that quarter. This is a lesson to me as to interference in family matters … I resolve for the future not to put gratitude to the test, & to have as little to do with mankind as possible, & therefore, if I do not demand that our correspondence may altogether cease, it must certainly undergo a very considerable pause.[41]

Through 1824 and 1825 Pinkerton's disease got worse. On 27 June 1825, he made his will. He disowned his two daughters and appointed as his heir and executor his "worthy" friend James Lewis Busson du Maurier, "Student of law residing at Paris." He left to him a few manuscripts he had failed to sell, his personal belongings, a bond upon Cadell and Davies and Longman's for £1,000 (payable "upon the fulfilment of certain conditions or events Specified in the said Bond the Amount of which must become due in length of time"), and a chest containing mostly letters he had received during the years 1776–1814. Du Maurier was to pay Pinkerton's debts, which amounted to 1,000 f. in Paris and £20 in England.[42] The two letters that survive from 1825 were dictated by Pinkerton and signed by him in a very broken hand. In June he wrote to Walter Scott, begging for money. For some weeks beforehand, he revealed, his disease had become unbearably severe, "having seized both my hands so as to render me utterly incapable of holding a book and much less a pen."[43] His last letter[44] was a note to Douce on 30 August to ask that the manuscript of "Literary Singularities" be passed to du Maurier when he required it. Douce was relieved of it on 1 January 1826.[45]

On 10 March 1826 Pinkerton died in his chambers at 10, Rue de Ponthieu, Paris.[46] His will specified that his funeral arrangements were to be "as oeconomical as possible." No lasting monument was placed on his grave.[47]

Pinkerton's daughters from his marriage to Henrietta Burgess had been well cared for during his declining years. Their mother died around 1820,[48] but Thomas Burgess looked after them "with an affection almost paternal."[49] The "Christian character" of the older sister, Mary Margery, especially endeared her to the bishop, and it was she who wrote to Douce in March 1826, asking for "the particulars of my father's death." She had seen a notice of his death in a newspaper.[50] In his reply Douce, while frank in some respects, was careful lest the association of his name with Pinkerton's reflect badly on his own reputation:

I am unable to give you any informn. concerning your fathers death not having had any intercourse with him for a considerable time & not since I was betrayed into an interview with the relatv of yours in the Strand which I shall not easily forget. I hope my letter to yr father on that occasion will be found among his papers, with other records of my benevolent feelings towards him and his misery and misfortunes, on the latter of which I forbear to make any comments: but I wish it to be understood (for it may perhaps have been misunderstood) that my connexion with him originated & continued with some intermission, on account only of the value I entertained for his literary talents.[51]

Mary Margery died of consumption in 1831.[52] Her sister Harriet Elizabeth married the Rev. Charles Buchanan Pearson, son of the Dean of Salisbury, in 1833, and a son was born to them in 1834.[53] One of their granddaughters, Mrs B.M.H. Riddell, was still living in 1934.[54] What eventually happened to Pinkerton's son John, and to his other children, is unclear. In 1835 John called on Dawson Turner, editor of Pinkerton's literary correspondence, "in very shabby dress & with much the look of a vagabond." He had led "a wandering life," motivated by "a love for science & a persuasion that he shall discover perpetual motion." He wanted to know the name of his mother.[55] He then disappears from view.

In the house of Scottish history and literature there are many mansions, and Pinkerton comes knocking. Not everyone will let him in. His disparagers focus on two issues: forgery and racism. That he was a literary forger is true, as already stated and explained; but so were quite a number of literary figures in the eighteenth century, some much more stubbornly persistent in their lies and hoaxes than he. Even Dr Johnson, recently declared by an American critic to be "the supreme truth-teller in English literature,"[56] was a forger, and didn't confess he was one until close to death. He fabricated speeches for members of Parliament and boasted that he made sure "the Whig dogs should not have the best of it" – surely a more serious deception than forging ballads. The forger Pinkerton should receive as much "rehabilitation"[57] as that given to more eminent counterfeiters.

As for Pinkerton's racist streak, the possible causes of this aberration have been examined in chapter 2, and it seems unnecessary to revert to it at length. It is found in his writing during a brief span of years, after which it is mostly missing. He felt it; he told what he felt; he suffered for it; he tried to explain and make amends for it; he got over it. His complete bibliography, not just that part of his work on Scotland where the racial bias against Celts is to be found, must be considered in determining his contribution to literature. Not that he should escape the whip of censure; not that he should be "rehabilitated" on that front too! Yet some slight abatement in the harsh language sometimes used of him might be in order. The truth is that while Pinkerton has had to answer again and again for his bigotry, the manifest bigotries of some of the greatest writers are often slipped over in critical discourse, as if high achievement compensated for them. We do not stop praising Trollope because the gallant Lord Lofton in *Framley Parsonage* delivers a sneer at "a Jew tailor" that is applauded by his virtuous mother and evokes in the author no sign of disagreement. Examples like this could be multiplied. Pinkerton gets no such *carte blanche*. If his prejudices mark him as a "vermicule,"[58] he has plenty of company in the manure pile.

There is, too, the unpalatable part – parts – of his private life. We need not repeat the details, except to say that neither his side of the story nor that of the women he lived with has been disclosed in sufficient detail to pass firm judgment. On the basis of what we do know, not much, if anything, can be advanced in his favour. It may also be said, though not to extenuate his apparent harsh treatment of women, that more is known, or surmised, about his sexual life than about that of other prominent men of his time, simply because he shocked public opinion in his books and so invited close observation and malicious gossip. One wonders how the bedroom and marital practices of the pious poohbahs among his contemporaries would stand up to the kind of scrutiny he endured.

So much, by way of conclusion, for his trespasses. As for his achievements, again extensive recapitulation seems unnecessary. A terse summary may serve, from an encyclopedia entry in 1840: "Pinkerton was no ordinary man; and his best performances, such as his Dissertation, his Enquiry, his History, and his edition of the Maitland Poems, with all their faults, not only all overflow with curious learning and research, but bear upon them the impression of a vigorous, an ingenious, and even an original mind."[59]

And one more. Historian Hugh Trevor-Roper called him "the greatest of Scottish antiquaries after Innes."[60]

In recent years his work has been drawn into the discussion of Scottish identity,[61] a subject that can here be broached only briefly. The historian Linda Colley has commented on the different ways Scots who "went south" in the eighteenth and early nineteenth centuries reacted to being uprooted. Some, she says, returned home quickly, "deeply alienated and disillusioned." Others "stayed on as foreign mercenaries, taking what advantage they could from their new surroundings while remaining fundamentally aloof." Still others, like Boswell, "were turned into perpetual exiles … feeling themselves too Scottish to settle comfortably in England, yet becoming too English ever to return to their native land." And some "were able to reconcile their Scottish past with their English present by the expedient of regarding themselves as British."[62] Pinkerton does not fit well into any of those categories. He did not worry about his own identity or the identity of Scotland. Few overt signs of alienation appear in his correspondence or published writings. He did not seem to have had any difficulty in becoming reconciled to the new, or pain in estrangement from the old. He was not an arguer against the 1707 union with England; indeed he appeared to have not the slightest misgiving about it. "From the English," he wrote in 1784, "the people of Scotland derived, and now derive, most of their improvements."[63] He wished for no separation of Scottish from English interests in politics. In 1794, amidst the ferment and uncertainty following the Terror in France, he said it was not Scotland's role to "take the lead" in reform; "on the contrary, she should strictly follow the fortunes of England."[64] To him, England, with France, stood highest of all nations "with regard to the arts and sciences, the general intelligence and enterprise of the people, and all the arts of civilized life."[65] He chose England as a place to live out most of his adult life, quitting Scotland when very young, as soon as he had the opportunity. As we have noted, his stance in *Modern Geography* in 1802 was that of a writer with an English perspective on world developments. In *Recollections of Paris* he wrote as an Englishman, even referring to himself obliquely as "an English patriot," and in 1817, defiantly answering critics, he wrote "I am still an Englishman and a heretic."[66] It was only the onset of failure and misery that drove him back to Scotland in 1812; had he succeeded financially as a writer in London, or had he not suffered financial losses unconnected to writing, he might never have returned. He didn't, until desperate, see Edinburgh or any other Scottish city or town as a suitable literary location for himself.[67] Scotland was "narrow"[68] and "remote"[69]; his eye was on London and, latterly, Paris. These were the milieux for a man of his talents and ambitions: so he thought. This fact, that he was an expatriate living permanently away from his native land, may account for some of the frankness of his views on Scottish topics, including racial questions, in his early work and his correspondence. He had gone for good: he could say what he liked, and not bother too much about what those he'd left behind in the north thought or wrote.

He started *The Treasury of Wit* with a series of Greek apophthegms, sometimes adding an explanatory comment, thus: "STRATONICUS, the harper teaching a Macedonian, who was very dull, once said in a passion, *I shall send you to Macedon.* Macedon was the Scotland of the Greeks."[70]

Macedon was a fine place, but it wasn't Greece. So was Scotland a fine spot, but it wasn't Broadway.

None of this is to say that Pinkerton lacked patriotic feeling for his native land. He did genuinely have such feeling, which was expressed often, in different ways, in his writing. In responding to the Earl of Buchan's remark that his *Enquiry* had created "violent disgust" among his countrymen, he said "if any Scotish man thinks that he is a warmer friend of his country than its author, he is mistaken."[71] Part of the motivation in his work was to rescue Scotland from misconceptions about the past. "The object of a true lover of his country," he said, "is to sow the seeds of future spirit."[72] As we have seen, he worked hard at bringing Middle Scots poetry out of obscurity into the light of day. From being a forger, he became an advocate for the pursuit of truth in history, though he fell into error, as subsequent historians of Scotland and other countries did, and still do. Nor did he forget Scotland in his later works. Perhaps the text that best illustrates his attitude towards his homeland was the section on Scotland in *Modern Geography* (1802).[73] It was long, indulgent, and affectionate. The clergy, lower classes, nobility and gentry, the education system – "perhaps, the best practical system pursued in any country in Europe" – cities and towns, the "remarkably pure and transparent" rivers, "remarkable" edifices, the "excellent and extensive" canal between the firths of Forth and Clyde, the New Town in Edinburgh, the University of Edinburgh with its "illustrious professors," the lakes, mountains, scenery, botany – all these and more features came in for their share of praise. He listed some authors too, because "Scotland can now produce able writers in almost every branch." As we've noted, he singled out Robert Burns when describing the activities of ordinary people, saying "Two exquisite poems of Mr. Burns, his *Halloween* and his *Cotter's Saturday Night,* will convey more information concerning the amusements, superstitions, and manners, of the Scotish peasantry, than the most long and animated detail." This was a generous stepping aside to acknowledge a writer of greater talent.

It is perhaps time, nearly two centuries after this pariah's death, to extend to him a modicum of pity and forgiveness, and, among scholars, to include him and his writing in the ongoing re-examination of eighteenth- and nineteenth-century literary history.

Notes

1. Youth, 1758–1781

1 David Douglas, *English Scholars 1660–1730*, 2nd ed. (London, 1951), 273.
2 See the Bibliography, II (2).
3 See below, ch. 3.
4 A.B. Friedman, *The Ballad Revival* (Chicago, 1961), 236.
5 Pinkerton (afterwards JP in notes) to David Steuart Erskine, the Earl of Buchan – afterwards normally Lord or Earl Buchan – 10 Dec. 1794; in JP, *The Literary Correspondence of John Pinkerton*, ed. Dawson Turner, 2 vols. (London, 1830), Vol. 1, 364.
6 JP to Earl Buchan, 10 Jan. 1795 (ibid., 378).
7 Phrase from Thomas Carlyle; in J.G. Lockhart, *The Life of Robert Burns*, ed. J.M. Sloan (London: n.d.), 243.
8 Variously spelt: Pinkerton, Pynkertoune, Pincartoun, Pynkerton.
9 See J.C. Pinkerton, "Pinkerton, East Lothian," *Notes and Queries*, Vol. 156 (3 June 1939), 388. A note by JP on its location is in Nat. Lib. Scot. MS 1716, f. 51. There he called it "the town of Pincarton, from which my name."
10 *Public Characters of 1800–1801* (London, 1801), 21. This biography of JP (20–31) contains information that could only have come from him personally. The *Critical Review* said the biography "is given with the minuteness of anecdote, which shows the author ... to be an intimate friend, and probably countryman" (2nd ser., Vol. 38 [1803], 82). A yeoman was a farmer; see Jane Austen's account of the "yeomanry" in *Emma*, ch. 4.
11 *Public Characters of 1800–1801*, 21. A hair-merchant supplied hair to wig-makers.

12 F.J. Grant, ed., *Register of Marriages in the City of Edinburgh, 1751–1800* (Edinburgh, 1922), 620.

13 *Public Characters of 1800–1801*, 21. She married Bowie in 1738. See Henry Paton, ed., *Register of Marriages For the Parish of Edinburgh, 1701–1750* (Edinburgh, 1908), 55. Her sons by him, JP's half-brothers, were James (b. 1739), slain in the battle of Minden (fought 1 Aug. 1759) and Robert (b. 1748). Robert Bowie (Sr) also had a third child. JP refers to (but doesn't name) "my step-sister" in a letter to James Gibson on 6 Aug. 1799 (Edin. Univ. Lib. MS La. II. 597/4), presumably – unless JP is using the term stepsister loosely, which is unlikely – Bowie's daughter by an earlier marriage.

14 Both to JP; 1 Apr. 1785, 3 Mar. 1790 (LWL MSS Vol. 82.1).

15 I assume she was a Scots speaker trying to write English. It is possible that her writing was affected by poor sight, illness, or advancing age; perhaps the letters were written out for her by a servant or relative.

16 *Public Characters of 1800–1801*, 22–3.

17 Her letters, written after James Pinkerton's death, were signed Mary Heron, which may indicate she called herself that during their marriage.

18 JP notes his birthday in *Ancient Scotish Poems*, 2 vols. (London: [1785]), Vol. 1, cxxviii. He is called the "only child of the marriage" between James Pinkerton and Mary Heron in a "Registered Commission," executed 6 May 1800, appointing a factor to sell property in Edinburgh (LWL MSS Vol. 82.4, [appendix]).

19 *Public Characters of 1800–1801*, 22. Elementary "English" schools were private schools "where the rudiments of reading, spelling, and counting" were taught. The Edinburgh Town Council established such schools in 1759, but many existed prior to then. See Alexander Law, *Education in Edinburgh in the Eighteenth Century* (London, 1965), 29, 47–54.

20 George Irving and Alexander Murray, *The Upper Ward of Lanarkshire*, 3 vols. (Glasgow, 1864), Vol. 2, 278. JP is said to have been "sent" to the school in Lanark at "About 1764" in *Public Characters of 1800–1801*, 22. He was six years old then.

21 Law, *Education in Edinburgh*, 58.

22 The Latin curriculum JP followed could have been close to that described in William Ferguson, *Scotland 1689 to the Present* (Edinburgh, 1968), 96; for the curricula at the Edinburgh High Schools, perhaps similar to that followed by JP, see Law, *Education in Edinburgh*, 74–81.

23 John Strong, *A History of Secondary Education in Scotland* (Oxford, 1909), 136. JP studied Greek in his final year at the school.

24 Nathaniel Hooke, *Roman History* (1739–71).

25 *Public Characters of 1800–1801*, 22.

26 Nat. Lib. Scot. MS 1711, ff. 25–6 (for Lithgow); JP, *Modern Geography*, 2 vols. (London, 1802), Vol. 1, 168; see also *Public Characters of 1800–1801*, 22.

27 In addition to James Pinkerton's connection, Robert Bowie, JP's half-brother, succeeded to an estate in Lanarkshire on the death of his father (ibid., 21); "capital grammar-school," 22.

28 Ibid., 22. Here it is said JP was "absolutely devoured by *mauvaise honte*," i.e., bashfulness.

29 JP, *Select Scottish Ballads*, 2 vols. (London, 1783), Vol. 2, 122. Title: "The Banks of Clyde." John Wilson had preceded JP as a praiser of the Clyde (George Eyre-Todd, ed., *Scottish Poetry of the Eighteenth Century*, 2 vols. [Glasgow, 1896], Vol. 1, 142–7), as indeed Lithgow had in 1617. Wordsworth succeeded him (*Poetical Works*, ed. Thomas Hutchinson [London, 1905], 299–300).

30 But see n. 93 below. His poem "Bothwell Bank" is also a love lament. Bothwell is on the Clyde.

31 JP, *Scottish Tragic Ballads* (London, 1781), 101; also *Ancient Scotish Poems*, 1, xxvii.

32 *Scottish Tragic Ballads*, 72–4.

33 *Public Characters of 1800–1801*, 23.

34 JP to Macvey Napier, 28 Jan. 1818 (B.L. Add. MS 34612, f. 172).

35 See *Public Characters of 1800–1801*, 23. In 1785 he complained that "The vast power that parents have over the happiness of their children, gives many depraved parents an unlimited tyranny" and suggested a "Domestic Tribunal" should be set up to settle disputes between father and son (*Letters of Literature* [London, 1785], 410–11). See John Dwyer, et al., eds., *New Perspectives on the Politics and Culture of Early Modern Scotland* (Edinburgh, c. 1982), 265 ff., esp. 279–80, for comment on child-parent relations at the time.

36 JP to Douce, 30 Mar. 1819 (Bodl. Lib. MS Douce d. 37, f. 28).

37 *Public Characters of 1800–1801*, 23.

38 He was able to cite Knox, Buchanan, and Goodall in notes, written in Sept. 1775, to *Craigmillar Castle* (Edinburgh, 1776).

39 As he termed him in *Letters of Literature*, 372.

40 I.e., "that higher class of Edinburgh attornies (among whom are often men of old family and good fortune) who are incorporated as an integral part of the 'college of justice,' deriving their title from their exclusive right to draw up all such law-papers as require an affix of the royal seal" (R.P. Gillies, *Recollections of Sir Walter Scott, Bart.* [London, 1837], 6).

41 Ferguson, *Scotland: 1689 to the Present*, 86; perhaps exaggerated – see
    Gillies, *Recollections of Sir Walter Scott*, 29–30. But for a description
    similar to Ferguson's, see D.R. Dean, *James Hutton and the History of
    Geology* (Ithaca, 1992), 3; also Law, *Education in Edinburgh*, 26–7. Nicholas
    Phillipson, "The Scottish Enlightenment," in Roy Porter and Mikuláš
    Teich, eds., *The Enlightenment in National Context* (Cambridge, 1981), 19.

42 He praised the *Weekly Magazine* as "a most useful periodical publication,
    and in which several valuable original pieces may be found" in *Ancient
    Scotish Poems*, Vol. 1, cxl – a tipoff that he had appeared there. David
    Macpherson commented on this statement in his copy of the work:
    "Certainly: for in that repository the great Pinkerton first permitted the
    public to be favoured with a specimen of the production of his muse."
    See W.A. Craigie, "Macpherson on Pinkerton," *Publications of the Modern
    Language Association of America (PMLA)*, Vol. 42 (1927), 440. I suggest
    that either of these five poems from the *Weekly Magazine*, being similar
    in style to his *Rimes* or *Tales in Verse*, may be his: 1. A poem in praise
    of Beattie by Urbanus, Vol. 28 (6 Apr. 1775), 47; 2. "Carlos *to* Sylvia: *A
    New-Year's Offering*" by Carlos, ibid., 48; 3. "Ode. *On* Contentment," by
    Lycidas, Vol. 37 (21 Aug. 1777), 184–5; 4 & 5. "Pandora" & "The Happy
    Fire-Side," by "J.P.," Vol. 40 (8 Apr. 1778), 39.

43 JP to Beattie, 20 Mar. 1784; in Margaret Forbes, *Beattie and his Friends*
    (Westminster, 1904), 199.

44 JP, *Literary Correspondence*, Vol. 1, 1–2.

45 Beattie to JP, 9 Mar. 1776 (ibid., 3–4). Beattie's kindness in dealing with
    the intrusive JP was typical of him; see E.H. King, *James Beattie* (Boston,
    1977), 20–1.

46 See Percy, *Reliques of Ancient English Poetry*, 3 vols. (London, 1765), Vol. 2,
    87–8; also George Chalmers' view, challenging JP's later opinion (in
    Ancient Scotish Poems, Vol. 1, cxxvii–cxxviii) that Sir John Bruce was the
    author, in Craigie, "Macpherson on Pinkerton," 436–7; George Eyre-
    Todd, ed., *Scottish Poetry of the Eighteenth Century*, Vol. 1, 15.

47 See Mel Kersey, "Ballads, Britishness and *Hardyknute*, 1719–1859,"
    Scottish Studies Review, 5.1 (2004), 43, 49–50.

48 Percy, *Reliques*, Vol. 2, 87–102; Ramsay, *The Ever Green*, 2 vols.
    (Edinburgh, 1724), Vol. 2, 247–64; Warton, *The Union* (Edinburgh, 1753),
    124–38.

49 As in *Hardyknute, A Fragment. Being the First Canto of an Epick Poem*
    (London, 1740). Foulis's edition was *Hardyknute. A Fragment of an Antient
    Scots Poem* (Glasgow, 1745). On *Hardyknute*'s "curious metamorphosis"
    from fragment into epic, see Kersey, "Ballads, Britishness and

Hardyknute," 40. Kersey sees an implied political theme in the poem; so did JP (*Ancient Scotish Poems,* Vol. 1, cxxvii).

50 See Gray to Walpole, Apr. 1760, in *Correspondence of Thomas Gray,* eds. Paget Toynbee and Leonard Whibley, 3 vols. (Oxford, 1935), Vol. 2, 665.

51 Thomas Warton, *Observations on the Faerie Queene* (London, 1754), 114.

52 He wrote his second part when he was seventeen, he told Douce on 30 Mar. 1819 (Bodl. Lib. MS Douce d. 37, f. 28). Perhaps stretching the truth (or memory) slightly.

53 "The Laird of Woodhouselie," "Lord Livingston," "The Death of Menteith," and the poem beginning "I wish I were where Helen lies," the first two lines of which were not his. But see n. 73 below.

54 Whether "forged" and "forgery" should apply to such titles as in the preceding note, or to *Hardyknute,* JP's completion of it, or James Macpherson's Ossian poems, has become an issue in contemporary literary discourse. While acknowledging the contribution of such discussion to our understanding of eighteenth-century literature, I will continue to use the words, though an amelioration in phraseology might sometimes be more appropriate. (For JP's own discussion of forgery, see below in this ch.) Most of this new discussion has centered on Macpherson. For a summary of the late-twentieth-century revised opinion of him, which has allegedly "nailed the legend that he was a complete charlatan," see Colin Kidd, *Subverting Scotland's Past* (Cambridge, 1993), 220 ff. Nick Groom, in *The Forger's Shadow* (London, 2002), 109, says Macpherson "had translated and edited, or written, or indeed forged, *depending on one's point of view* ... the lost epics of a third-century Celtic bard" (my italics). But the "introductory survey of scholarship" in T.M. Curley's *Samuel Johnson, the Ossian Fraud, and the Celtic Revival* (Cambridge, 2009), 1–21, may show opinion swinging back towards a more traditional view. He says of the Ossian epics that "This was a work of almost complete fabrication, composed rather quickly over several years and based on very meager original materials or, mostly, none at all." Jack Lynch also, in *Deception and Detection in Eighteenth-Century Britain* (Aldershot, 2008), 10, says the eighteenth-century critics of Ossian "were mostly right that Macpherson had doctored his evidence."

55 F.J. Child accepted seven couplets of JP's "Binnorie" as traditional and printed it as version N of the ballad "The Twa Sisters." See Child, *The English and Scottish Popular Ballads,* 5 vols. (Boston and New York, 1882–98), Vol. 1, 134.

56 Percy, *Reliques,* Vol. 1, 90. "Sir Cauline" was treated similarly (ibid., 35).

57 Percy, *Bishop Percy's Folio Manuscript*, eds. J.W. Hales and F.J. Furnivall, 3 vols. (London, 1867–8), Vol. 1, xvii.

58 JP, *Scottish Tragic Ballads*, xxxv.

59 Child, *English and Scottish Popular Ballads*, Vol. 4, 159.

60 The MS contained "a collection of Poems by different hands from the reign of Queen Elizabeth to the end of the last century" (*Scottish Tragic Ballads*, 113). From it he also printed two previously unpublished poems of Sir Robert Aytoun (117–18). At the sale of JP's library in 1812 the MS was bought by Heber. It is mentioned in James Watson, *Watson's Collection of Comic and Serious Scots Poems* (Glasgow, 1869), xxv, as being in David Laing's possession.

61 Dodsley to JP, 10 Jan. 1778 (LWL MSS Vol. 82.1).

62 JP to Dodsley, 6 Feb. 1788 (Edin. Univ. Lib. MS La. II. 266).

63 But was probably just a congenial, like-minded group. Young, a writer (in the legal sense) like JP, was one of this Edinburgh clique; comments in some of his letters to JP (12 survive; one each in 1782, 1784, and 1785; the rest from 1807 to 1815) shed occasional light on this aspect of JP's early life. They are not sufficiently detailed to give an idea of the group's influence on him. The letters are in LWL MSS 82.1, 3, and 4. Young names G. (i.e., George) Clapperton and Dr Boyd as members of the group, and refers to others. A letter from JP in London to Young, dated 10 July 1784 (Nat. Lib. Scot. MS 585, f. 44) is irreverent, playful, and intimate. In it he sends his compliments to Clapperton, and to no one else. A letter from Clapperton to JP, 26 Apr. 1808, in LWL MSS Vol. 82, has been lost. The "surly dictator" reference is from the *European Magazine*, Vol. 2 (July 1782), 49, and could be just a bilious insult.

64 See his account of being duped, in *Enquiry into the History of Scotland*, 2 vols. (London, 1789), Vol. 2, 78. Afterwards (often) *Enquiry*. In *Scottish Tragic Ballads*, xviii, he wrote: "How should we have been affected by hearing a composition of Homer or Ossian, sung and played by these immortal masters themselves! ... I suppose that Ossian's poetry is still recited to its original cadence and to appropriated tunes."

65 He wrote comments on these works in a notebook (Nat. Lib. Scot. MS 1721, f. 9).

66 Percy to JP, 27 Nov. 1778 (JP, *Literary Correspondence*, Vol. 1, 14). JP's and Percy's letters have been collected in *The Correspondence of Thomas Percy & John Pinkerton*, ed. H.H. Wood (New Haven and London, 1985); but I had access to MS and printed texts of most relevant letters in my work in the 1960s, and made my own copies, where needed. I choose to use those copies.

67 As in n. 65, ff. 7, 13.

68 Two MSS of which survive; it will figure later in this biography. JP told Douce twice he had written it at 18 (11 Mar. 1798, Bodl Lib. MS Douce d. 37, f. 3; 30 Mar. 1819, ibid., f. 28).

69 A 3½ pp. outline, titled "Disposition of the Plot, &c. of The British Princess, a Tragedy in Three Acts," is Nat. Lib. Scot. MS 1712, ff. 150–1. It is undated, but the handwriting is early, probably before 1778.

70 JP to Beattie, 12 Aug. 1778 (Nat. Lib. Scot. MS 1713, f. 4 [rough draft]).

71 JP to Beattie, 18 Sept. 1778 (ibid., ff. 5–6). Two pages of a MS collection of this nature, containing transcripts of early Italian poetry, are in Nat. Lib. Scot. MS 1711, f. 230, in JP's handwriting, numbered 265 and 266.

72 In Hugo Arnot, *History of Edinburgh* (Edinburgh, 1779), 639.

73 On David Dalrymple's saying that all of JP's version except the first line was modern, JP responded 10 June 1783: "Had your lordship said *the three first lines*, you would have been right" (JP, *Literary Correspondence*, Vol. 1, 36).

74 Nat. Lib. Scot. MS 1713, f. 7; *Scottish Tragic Ballads*, 79.

75 JP to Percy, c. May 1779 (Nat. Lib. Scot. MS 1713, ff. 8–9).

76 Scott, *Poetical Works*, ed. J.G. Lockhart, 12 vols. (Edinburgh, 1851), Vol. 3, 99; Burns, *Letters*, ed. J. De Lancey Ferguson, 2 vols. (Oxford, 1931), Vol. 2, 181.

77 Hume, "My Own Life," in his *History of England*, 6 vols. (Boston, 1850), Vol. 1, iv; Johannes Voet, Arnold Vinnius, famed jurists.

78 Aytoun chided him "for poring over Copernicus, when he ought to have been reading Dallas's Styles" (*Public Characters of 1800–1801*, 23–4). The reference is to George Dallas, *System of Styles* (1697), a legal handbook for Scotland.

79 Thomas Crawford, in *The History of Scottish Literature*, ed. Craig Cairns, 4 vols. (Aberdeen, 1987–8), Vol. 2, 123.

80 JP, *Letters of Literature*, 383.

81 In *Ancient Scotish Poems*, Vol. 1, cxxviii–cxxxi.

82 To whom he sent his tragedy "Malvine"; see JP's letter to Percy, c. Oct. 1778 (Nat. Lib. Scot. MS 1713, f. 9). JP said Blair was partly responsible for his acceptance of Ossian's authenticity. In *Letters of Literature*, 187, he intimated that Blair was a plagiarist.

83 Nat. Lib. Scot. MS 1713, f. 7.

84 Perhaps taken in part from a medical handbook of some sort.

85 Nat. Lib. Scot. MS 1721, ff. 5–7.

86 John Gibson Lockhart in 1843 (see *Notes and Queries*, Vol. 187 [23 Sept. 1944], 135) and the obituary in the *Gentleman's Magazine*, Vol. 96,

pt. i (May 1826), 472, both mention scars on JP's face and both attribute them to smallpox. I assume this was a childhood illness. In discussing round crystals of feldspar in *Petralogy: A Treatise on Rocks*, 2 vols. (London, 1811), Vol. 1, 133, JP says they have "a faint resemblance to the pustules of the small-pox."

87 JP, *Ancient Scotish Poems*, Vol. 1, cxxvi; Beattie referred to his own "bad eyes" in a letter to JP on 7 Feb. 1782, adding "I am sorry that yours should resemble mine" (JP, *Literary Correspondence*, Vol. 1, 22). But in 1796 JP told Earl Buchan that "cold water is the best eye-water; and I always use an eye-cup, so that I never had any thing the matter with my eyes, though I use them so much" (ibid., 413).

88 Horace Walpole, *Horace Walpole's Correspondence*, eds. W.S. Lewis et al., 48 vols. (London and New Haven, 1937–83), Vol. 29, 359. The remark, by George Simon Harcourt, may be jocular and derisive rather than accurate.

89 JP explained the difference between writer and Writer to the Signet in a letter to Grímur Thorkelín on 16 Jan. 1792 (Edin. Univ. Lib. MS La. III. 379/665). In the Deed of Factory between him and William Buchan, quoted below, JP describes himself as a writer but not as a Writer to the Signet. In his father's testament he is also termed a writer.

90 *Scots Magazine*, Vol. 42 (May 1780), 280. He died 10 May.

91 "In this line I myself studied for five years, the term of apprenticeship" (JP to Thorkelín, 16 Jan. 1792, as in n. 89). On 28 Jan. 1818 he told Macvey Napier "My father and Mr. Aytoun died about the same time and I doubt if even my apprenticeship had expired" (B.L. Add. MS 34612, f. 172). There is, then, some doubt whether he was at the exact termination of his five-year term, but the statement to Thorkelín, and his being called a writer in other sources, indicate he was close to being finished, if not actually finished. The acct. of JP in *Public Characters of 1800–1801*, 25, says his apprenticeship had "expired."

92 See the quotation from the letter to Napier in n. 91; also *Public Characters of 1800–1801*, 25.

93 Info. from James Pinkerton's testament, General Register House, Edinburgh (CC8/8/125/1236-1238). Readers of Scott's *The Antiquary* will note resemblances between JP's early career and that of Jonathan Oldbuck, as described in the brief "recapitulation" of Oldbuck's youth early in the novel. Scott knew "the learned Pinkerton," having met him during JP's trip to Scotland in 1812–14. He wrote the novel in 1815–16 and it is possible that, while writing it, he recalled and echoed episodes from JP's life that he'd heard from JP himself. I am aware that Scott said his model for Oldbuck's "character"

was the lawyer George Constable. But I doubt if details supplied for Oldbuck such as his "halting upon the threshold" of becoming a writer (in the legal sense), and not practising in "the hated drudgery of the law" apply to Constable. Other details thrown out in the "recapitulation," e.g., misogyny deriving "from an early disappointment in love," and his father's destining him "to a share in a substantial mercantile concern, carried on by some of his maternal relations," might spring from JP too. All this is, of course, speculative.

94 JP, *Literary Correspondence*, Vol. 1, 18 n.

95 The MS of *Rimes* is the first 100 pp. of Nat. Lib. Scot. MS 1715. On p. 100 is this note: "Revised and approved as correct. 20 June 1780. J P."

96 "… a distant relation through marriage" (B.H. Davis, *Thomas Percy: A Scholar-Critic in the Age of Johnson* [Philadelphia, 1989], 244). Percy "turned … over" the MS of *Scottish Tragic Ballads* to Nichols (ibid., 245).

97 The proofs with Percy's corrections are Nat. Lib. Scot. MS 1711, ff. 11–18.

98 Kersey's term for JP's "continuation" of the poem, presumably more acceptable than "forgery"; see "Ballads, Britishness and *Hardyknute*," 49. Dr Johnson used the word "imposture" of the Ossian poems in his famous letter to Macpherson on 20 Jan. 1775.

99 *The Mirror*, No. 110 (27 May 1780), 439.

100 General Register House, Edinburgh (CC8/8/125/1236–1238).

101 Again, writer in the legal sense.

102 General Register House, Edinburgh, Register of Deeds, Second Series, McKen. Office, Vol. 230, 873–4.

103 JP to John Young, 10 July 1784, Nat. Lib. Scot. MS 585, f. 44; a magazine writer in 1782 estimated JP's fortune at "about six thousand pounds," but this was probably just a guess; other supposed facts in the piece are errors; see *European Magazine*, Vol. 2 (July 1782), 49.

104 JP's address in the early 1780s, as well as his mother's in the 1780s and in 1790.

105 This is made clear by an "Accompt" of transactions conducted by William Buchan on JP's behalf from 1 Jan. 1788 to 15 May 1790 (LWL MSS Vol. 82.4 [appendix]).

106 Sold for him by John Macintosh, stockbroker. JP's receipt, Nat. Lib. Scot. MS 709, f. 1.

107 From there he wrote a letter to John Bell, the Edinburgh bookseller (Nat. Lib. Scot. MS 583, f. 160). A "Mr. Pinkerton" was minister at Markinch in 1760. See Thomas Somerville, *My Own Life and Times 1741–1814* (Edinburgh, [1861]), 49.

108 As he described himself on 10 June 1783 (JP, *Literary Correspondence*, Vol. 1, 38).
109 "Knightsbridge near London": thus he would sometimes give his address in letters.

2. Finding His Way, 1782–1789

1 He said (much later) that a poet "required book-learning most especially, and ought to have all Dante and Petrarch, and countless others, at his fingers' ends, before he presumed to … compose a single sonnet" (R.P. Gillies, *Memoirs of a Literary Veteran*, 3 vols. [London, 1851], Vol. 2, 131).
2 Sir William Forbes, *An Account of the Life and Writings of James Beattie, LL.D.*, 2 vols. (Edinburgh, 1806), Vol. 2, 16–17. For some of these Scottish poets English was "not even a second but actually a third language," the second being Latin (James Kinsley, ed., *Scottish Poetry: A Critical Survey* [London, 1955], 121). JP, later writing about comedy, made a somewhat similar point to Beattie's. He said Scots were "strangers to the English humour and wit" since these "depend so much on 'thinking in a language,' a perfect use of all its delicate lights and shades … it may be reasonably inferred that, when the nations are blended into the same speech and pronunciation, we may aspire to comic fame, especially as in our own dialect, written and spoken, much humour at times appears" (*Scotish Poems*, 3 vols. [London, 1792], Vol. 1, xxii).
3 JP, *Rimes* (London, 1781), 94; this is the poem with thee lines from "Il Penseroso"; quote from "To the Lark," 106.
4 Wordsworth on "poetic diction" in 2nd edition of *Lyrical Ballads*; also Coleridge, *Selected Poetry and Prose*, ed. Elisabeth Schneider, 2nd ed. (San Francisco, 1971), 204–5 (in ch. 1 of *Biographia Literaria*); "To the Lark," *Rimes*, 106; "… perfectly congenial to the mind," ibid., 50 n.
5 *Critical Review*, Vol. 51 (Mar. 1781), 217–19.
6 *Monthly Review*, Vol. 65 (July 1781), 15. Cartwright was the inventor of the power loom. Identified in B.C. Nangle, *The Monthly Review First Series* (Oxford, 1934), 188. Later identifications of *Monthly* reviewers are from this work or Nangle, *The Monthly Review Second Series 1790–1815* (Oxford, 1955).
7 Gillies, *Memoirs of a Literary Veteran*, Vol. 2, 124.
8 Nat. Lib. Scot. MS. 1715, ff. 9–10.
9 *European Magazine*, Vol. 2 (July 1782), 46, 48–9. The last phrase referring to gods, scenes, etc. from the past, as in the historical ode "To the Lyre," *Rimes*, 45–59.
10 See ch. 1, esp. n. 63.

11 Douce to JP, 11 Nov. 1808 (Bodl. Lib. MS Douce d. 37, ff. 10–11).

12 Beattie's description in 1798, recalling JP of 1784; in Forbes, *Account of the Life and Writings of James Beattie*, Vol. 2, 321.

13 On anti-Scot feeling see Murray Pittock, *Celtic Identity and the British Image* (Manchester, 1999), 30–4; Pat Rogers, "Johnson, Boswell, and Anti-Scottish Sentiment" in his *Johnson and Boswell: The Transit of Caledonia* (Oxford, 1995), 192–215; P. O'Flaherty, "Johnson in the Hebrides: Philosopher Becalmed," *Studies in Burke and his Time*, Vol. 13 (1971), 1988–9. Mrs Thrale termed Johnson's attitude to Scots "hatred."

14 JP, *Tales in Verse* (London, 1782), 25.

15 I.e., revealing an interest in macabre elements of medieval society, as on display in Horace Walpole's *The Castle of Otranto* (1764). In ch. 2 of Walpole's book the character Manfred is warned not to pursue an "incestuous design" against his "contracted daughter."

16 In the Advertisement to *Other Juvenile Poems* [1798].

17 *Critical Review*, Vol. 53 (May, 1782), 343–4.

18 *Monthly Review*, Vol. 67 (Aug., 1782), 110–11.

19 E.g., *Gentleman's Magazine*, Vol. 52 (Mar., 1782), 131; but see a rev. of *Tales in Verse*, ibid. (May 1782), 243.

20 In a note in his *Dissertations Moral and Critical* (London, 1783), 552 n. JP had asked for the puff c. Jan. 1782 (Margaret Forbes, *Beattie and his Friends*, 181); see Beattie to JP, 7 Feb. 1782 (JP, *Literary Correspondence*, Vol. 1, 21–2). On the practice of puffing, see Richard B. Sher, *The Enlightenment & the Book* (Chicago, 2006), 137–8.

21 *Monthly Review*, Vol. 68 (Apr. 1783), 355.

22 In Eyre-Todd, ed., *Scottish Poetry of the Eighteenth Century*, Vol. 2, 149–50; D. McCordick, ed., *Scottish Literature: An Anthology*, 2 vols. (New York, 1996), Vol. 2, 77; see JP, *Select Scotish Ballads*, Vol. 2, 131–2.

23 See his comment on *Letters of Literature*, in *Gentleman's Magazine*, Vol. 56, pt. i (Feb., 1786), 95.

24 *European Magazine*, Vol. 2 (July, 1782), 50. The play does not survive.

25 The "General Argument" of this poem is Nat. Lib. Scot. MS 1712, ff. 120–2. It is an expanded version of a sketch for "The Mantle Or The Ribbon an epic poem in 20 books," written 12 Oct., 1781 (ibid., MS 1719, f. 3). The incidents and characters of "Ratho" are similar to those of "A Northern Tale," noted earlier.

26 Cartwright said he was possessed by one; *Monthly Review*, Vol. 68 (Apr. 1783), 355.

27 *Gentleman's Magazine*, Vol. 51 (June 1781), 279–80; *Critical Review*, Vol. 52 (Sept. 1781), 205–8.

28 JP to John Nichols, 28 Nov. 1782 (JP *Literary Correspondence*, Vol. 1, 28 n.).
29 I.e., the language of Lowland Scotland from (roughly) the fifteenth to the seventeenth century; used by the makars (makers), as the poets in this language were and (often) still are termed.
30 JP, *Literary Correspondence*, Vol. 1, 30–2.
31 Dalrymple, ed., *Ancient Scottish Poems. Published from the MS of George Bannatyne, MDLXVIII* (Edinburgh, 1770).
32 JP to Dalrymple, 10 June 1783 (JP, *Literary Correspondence*, Vol. 1, 39).
33 The architect William Porden knew the truth, and Beattie suspected it (ibid., 25, 82); Cartwright in the *Monthly Review*, Vol. 66 (Apr., 1782), 292, said he had "many doubts" about its "authenticity." Joseph Ritson was another skeptic – see below. Dalrymple's comments on *Select Scotish Ballads*, dated 26, 27 May 1783, are in LWL MSS Vol. 82.1.
34 As in n. 32, 37–8.
35 JP, *Select Scotish Ballads*, Vol. 2, 188.
36 JP to John Nichols, 27 Oct. 1783 (Bodl. Lib. MS Montagu d. 5, f. 135); for the proposal to Nichols, 3 Oct. 1783, see JP, *Literary Correspondence*, Vol. 1, 35 n.
37 Johnson, *The Rambler*, eds. W.J. Bate and Albrecht B. Strauss, 3 vols. (New Haven and London, 1969), Vol. 2, 64–70.
38 For the latter point and an explanation of "the ridicule attached to antiquarianism" in the Scottish intellectual context, see Susan Manning, "Antiquarianism, the Scottish Science of Man, and the emergence of modern disciplinarity," in Leith Davis et al., eds., *Scotland and the Borders of Romanticism* (Cambridge, 2004), 57–76; see also Douglas, *English Scholars*, 272–84, esp. 275.
39 Vol. 2, 186; also mocking comments in *Ancient Scotish Poems*, Vol. 1, xxxi–xxxv.
40 As in *A Dissertation on the Origin and Progress of the Scythians or Goths* (London, 1787), 18.
41 See his self-description as historian in 1794; JP, *Literary Correspondence*, Vol. 1, 365–6.
42 See John Nichols to JP, 22 Jan. 1784 (ibid., 43–4).
43 The term "medals" as used by collectors in this period often included coins. See *OED*. JP has, however, a Section (*An Essay on Medals* [London, 1784], 96–9), on "Medallions" (what is termed medals today), and says, 131, that while up to that point in the book he has been using the terms medal and coin as synonyms, meaning coins only, from then on the reader "will be pleased to remark, that, in treating of modern coinage,

the word Coin only is used in speaking of common cash; and that of Medal supplies the place of the term Medallion." Yet confusion over whether he means medal or coin sometimes occurs afterwards.

44 The work "is just published," JP told Percy on 14 June 1784 (MS Historical Society of Pennsylvania).

45 JP, *An Essay on Medals*, xiii.

46 *Gentleman's Magazine*, Vol. 54, pt. ii (July 1784), 521.

47 Southgate to JP, probably in 1784 (Nat. Lib. Scot. MS 1709, f. 185); JP's tribute to Southgate, *An Essay on Medals*, 148.

48 Ibid., 13. See 14–18 on the usefulness to history of studying medals. Rosemary Sweet, in *Antiquaries: The Discovery of the Past in Eighteenth-Century Britain* (London and New York, 2004), xiv–xviii, clarifies the differing (but related) roles of antiquaries and historians.

49 JP, *Letters of Literature*, 335.

50 JP, *An Essay on Medals*, 113. Virgil's much discussed use of material from other writers is usually referred to as borrowing rather than stealing. The fifth-century author Macrobius treated this aspect of the poet's work in his *Saturnalia*, and he wasn't the first to take note of it. See T.R. Glover, *Virgil* (London, 1930), 41–66; W. Lucas Collins, *Virgil* (Edinburgh, 1890), 13, 15, passim. In his note to "Virginia" in *Lays of Ancient Rome* Macaulay said that in their works apart from satire the Latin poets were "mere imitators of foreign models."

51 JP, *Literary Correspondence*, Vol. 2, 353.

52 Hack work, as I understand the term, is work undertaken, not principally, but solely, for money. JP rarely wrote on subjects that did not interest him. Yet at times the word hack fits him.

53 JP to John Young, 10 July 1784 (Nat. Lib. Scot. MS 585, f. 44).

54 As he told Thorkelín in a letter on 3 May 1792 (Edin. Univ. Lib. MS La. III. 379/670).

55 See JP, *Select Scotish Ballads*, Vol. 2, xv.

56 "It is certainly the most transcendent pleasure to be agreeably surprized with the confession of love from an adored mistress," he wrote (*The Treasury of Wit*, 2 vols. [London, 1786], Vol. 2, 109).

57 Bodl. Lib. MS Douce d. 37, f. 47. JP is slightly off about the length of their friendship. In 1808 he wrote that their acquaintance "dates from 1784 or 5" (ibid., ff. 10–11). Douce helped JP in the preparation of *An Essay on Medals*; 1784 is the correct date.

58 Boswell, *Boswell's Life of Johnson*, ed. G.B. Hill, rev. L.F. Powell, 6 vols. (Oxford, 1934–50), Vol. 4, 330. Of the Scot Vicesimus Knox, Boswell says

he "perhaps … formed the notion of [the dinner] which he has exhibited in his *Winter Evenings*." He may allude to ch. 5 of Book the Ninth of that work in the edition of 1788, Vol. 3, 245–52; no mention there of JP.

59 See the ungracious remark in *Ancient Scotish Poems*, Vol. 2, 403–4.

60 A formally worded short note from Boswell to JP, dated 23 May 1785, is in his *Letters*, ed. Chauncey Tinker, 2 vols. (Oxford, 1924), Vol. 2, 327.

61 JP, *Walpoliana*, 2 vols. (London, 1799), Vol. 1, xli.

62 *Horace Walpole's Correspondence*, Vol. 16, 263 n.

63 Ibid., Vol. 29, 357–9.

64 See JP, *Literary Correspondence*, Vol. 1, 78.

65 "A lady to whom I refuse nothing begs me to enquire if Your Lordship has any acquaintance with Mr Colman," he lied to Percy on 14 June 1784 (MS, Historical Society of Pennsylvania).

66 Walpole to JP, 27 Sept. 1784 (*Horace Walpole's Correspondence*, Vol. 16, 253–4).

67 For JP's views on comedy, see *Letters of Literature*, 42–50.

68 *Horace Walpole's Correspondence*, Vol. 16, 265; JP then tried to court Colman by flattery in *Letters of Literature*, 47.

69 *Horace Walpole's Correspondence*, Vol. 16, 261, 275–7.

70 Ketton-Cremer, *Horace Walpole* (London, 1946), 291.

71 Walpole to JP, 19 Aug. 1789 (*Horace Walpole's Correspondence*, Vol. 16, 308); clarifying his letter to JP, 14 Aug. 1789 (ibid., Vol. 43, 228–9).

72 Ibid., Vol. 16, 327.

73 JP's explanation is in a letter to Sir Egerton Brydges, 5 Aug. 1814, where he refers to "Strawberry Hill where I was intimate for 15 years till the good old Earl fell into the hands of interested flatterers a trade quite out of my way" (Misc. MS, LWL).

74 Peter Peckard to JP, 10 Nov. 1784 (JP, *Literary Correspondence*, Vol. 1, 62). JP had applied for leave to copy the MS early in 1784, but permission was slow to come (JP to Thomas Warton, 2 Apr. 1784 [B.L. Add. MS 42561, f. 139]). Samuel Knight, in a letter to JP on 12 Apr. 1784, said he had got a note from "Mr. Hey" (Samuel Hey, President of the College) saying JP had been refused "the liberty of copying the Maitland MSS" (LWL MSS Vol. 82.1). (I thank Phillipa Grimstone, Sub-Librarian at the College, for information about Hey.) Percy's intervention, in July, appears to have cleared a path for him; see *Correspondence of Thomas Percy & John Pinkerton*, ed. Wood, 48–9.

75 See JP, *Literary Correspondence*, Vol. 1, 62; *Ancient Scotish Poems*, Vol. 1, viii, Vol. 2, 433 (JP's poem to the robin).

76 The two MSS are described in his Preface, ibid., Vol. 1, v–vii.

77 Craigie, "Macpherson on Pinkerton," 440; also in Craigie, ed., *The Maitland Folio Manuscript*, 2 vols. (Edinburgh, 1919–27), Vol. 2, 26 n.

78 See Peckard to JP, 14 Sept. 1784 (JP, *Literary Correspondence*, Vol. 1, 47–9).

79 Bywater to JP, 7 Dec. 1785 (Edin. Univ. Lib. MS La. II. 647/62); Peckard to JP, 9 Dec. 1785 (JP, *Literary Correspondence*, Vol. 1, 105–6).

80 Craigie, "Macpherson on Pinkerton," 441. "This story told me by Mr. Ritson" (ibid.).

81 *Gentleman's Magazine*, Vol. 54, pt. ii (Nov. 1784), 812; full letter, 812–14.

82 Walter Scott, who admired Ritson, said he arraigned "each trivial inaccuracy as a gross fraud, and every deduction which he considered to be erroneous as a wilful untruth" (*Supplement to the fourth, fifth, and sixth editions of the Encyclopædia Britannica*, 6 vols. [Edinburgh, 1824], Vol. 6, 441).

83 John Nichols & J.B. Nichols – hereafter Nichols – *Illustrations of the Literary History of the Eighteenth Century*, 8 vols. (London, 1817–58), Vol. 8, 103 n. This says the Notes and Introduction were "probably" written by JP; but there is no doubting they are his. John Nichols, who was influential in running the magazine, was likely responsible for the way JP was treated.

84 *Gentleman's Magazine*, Vol. 55, pt. i (Mar. 1785), 164–5. See B.H. Bronson's arguments for JP's authorship of the letter in *Joseph Ritson: Scholar at Arms*, 2 vols. (Berkeley, 1938), Vol. 1, 117–20.

85 *St. James's Chronicle*, 25/27 Aug. 1785. Copies were circulating prior to then. Walpole had finished reading his copy by June 22 (*Horace Walpole's Correspondence*, Vol. 16, 263).

86 JP, *Letters of Literature*, 231.

87 Ibid., 401.

88 See his defence of his method, *Gentleman's Magazine*, Vol. 56, pt. ii (Dec. 1786), 1022.

89 Walpole to JP, 22 June 1785 (*Horace Walpole's Correspondence*, Vol. 16, 263).

90 *New Review*, Vol. 8 (July 1785), 34–5.

91 *Gentleman's Magazine*, Vol. 55, pt. ii (July 1785), 544.

92 JP, *Letters of Literature*, 188.

93 Ibid., 186–7. For scholarly controversy over the Longinus quote from Genesis, see W. Rhys Roberts, "The Quotation from *Genesis* in the *De Sublimitate*," *Classical Review*, 11, 9 (1897), 431–6. JP's remarks are without foundation.

94 JP, *Letters of Literature*, 372–3.

95 Ibid., 141, 326.

96 Chs. 3 and 6 below deal further with his religious views.

97 He explains what he sees as the difference between whig and tory in
    his letter on Hume. A tory is one who exalts and wishes to extend the
    "prerogative," i.e., the power of the king; a whig argues that the king is
    rather "by the present constitution … only elective chief magistrate of the
    country," deriving his power "only from the confidence of the nation."
    To a whig "public spirit [is] the most laudable principle of society"
    (*Letters of Literature*, 365–72).

98 Ibid., 370.

99 Ibid., 306, 339, 64–5, 150, 40–1.

100 Ibid., 254. The vicar William Tremayne, of Helston, Cornwall, wrote a
    long appreciative comment on JP's system; JP, *Literary Correspondence*,
    Vol. 1, 83–6.

101 Shakespeare, *Plays*, eds. Samuel Johnson and George Steevens, 2nd ed.,
    10 vols. (London, 1778), Vol. 10, 292 n.

102 Fifteenth-century Italian scholar, normally called Politian.

103 *Letters of Literature*, 312.

104 Ibid., 318–19.

105 Kathleen Wilson points to "evidence of a rise in the expression of
    English hostility to sodomy in the last quarter of the eighteenth century,
    a hostility that exceeded perhaps anywhere else in Europe" (*The Island
    Race: Englishness, Empire and Gender in the Eighteenth Century* [London and
    New York, 2003], 190). Sodomy was a capital crime in England – a man
    was executed for it in 1772 – but it was normally punished by the pillory;
    see Lawrence Stone, *The Family, Sex and Marriage In England 1508–1800*
    (London, 1977), 540–1; Lucy Worsley, *If Walls Could Talk* (London, 2011),
    149, 78; Tim Hitchcock, *English Sexualities, 1700–1800* (Houndmills,
    UK, 1997), 60–7; Randolph Trumbach, "Erotic Fantasy and Male
    Libertinism in Enlightenment England," in Lynn Hunt, ed., *The Invention
    of Pornography: Obscenity and the Origins of Modernity, 1500–1800* (New
    York, 1993), 255–7. JP was more circumspect in a later reference to the
    "abominable vices" among the "savages" of America and the southern
    Pacific islands (*Recollections of Paris*, 2 vols. [London, 1806], Vol. 1, 225).

106 *European Magazine*, Vol. 8 (Oct. 1785), 293; attr. to William Julius Mickle,
    the Scottish poet and translator of the *Lusiad*, in Arthur Sherbo, "Isaac
    Reed and the *European Magazine*," *Studies in Bibliography*, Vol. 37 (1984),
    224 n. Sherbo here also attr. to Mickle a severe letter signed "Common
    Sense," in *European Mag.*, Vol. 9 (Feb. 1786), 87–9. Note the word "talks."
    Mickle had evidently been in JP's company.

107 *Gentleman's Magazine*, Vol. 55, pt. ii (Sept. 1785), 719.

108 *Edinburgh Magazine*, Vol. 2 (Sept. 1785), 161.

109 *Critical Review*, Vol. 60 (Dec. 1785), 405–13; *Monthly Review*, Vol. 74 (Mar. 1786), 175–82.

110 *Gentleman's Magazine*, Vol. 55, pt. ii (Aug., 1785), 579–81 (see Nichols, *Illustrations*, Vol. 8, 108–10); *Gentleman's Mag.*, Vol. 55, p. ii (Sept. 1785), 681; ibid. (Oct. 1785), 784–5; ibid. (Dec. 1785), 949–50 (see Nichols, Vol. 8, 110–12); *Gentleman's Mag.*, Vol. 56, pt. i (Jan. 1786), 16–17; *European Mag.*, Vol. 9 (Feb. 1786), 87–9; *Gentleman's Mag.*, Vol. 56, pt. i (Mar. 1786), 214, 221, 252; ibid. (Apr. 1786), 280, 284 (see Nichols, Vol. 8, 126–7); *Gentleman's Mag.*, Vol. 56, pt. ii (Dec. 1786), 1040–1 (see Nichols, Vol. 8, 132–3); *Gentleman's Mag.*, Vol. 56, pt. ii (Supp. 1786), 1116, 1128; ibid., Vol. 57, pt. i (Apr. 1787), 296–7 (see Nichols, Vol. 8, 131); *Gentleman's Mag.*, Vol. 57, pt. i (Feb. 1787), 130–1. See also n. 106 above.

111 *A Letter to Robert Heron, Esq. Containing a Few Brief Remarks on his Letters of Literature by one of the Barbarous Blockheads of the Lowest Mob, Who is a true Friend to Religion and a sincere Lover of Mankind* (London, 1786).

112 *Gentleman's Magazine*, Vol. 55, pt. ii (Nov. 1785), 88. See *Horace Walpole's Correspondence*, Vol. 16, 278–80; Vol. 43, 225–6.

113 He'd mentioned *Letters of Literature* to Percy as early as 1781.

114 *Gentleman's Magazine*, Vol. 56, pt. i (Feb. 1786), 95.

115 Ibid., Vol. 56, pt. ii (Nov., Dec. 1786), 942–4, 1021–3. See Nichols, *Illustrations*, Vol. 8, 127–31; also Joseph Towers in *Gentleman's Magazine*, Vol. 56, pt. ii (Supp. 1786), 1128.

116 William Zachs, *The First John Murray and the Late Eighteenth Century Book Trade* (Oxford, 1998), 210.

117 *English Review*, Vol. 7 (Jan. 1786), 37.

118 *Gentleman's Magazine*, Vol. 56, pt. ii (Supp. 1786), 1128.

119 Ibid., Vol. 57, pt. i (Feb. 1787, 120–1, 397–9; I also attribute to JP another defence of the book, signed "Hint," Ibid., Vol. 56, pt. i (May 1786), 390. This is in the vein of his other letters and expresses one of his repeated ideas: "an apparent friend is the worst of enemies." A friendly letter on the *Letters*, signed "Philo-Pinky," is in the *St. James's Chronicle*, Oct. 6/8, 1789. He would, in 1812–14, be nicknamed "Pinky" on his return to Edinburgh.

120 Robert Lowth's Latin lectures on the sacred poetry of the Hebrews had appeared in an English translation in 1787. There were to be 29 letters in JP's new volume. Another was to be on the "Art of bearing abuse" (Nat. Lib. Scot. MS 1719, f. 19).

121 JP, *Walpoliana*, Vol. 1, p. 78 n. In 1806 he fancied bringing out a new edition, "omitting one third part of crude matter" (JP, *Literary Correspondence*, Vol. 2, 345); but he didn't.

122   De Quincey, *Collected Writings*, ed. D. Masson, 14 vols. (London, 1896–7), Vol. 11, 443.

123   Cowper, *Poetical Works*, ed. Robert Southey (London, 1854), 277.

124   JP, *Letters of Literature*, 97, 150; see n. 50 above.

125   Ibid., 207, 211.

126   A.F.B. Clark, *Boileau and the French Classical Critics in England* (Paris, 1925), 54.

127   JP, *Letters of Literature*, 356.

128   Aisso Bosker, *Literary Criticism in the Age of Johnson* (Groningen, 1930), 251.

129   JP, *Letters of Literature*, 361, 212.

130   See *Horace Walpole's Correspondence*, Vol. 16, 288 n. The title-page incorrectly gives the date as MDCCLXXXVI.

131   The term makar, meaning maker, was applied to poets writing in Middle Scots; see note 29.

132   Craigie, ed., *The Maitland Folio Manuscript*, Vol. 2, 11. He then printed the whole of JP's preface (13–26).

133   Peter Peckard to JP, 14 Sept. 1784 (JP, *Literary Correspondence*, Vol. 1, 49).

134   JP, *Ancient Scotish Poems*, Vol. 1, cxxviii.

135   Ibid., Vol. 2, 444–9; commentary, 443, 449–51.

136   Agnes M. Mackenzie, *An Historical Survey of Scottish Literature to 1714* (London, 1933), 85.

137   Adam de Cardonnel in a letter to JP, 27 Sept. 1790 (LWL MSS Vol. 82.1) tells of getting access to the imprints, and asks what JP wants copied.

138   For doubt about Douglas's authorship, see F.H. Ridley, "Did Gavin Douglas write *King Hart?*" *Speculum*, Vol. 34 (July, 1959), 402–12.

139   It is not by him. See Priscilla Bawcutt, *Dunbar the Makar* (Oxford, 1992), 10.

140   Kinsley, ed., *Scottish Poetry*, 15.

141   See discussion in Alexander Montgomerie, *Poems*, ed. James Cranstoun (Edinburgh, 1887), xli–xliv.

142   Vol. 1, vi.

143   JP, *Ancient Scotish Poems*, Vol. 1, lxxxii, lxxxix, xciv.

144   Ibid., cxxv–cxxvi; his plan for the "general Glossary" is stated in Vol. 2, 520.

145   Ibid., 542–3. The copy of the *Complaynt* described by JP is B.L. C. 21. a. 56. The title-page is in JP's handwriting and the pages are numbered by him.

146   JP, *Ancient Scotish Poems*, Vol. 1, xvi.

147   *Critical Review*, 2nd ser., Vol. 13 (Jan. 1795), 56. See ch. 4 for JP's connection with the journal then.

148   JP, *Ancient Scotish Poems*, Vol. 1, xviii.

149   Craigie, ed., *The Maitland Folio Manuscript*, Vol. 1, 258.

150   JP, *Ancient Scotish Poems*, Vol. 1, 8.

151   A.S.G. Edwards, "Editing Dunbar: The Tradition," in Sally Mapstone, ed., *William Dunbar, 'The Nobill Poyet'* (East Linton, Scot., 2001), 56.

152   Chalmers gives correct meanings of "bait," "digest," "fre," "lurdans," "mensit," "pallet," and "break your pallet"; he points out that "fackless," "smattis," and "walter cail" are misreadings by JP. But JP, contrary to what Chalmers says, gives correct meanings of "cleuchs," "dispense," "droup," "kappis," "sornars," and "bans." Chalmers mistakenly thought JP's "buist" and "south" were misreadings. Both are wrong in their meanings of "pak," "morgeons," and "preis." I use Chalmers' spellings, which are not always the same as JP's. Cf. Sir David Lyndsay, *Poetical Works*, ed. Chalmers, 3 vols. (London, 1806), Vol. 3, 195–6 n.; JP, *Ancient Scotish Poems*, Vol. 2, 520–34; Craigie, *The Maitland Folio Manuscript*, Vol. 2, 135–82. Craigie is assumed to give correct meanings.

153   See Bawcutt's comments, *Dunbar the Makar*, 131–2, 186.

154   JP, *Ancient Scotish Poems*, Vol. 2, 455–6, 460, 462.

155   Ibid., 383.

156   Ibid., Vol. 1, cxi.

157   Ibid., x.

158   These are ll. 139–46. See Craigie, *The Maitland Folio Manuscript*, Vol. 2, 23 n.

159   Percy, *Reliques*, Vol. 1, xiv.

160   Dalrymple, *Ancient Scottish Poems*, 302.

161   Dalrymple to JP, 26 Dec. 1785 (JP, *Literary Correspondence*, Vol. 1, 107).

162   Leyden in the *Scots Magazine*, Vol. 64 (July, 1802), 571, forgetting that JP drew a distinction between immodesty and obscenity; see *Ancient Scotish Poems*, Vol. 2, 383; Scott in *Poetical Works*, Vol. 1, 75.

163   Craigie, "Macpherson on Pinkerton," 434.

164   Dunbar, *Poems*, ed. James Kinsley (Oxford, 1979), vii.

165   JP, *Ancient Scotish Poems*, Vol. 1, cxxxv–cxxxvi, cxxxiii–cxxxiv, cxiii–cxiv (n.), xlv–lii.

166   Ibid., xxiv–xxv.

167   Hailes to JP, 26 Dec. 1785 (JP, *Literary Correspondence*, 108). Hailes had been embittered by the poor reception of his *Ancient Scottish Poems* and *Annals of Scotland* and had resolved "to stop short" (John Ramsay, *Scotland and Scotsmen in the Eighteenth Century*, 2 vols. [Edinburgh, 1888], Vol. 1, 404). The reception of the *Annals* led him to abandon the idea of writing a history of the Stuart kings of Scotland; see his comment to William Smellie, 21 Jan. 1779 in Robert Kerr, *Memoirs of William Smellie*, 2 vols. (London, 1811), Vol. 2, 196.

168   Letter signed "Z Y," Nat. Lib. Scot. MS 1709, f.18. JP recognized the handwriting.

169   JP, *Ancient Scotish Poems*, Vol. 1, cxxv; see also his tribute, iv–v. He'd written to Tytler to praise his work and ask for help in 1783; Tytler's reply, 27 Dec. 1783, is in JP, *Literary Correspondence*, Vol. 1, 40–2.

170   JP to Earl Buchan, 24 Nov. 1785 (ibid., 98). Four-page "Proposals" for the work were printed separately in November, 1787 (LWL MSS Vol. 82.1).

171   JP, *Ancient Scotish Poems*, Vol. 2, 500–19.

172   A letter from William Buchan, his factor, 17 June 1786, said he'd been "plagued with your tenants about repairs"; roofs were leaking, walls were damp, &c. (LWL MSS Vol. 82.1). Mary Heron to JP, 29 June 1785, after acknowledging one from him that "gave me great pleasure," says "I will taike itt kind if you writt me a long letter as you have time two or three for my one for I have nothing to write that you will care for" (ibid.)

173   JP to Earl Buchan, 24 Nov. 1785 (JP, *Literary Correspondence*, Vol. 1, 95–9).

174   See JP to Earl Buchan, 3 Apr. 1786 (ibid., 124–6, and 15 Oct. 1787, 164–5).

175   Percy to JP, 23 Mar. 1786 (ibid., 117; also 147, 176).

176   JP to Percy, 19 Nov. 1785; *Gentleman's Magazine*, Vol. 102, pt. ii (Aug. 1832), 121–2.

177   The most significant notebook for the *Enquiry* and *Dissertation on the Scythians or Goths* is Nat. Lib. Scot. MS 1718. It contains notes, in small, dense handwriting, on the numerous books he read while preparing these works. Two other related notebooks survive, as well as various scraps and materials.

178   Nat. Lib. Scot. MS 1712, ff. 154–67.

179   He didn't acknowledge he wrote it in any letter I've seen; but it is unmistakably his work. The fact that various anecdotes in it were repeated in *Walpoliana* (see ch. 4, n. 28) is enough to prove it is his, but other elements in it point to him as well, in particular the attitude towards Celts, esp. the Scottish Highlanders, who get the same harsh treatment here, in similar language, to what they receive in the two works discussed immediately below. See esp. *The Treasury*, Vol. 2, xxxvi–ix. See his statements about the need for "A book of Apothemes" in *Letters of Literature*, 340–1.

180   Nat. Lib. Scot. MS 1712., ff. 123–49.

181   The fragment, called "Nordymra," was published in Thorkelín's *Fragments of English and Irish History* (London, 1788), 2–31. For "the English translation of Nordymra, written shortly after my arrival in England in the year 1786, I am indebted to my friend the truly ingenious

and learned Mr. John Pinkerton" (xi). Thorkelín's first letter to JP, on 2 Sept. 1786, is in JP, *Literary Correspondence*, Vol. 1, 139.

182 JP wrote the rev. of D. Antonio Perabo's tragedy *Valsei* in the *Gentleman's Magazine*, Vol. 57, pt. i (Mar. 1787), 242–4 (see JP, *Literary Correspondence*, Vol. 1, 151–2). I suspect he wrote the puff of Thorkelín's *Diplomatarium Arna-Magnaeum, exhibens Monumenta Diplomatica*, ibid., Vol. 56, pt. ii (Sept. 1786), 773–4.

183 JP to Charles O'Conor, 13 Mar. 1786 (MS Henry E. Huntington Library).

184 In Irish "duan" means poem, "Albain" is Scotland.

185 O'Conor to JP, 4 Apr. 1786, in *The Letters of Charles O'Conor of Belangare*, eds. C.C. Ward and R.E. Ward (Ann Arbor, Mich., 1980), 245.

186 *St. James's Chronicle*, 19/21 June 1787.

187 Which JP distrusted; see *Dissertation on the Scythians or Goths*, 35; but he wasn't averse to using it when it suited his purposes.

188 Ibid., 27.

189 Ibid., 33.

190 Ibid., 48.

191 Ibid., 51.

192 Ibid., 161.

193 Ibid., 176.

194 Ibid., 158, 177. His explanation for using "Piks" instead of Picts, 23 n.

195 Ibid., 23.

196 Ibid., 33 n. On p. 26 he says: "The flood is now generally reputed a local event." On the Flood and "old outlooks and theories" vs. eighteenth-century science, see Roy Porter, *The Making of Geology: Earth Science in Britain 1660–1815* (Cambridge, 1977), 158–60.

197 JP, *Dissertation on the Scythians or Goths*, 33–4; also 40.

198 Pliny, *Natural History*, trans. J. Bostock and H.T. Riley, 6 vols. (London, 1855–7), Vol. 2, 330.

199 JP, *Dissertation on the Scythians or Goths*, 92.

200 Ibid., 69.

201 Ibid., 69 n., 123, 152 n.

202 *Quarterly Review*, Vol. 46 (July 1829), 135.

203 Part of a poem called "The Harp of Ossian."

204 JP, *Rimes*, 27, A4v.

205 JP, *Ancient Scotish Poems*, Vol. 1, xlvi–ii.

206 JP, *Dissertation on the Scythians or Goths*, 69 n.

207 JP, *Enquiry*, Vol. 2, 78; in *The Treasury of Wit*, Vol. 2, 145, "an English gentleman" argues that "it was an insult to common sense to believe so rank a forgery."

208 Nichols, *Illustrations*, Vol. 7, 770–1. He would exclaim against "the riff-raff of O's and Mac's" in conversation, saying "Show me a great O, and I am done." See *Gentleman's Magazine*, Vol. 96, pt. i (May 1826), 471–2.

209 On this point see Hugh Trevor-Roper, *The Invention of Scotland: Myth and History* (New Haven and London, 2008), 83–4, 110; Murray Pittock, in *Inventing and Resisting Britain* (New York, 1997), 119, says "The idea that the Scots Highlanders were cannibals was a real fear in the 1745 Rising."

210 JP, *Modern Geography* (London, 1811), 68.

211 Pittock, *Celtic identity and British image*, 31; also Linda Colley, *Britons: Forging the Nation 1707–1837* (London, 2003), 113–17.

212 Us meaning here, the English; for JP's identifying himself with the English later, see Ch. 7; also disc. of *Modern Geography*, chs. 4 & 5.

213 JP, *The Treasury of Wit*, Vol. 2, 145, 189.

214 JP, *A New System of Religion* (?Amsterdam, 1790), 38. See the Bibliography, I (16), for a note on the authorship of this work.

215 Not completely; e.g., in *Recollections of Paris* (1806), Vol. 1, 102, we find a reference to an antiquary named Cambri, "a warm admirer of the Celts" who "affects to regard" them "as the most ancient civilized people in the world." JP comments: "he is puzzled when he asked for proofs of this civilization." See also a comment on superstitions, Vol. 2, 142, and a harsh description of Bretons, 229. A leaven of dislike and suspicion stayed with him, which he normally kept well hidden.

216 See *Analytical Review* on his "*insolence*," Vol. 4 (Aug. 1789), 404; *Gentleman's Magazine*, Vol. 59, pt. ii (Sept., 1789), 837–8.

217 JP to Earl Buchan, 16 Mar. 1789 (JP, *Literary Correspondence*, Vol. 1, 215).

218 He told Buchan, 16 Mar. 1789, it "will be ready in two months" (ibid., 214).

219 Ritson, *Letters*, ed. Harris Nicolas, 2 vols. (London, 1833), Vol. 1, 96.

220 See JP, *Literary Correspondence*, Vol. 1, 141–2.

221 Lorimer's quote from Vallancey's letter to him of 26 Sept. 1786. See *Gentleman's Magazine*, Vol. 59, pt. ii (Aug. 1789), 679.

222 A letter from Lorimer to JP, 28 Feb. 1788, is in JP, *Literary Correspondence*, Vol. 1, 180–3. JP had consulted him on a number of matters, including one relating to the Duan. Thorkelín refers to "our friend Dr. Lorimer" in a letter in 1787 (ibid., 161). A letter from Lorimer to JP, 31 May 1788, is in LWL MSS Vol. 82.1.

223 "I should be sorry if you lose by that large work, my 'Enquiry into the History of Scotland.' I wish you would use any means you please … to promote the sale." JP to Nichols, 28 Jan. 1794 (Nichols, *Illustrations*, Vol. 5,

677). In 1794 the unsold copies were reissued with the most offensive sheets cancelled. The book was still on sale in Nichols' shop in Red Lion Passage in 1803. See also corres. with Earl Buchan in Oct. and Nov., 1789, JP, *Literary Correspondence*, Vol. 1, 235–7.

224   *Gentleman's Magazine*, Vol. 59, pt. ii (July 1789), 583.

225   Ibid. (Aug., 1789), 679–80; (Sept., 1789), 801–2; (Oct. 1789), 905–6; (Nov. 1789), 984; (Dec. 1789), 1066 – Lorimer having the last word.

226   Fordun has regained respectability; see William Ferguson, *The Identity of the Scottish Nation* (Edinburgh, 1998), 43–51; but also Trevor-Roper, *The Invention of Scotland*, 15–20. And Steve Boardman, "Late Medieval Scotland and the Matter of Britain," in E.J. Cowan and R.J. Finlay, eds., *Scottish History: The Power of the Past* (Edinburgh, 2002), 47–72.

227   Thus James Wallace wrote that Scotland *"has been a continued Monarchy under one Race of Kings for the Space of* 2037 *Years"* (*History of the Kingdom of Scotland* [Dublin, 1724], Cr).

228   This name occurs in early Roman references to the peoples of northern Britain, notably that of Tacitus. W.A. Cummins assumes Caledonia was a general name given to all the "tribes" of early Scotland (*The Age of the Picts* [Phoenix Mill, Glos., UK, 1995], 28 ff.)

229   Maitland, *The History and Antiquities of Scotland*, 2 vols. (London, 1757), Vol. 1, 23.

230   John Macpherson, *Critical Dissertations on the Origin, Antiquities, Language, Government, Manners, and Religion, of the Ancient Caledonians* (London, 1768), 95–6.

231   Ibid., 76.

232   Gibbon, *The History of the Decline and Fall of the Roman Empire*, ed. J.B. Bury, 7 vols. (London, 1896–1900), Vol. 3, 41 n, 42–3 n.

233   Still being asked in 1987; see Marjorie O. Anderson, "Picts – the Name and the People," in *The Picts: a new look at old problems*, ed. Alan Small (Dundee, 1987), 13; and in 2002, see Isabel Henderson, "The Problem of the Picts," in Gordon Menzies, ed., *Who are the Scots? and The Scottish Nation* (Edinburgh, 2002), 20–35.

234   Bede, *A History of the English Church and People*, trans. Leo-Sherley-Price, rev. R.E. Latham (London, 1970), 38–9.

235   A.O. Anderson, *Early Sources of Scottish History*, 2 vols. (Edinburgh, 1922), Vol. 1, 270–4.

236   Henry of Huntingdon, *Chronicle*, trans. Thomas Forester (London, 1853), 8–9.

237   Sibbald, *The History, Ancient and Modern, of the Sheriffdoms of Fife and Kinross* (Edinburgh, 1710), 7.

238 Innes, *Critical Essay*, 105.
239 Sibbald, *History of Fife and Kinross*, 18.
240 Who said of the "Caledonians," his single name for the northern "barbarians" in Britain, that their "reddish hair and large limbs … proclaim a Germanic origin" (Tacitus, *Agricola; Germania*, trans. Harold Mattingly, rev. J.B. Rives [London, 2009], 9).
241 Innes, *Critical Essay*, 55–7.
242 John Macpherson, *Critical Dissertations*, 47.
243 Ibid, xx (James Macpherson's preface).
244 JP, *Enquiry*, Vol. 2, 77–86.
245 Ibid., Vol. 1, 123.
246 Ibid., 235.
247 Ibid., Vol. 2, 132.
248 Innes, *Critical Essay*, 93.
249 Flann Mainistrech was an eleventh century Irish poet whose *Synchronisms* contained information on Scottish as well as Irish kings.
250 Nennius was a medieval Welsh monk, allegedly the author of a history of Britons, a work that went through a number of revisions.
251 A twelfth-century chronicle of Ireland.
252 John of Fordun, *Chronicle of The Scottish Nation*, ed. Skene (Edinburgh, 1872), xxxvi.
253 Dalriada being the name given to territory in western Scotland and northeastern Ireland occupied by the Gaels in the sixth and seventh centuries.
254 Ref. to *Rerum Hibernicarum Scriptores Veteres*, four-vol. work, ed. Charles O'Conor (the younger), pub. 1814–26.
255 A valuable fifteenth- and sixteenth-century Irish compilation.
256 James Johnstone, *Antiquitates Celto-Normanicae* (Copenhagen, 1786), 56–71; see JP's comments on the "errors" in Johnstone's text, *Enquiry*, Vol. 2, 307 n. In July 1785, he'd approached Johnstone for help on a number of historical points; the polite reply, 30 Mar. 1786, from Copenhagen is in JP, *Literary Correspondence*, Vol. 1, 118–20.
257 John of Fordun, *Chronicle*, ed. Skene, lxx–lxxii.
258 See *Enquiry*, Vol. 1, 413–5; also discussion in Sweet, *Antiquaries*, 138–9, 226.
259 A fourteenth-century monk. The forgery by Charles Bertram (1723–65) was published at Copenhagen in 1757. It deceived, among others, William Stukeley, Ritson, Chalmers, William Roy, and JP. The work dealt with British antiquities and especially with the Roman walls. B.B. Woodward exposed it as a forgery in the *Gentleman's Magazine* in

1866–7. JP had been fooled by it in his prefatory "Essay on the Origin of Scotish Poetry" in *Ancient Scotish Poems*, though noting that Richard's authority, since he wrote in the fourteenth century "is very small" (Vol. 1, xl). In the *Enquiry* he used the work, again hesitantly, saying "it must be used with much caution" (Vol. 1, 11–12). His chapter on the province of "Vespasiana" is derived from this source.

260   JP, *Enquiry*, Vol. 2, [307] –20 (*Annals of Ulster*); 321–6 (*Albanic Duan*); "free translation" of the *Duan* by O'Conor (the elder), 327–8; remarks on it by O'C. and JP, 328–9. For disc. of JP's version and where it stands in relation to other texts of the poem, see Kenneth Jackson, "The poem *A Eolcha Alban Uile*," *Celtica*, Vol. 3 (1956), 149–67. When Jackson wrote, the MS from which JP's text was made was untraced; JP's text therefore was a valuable source of readings. Jackson said O'Conor's trans. was "very bad" (152).

261   JP, *Enquiry*, Vol. 1, 60.

262   These were St. Ailred's "Life of St. Ninian" and "Eulogium of St. David." Parts of Jocelin's "Life of St. Kentigern," which JP printed in full, had been published before 1789.

263   If "fact" can be used of this period of Scottish history, which has been called "a region of almost Balkan complexity" (David Miles, *The Tribes of Britain* [London, 2006], 180).

264   Burton, *History of Scotland*, 2nd ed., 8 vols. (Edinburgh, 1873), Vol. 1, 194.

265   Mary-Ann Constantine, *The Truth Against the World: Iolo Morganwg and Romantic Forgery* (Cardiff, 2007), 99–100, 130–1.

266   JP, *Enquiry*, Vol. 1, 340; also 341.

267   James Tytler, *A Dissertation on the Origin and Antiquity of the Scottish Nation* (London, 1795), 5.

268   George Buchanan, *Dialogue concerning the Rights of the Crown of Scotland* (London, 1799), 47, 51.

269   As JP and those who thought like him are termed in Kidd, *Subverting Scotland's Past*, 253. See also Kidd's "The Ideological Uses of the Picts," in Cowan and Finlay, eds., *Scottish History*, 169–90; on JP, 174–7.

270   His authorship was known by the mid-nineteenth century; see, e.g., Sharon Turner, *History of the Anglo-Saxons*, 3 vols. (London 1852), Vol. 1, 83 n. Coxe's letters to W.O. Pughe in 1802 (esp. that of 23 Dec.) prove he wrote it (Nat. Lib. of Wales, MS 132248, Vol. IV, f. 233).

271   Coxe, *A Vindication of the Celts* (London, 1803), 19.

272   Ritson, *Scotish Song*, 2 vols. (London, 1794), Vol. 1, xv (n).

273   As in Murray Pittock's *Scottish and Irish Romanticism* (Oxford, 2008), 64, where JP is lumped in, not unreasonably, with some other practitioners

of "pseudo-science," while in the 10 pp. on James Macpherson (71–81) the word "pseudo-epics" makes no appearance.

274   The idea that the Picts were Gothic or Teutonic has been completely dispensed with; a consensus has emerged – they were Celtic. (Which was what Innes said.) This is phrased differently by various scholars. A few quotes must suffice here. W.A. Cummins, while acknowledging that the Picts' "racial origin or origins" remained obscure, says their language was "a hybrid speech" of Goidelic and Brittonic elements that evolved into something "distinctive" over the centuries (*The Age of the Picts*, 47, 49). "It is most likely," writes Murray G.H. Pittock, "that the name [Picts] indicates a collective identification of a group of Celtic tribes." But Pittock states further that the extent to which the Picts "differed ethnically or culturally from the Welsh-speaking Britons … or the Gaelic-speaking Celts … is a moot point" (*A New History of Scotland* [Phoenix Mill, Glos., UK, 2003], 19). David Miles denies that the Picts spoke "a non-Indo-European language" – traces of which were detected in Scotland by Kenneth Jackson – saying instead that they were akin to the British and Welsh (*The Tribes of Britain*, 181). T.O. Clancy and B.E. Crawford say they sprang from "solidly British roots"; their language had only "dialectal difference" from that of Britons and was "demonstrably Celtic" (R.A. Houston and W.W.J. Knox, eds., *The New Penguin General History of Scotland* [London, 2001], 32–3, 36.) For JP's vigorously argued contrary beliefs, see the *Enquiry*, Vol. 1, 121 ff.

275   Mimicking Pinkerton's efforts at etymology.

276   "Peanfahel," a place-name, preserved by Bede (*A History of the Church and People*, 52). Cummins, citing W.F.H. Nicolaisen's *Scottish Place-names* (1976), says the word was "a British-Gaelic hybrid meaning 'the end of the wall'" (*The Age of the Picts*, 44, 154). JP says it meant "the extent, or end of the wall," deriving it, however, from "broad Gothic" (*Enquiry*, Vol. 1, 358).

277   Presumably the antiquary Sir Robert Gordon of Straloch; for a bibliography and brief biography, see JP, *The Scotish Gallery* (London, 1799), 112–13.

278   Scott, *The Antiquary*, ch. 6. In his *History of Scotland*, 2 vols. (London, 1830), Vol. I, 7, Scott says there may have been a Gothic strain in the blood of the Picts. But for his real opinion of the controversy, see his review of Ritson's *Annals* in the *Quarterly Review*, Vol. 41 (July 1829), 120–62.

279   *Gentleman's Magazine*, Vol. 59, pt. ii (Nov. 1789), 980–1; see *Analytical Review*, Vol. 5 (Sept. 1789), 1–13.

280 George Dempster, a persistent correspondent in 1789 who knew something of JP's personal life and financial situation, advised against Switzerland as a domicile in a letter, 25 Nov. (JP, *Literary Correspondence*, Vol. I, 239–40), suggesting instead that he become a "lodger and boarder" near the British Museum. JP had asked Dempster about securing an annuity. He got from him "advice respecting oeconomy" (letter to JP, 12 Dec. 1789, LWL MSS Vol. 82.1). JP had already asked Walpole to help get him a position at the Museum.

281 His address was Mansfield Place, Kentish Town. Kentish Town was a suburb of London, to the northeast. He was at his new address in July 1789.

3. The Great Work, 1790–1797

1 In a letter, 23 Oct. 1789 (JP, *Literary Correspondence*, Vol. 1, 236).

2 As John Leyden noted in 1802, *Scots Magazine,* Vol. 54 (July 1802), 568.

3 As when he wrote of the lives of the Scottish saints, "As Protestants, and as men of sense, let us laugh at their ridiculous miracles," *Ancient Scotish Poems*, Vol. 2, 556.

4 The unfinished MS, entitled *"Ye shall know them by their fruits,"* is Nat. Lib. Scot. MS 1711, ff. 253–5.

5 On the authorship of this work see a note in the Bibliography, I (16).

6 JP, *The Spirit of all Religions* (?Amsterdam, 1790), 6, 21, 27, 30, 33–4.

7 JP, *Modern Geography* (1802), Vol. 2, 122; he no doubt knew Hume's lengthy reflections on polytheism in *The Natural History of Religion* (1757).

8 JP to Macvey Napier, 28 Jan. 1818 (B.L. Add. MS 34612, f. 173).

9 JP, *The Spirit of all Religions*, 5, 14; "particle of littleness," 4, "shadow of nothing," 6, "dust," 12.

10 Ibid., 40, 45.

11 On the authorship of this work see a note in the Bibliography, I (15).

12 JP, *A New Tale of a Tub* (London, 1790), 114.

13 Ibid., 119–20.

14 Ibid., 23–4.

15 JP, *Modern Geography*, 2 vols. (London, 1817), Vol. 1, 762 [782].

16 Meaning the Glorious Revolution of 1688.

17 *Horace Walpole's Correspondence*, Vol. 16, 299; also notes, 299–302.

18 *Monthly Review,* 2nd ser., Vol. 3 (Nov. 1790), 289.

19 Modernized versions of the poem were in print earlier; Thomas Warton made use of one in his *History of English Poetry*, 3 vols. (London, 1774–81),

Vol. 2, 334–5. JP said the poem "has already gone thro' about twenty editions in Scotland since the year 1616 … all … modernized" (*The Bruce; or, the History of Robert I. King of Scotland. Written in Scotish Verse By John Barbour*, 3 vols. [London, 1790], Vol. 1, vii).

20 Earl Buchan to JP, 27 Sept. 1787 (Nat. Lib. Scot. MS 1711, f. 29).

21 JP, *The Bruce*, Vol. 1, ix–x.

22 Ibid., xxiii.

23 *The Metrical History of Sir William Wallace, Knight of Ellerslie*, 3 vols. (Perth, 1790), Vol. I, 3.

24 JP to William Herbert, 22 Feb. 1790 (JP, *Literary Correspondence*, Vol. 1, 244).

25 JP to Earl Buchan, 19 July 1790 (ibid., 256).

26 Gervais de La Rue (1751–1835).

27 Isaac Disraeli, *The Illustrator Illustrated* (London, 1838), 5.

28 See, e.g., Dalrymple to JP, 28 Sept. 1790 (JP, *Literary Correspondence*, Vol. 1, 260).

29 Dalrymple to Buchan, 10 May 1791 (Edin. Univ. Lib. MS La. II. 588).

30 JP to Thorkelín, 24 Oct. 1791 (ibid., MS La. III. 379/666).

31 He described his *Enquiry* as an "Antiquarian dissertation" in the *Gentleman's Magazine*, Vol. 49, pt. ii (Nov. 1789), 980.

32 JP to Earl Buchan, 11 Mar. 1792 (Edin. Univ. Lib. MS La. II. 588).

33 JP, *Enquiry*, Vol. 1, xxiv.

34 Dilly to JP, 26 Jan. 1797 (JP, *Literary Correspondence*, Vol. 1, 438).

35 See William Zachs, *Without Regard to Good Manners: A Biography of Gilbert Stuart 1743–1786* (Edinburgh, 1992), 173 ff.

36 See a description, dated 31 Jan. 1804, of the fire and its effects, *Critical Review*, 3rd ser., Vol. 1 (1804), Preface. The fire destroyed the home of the proprietor, his entire printing office, and all remaining copies of the review.

37 Cf. Thorkelín to JP, 29 Aug. 1791 (JP, *Literary Correspondence*, Vol. 1, 266–8), and *Critical Review*, 2nd ser., Vol. 3 (Appendix, 1791), 577–8. For a similar transplantation in 1792, cf. *Literary Correspondence*, Vol. 1, 299–301, and *Critical Review*, 2nd ser., Vol. 4 (Appendix, 1792), 573–4. Thorkelín thought JP was printing his news in the *London Chronicle* and asked him to make use of it there (letter to JP, 23 Feb. 1792, LWL MSS Vol. 82.1). JP answered 3 May 1792: "Your political intelligence i in fact intend for, and have occasionally inserted, in, a work of great superiority to a News Paper; but its title i am not at liberty to declare" (Edin. Univ. MS La. III. 379/670).

38 See Thorkelín to JP, 2 Sept. 1786 (JP, *Literary Correspondence*, Vol. 1, 139); also Magnús Fjalldal, "To Fall by Ambition – Grímur Thorkelín and his *Beowulf* Edition," *Neophilologus*, Vol. 92 (2007), 322–3.

39 JP to Thorkelín, 24 Oct. 1791 (Edin. Univ. Lib. MS La. III. 379/666).

40 See below for a discussion of the gap in his work for the *Critical*.

41 Owing to which I have decided not to make direct use of them in discussing JP's views in the text of this book.

42 Roper, *Reviewing before the Edinburgh 1788–1802* (London, 1978), 176–7.

43 *Critical Review*, 2nd ser., Vol. 1 (Apr. 1791), 425–30. The book was published anonymously. JP had helped Thorkelín prepare it for the press. Thorkelín told JP on 29 Aug. 1791 that "I esteem your encomium on my Sketches as the most illustrious mark of your friendship toward me and my native country" (JP, *Literary Correspondence*, Vol. 1, 265). Through 1791–3 the *Critical's* "Retrospect" and "Review" devoted much space to Denmark and its prince.

44 See, e.g., JP, *The Scotish Gallery*, 27 n.

45 *Critical Review*, 2nd ser., Vol. 1 (Apr. 1791), 433–4. Cf. the reviewer's attitude towards Dalrymple's defence of Christianity with that of JP in his letter to Earl Buchan, 5 Dec. 1787 (JP, *Literary Correspondence*, Vol. 1, 172). That "the worst kind of enmity is a weak defence" was a notion JP expressed often.

46 JP to Earl Buchan, 10 Nov. 1789 (ibid., 237).

47 JP to Earl Buchan, 27 Mar. 1790 (ibid., 247).

48 JP to Earl Buchan, 30 July 1787 (ibid., 158).

49 *Critical Review*, 2nd series, Vol. 1 (Apr. 1790), 430–3. Cf. the attack on the editorial procedure (431–2) with JP's advice to Earl Buchan in 1787 on how to copy MSS (JP, *Literary Correspondence*, Vol. 1, 158).

50 Letter to JP from the Society, 15 Sept. 1789 (LWL MSS Vol. 82.1). JP's name isn't mentioned in *Transactions of the Literary and Antiquarian Society of Perth* (Perth, 1827), but the papers of the society from the period around 1790 had been lost before 1827. The "Perth Academy," he told Earl Buchan on 10 Nov. 1789, had expressed "approbation" of his *Enquiry* (JP, *Literary Correspondence*, Vol. 1, 237). "F.S.A. Perth" is first given after his name on the title-page of *Scotish Poems*, 3 vols. (London, 1792).

51 JP was elected to the first probably in Nov. 1791; to the second, in Mar. 1792. See JP, *Literary Correspondence*, Vol. 1, 274, 308. Thorkelín secured these honours for him.

52 JP to Earl Buchan, 19 Dec. 1791 (JP, *Literary Correspondence*, Vol. 1, 277).

53 As he explained in the *Gentleman's Magazine*, Vol. 45, pt. i (Feb. 1795), 101.

54 Fjalldal, "To Fall by Ambition," 324, 326.

55 For the full assault, see *Critical Review*, 2nd ser., Vol. 5 (Aug. 1792), 402–10; (Appendix, 1792), 561–70. The indented quote is p. 566. The information on the state of the MS collections in the Society had been given to JP in Herd's comments on *Ancient Scotish Poems* (Nat. Lib.

Scot. MS 1711, f. 9). Herd told him Drummond's MSS instead of being "arranged & bound up in Volumes" remained in "large bundles." In May 1788, JP told Lord Buchan that a "traveller" informed him that the MS collections of the Society were handled "shockingly," and "in particular, that Drummond of Hawthornden's are all in confused heaps" (JP, *Literary Correspondence*, Vol. 1, 198). "It is with great regret, however," the *Critical* reviewer noted (p. 403), "we learn, that the collections of the Society are … of very little use to the public; for instance, the curious papers of Drummond of Hawthornden cannot be consulted, being left in a mass of confusion, instead of being arranged and bound up into volumes."

56   *Critical Review*, 2nd ser., Vol. 6 (Oct. 1792), 138–9; full review, 129–41. Cf. the review with JP's comments on Webb in the 1794 reissue of the *Enquiry*, Vol. 1, 2–3 (Advertisement). Also, cf. the indented passage quoted from the review here with this comment, ibid., p. 9 (Advertisement): "The attacks on the Celts partly arose from the extravagant praises, bestowed on them by some writers, who have at the same time exerted every art to calumniate our Gothic ancestors."

57   *Critical Review*, 2nd ser., Vol. 4 (Jan. 1792), 55–8; the letter from Ritson that ended up in JP's papers seems a sufficient indication of his authorship.

58   Nat. Lib. Scot. MS 1709, f. 44. See also Ritson's *Robin Hood*, 2 vols. (London, 1795), Vol. 1, ii.

59   Ritson, *Letters from Joseph Ritson, Esq. to Mr. George Paton* (Edinburgh, 1829), 7.

60   Percy to JP, 21 July 1792 (Bodl. Lib. MS Eng. Misc. d. 244).

61   Percy to JP, 28 July 1792 (JP, *Literary Correspondence*, Vol. 1, 317–18).

62   Part of Percy's letter to JP on 21 July 1792 (as in n. 60): "If you have ye Critical Review of Jany. last wch. contains the Critique on Mr. R.s *Tom Thumb* &c pray bring it with you." He was referring to the review of *Pieces of Ancient Popular Poetry*. As is made clear in this paragraph, he knew of Pinkerton's connection by 1794.

63   *Critical Review*, 2 ser., Vol. 6 (Nov. 1792), 283–4; full review, 283–93. JP's letter to Percy on 4 Sept. 1794, in *Gentleman's Magazine*, Vol. 102, pt. ii (Aug. 1832), 125, amounts to an admission that he wrote the review.

64   JP to Percy, 4 Sept. 1794 (ibid.). JP thereby reversed the opinion expressed in *Letters of Literature*, p. 76, where he'd said "the itinerant minstrels … were poets."

65   See also Davis, *Thomas Percy*, 286; Arthur Johnston, *Enchanted Ground: The Study of Medieval Romance in the Eighteenth Century* (London, 1964), 96.

66 Announced in *St. James's Chronicle*, 29 Nov./1 Dec. 1792; also JP, *Literary Correspondence*, Vol. 1, 320.
67 JP, *Scotish Poems*, Vol. 1, xiv.
68 *Select Works of Gawin Douglass, Bishop of Dunkeld* (Perth, 1787).
69 JP, *Scotish Poems*, Vol. 1, xvi–xvii.
70 He printed seven of these "ballads" – he included one, Lydgate's *Rhyme without accord*, at the end of *Gologros and Gawaine*, with which it had been bound. The others, following JP's order, are: a. unidentified; b. Henryson's *The Praise of Age*; c. unidentified; d. Henryson's *The Want of Wise Men*; e. Dunbar's *The ballade of Lord Bernard Stewart*; f. Dunbar's '*Kynd Kittok.*' The last is not a Chepman item but is bound up with the others. Pinkerton identified none of these. Cf. JP, *Scotish Poems*, Vol. 3, 124–42, and William Beattie, *The Chepman and Myllar Prints* (Edinburgh, 1957), vii–viii.
71 JP, *Scotish Poems*, Vol. 1, xxx.
72 *Gentleman's Magazine*, Vol. 63, pt. i (Jan. 1793), 32–3; also Ritson's comments in his edition of Laurence Minot's *Poems* (London, 1795), 156, and his letter to Paton, 8 Jan. 1793, in Ritson, *Letters to Paton*, 7–8.
73 Bodl. Lib. MS Douce 324, f. i. On his transcript of the MS, Douce noted that Pinkerton had "unwarrantedly & treacherously" printed the poem without his or Ritson's permission (ibid., MS Douce 309, f. vi). On this matter see Bronson, *Joseph Ritson*, Vol. 1, 188–90.
74 Ritson to George Paton, 8 Jan. 1793 (*Letters to Paton*, 8).
75 Barbour, *The Bruce*, ed. W.W. Skeat, 2 vols. (Edinburgh, 1894), Vol. 1, lxxxiii.
76 See Douglas Hamer, "The Bibliography of Sir David Lyndsay," *The Library*, 4th ser., Vol. 10 (June 1929), 35.
77 *The Thre Prestis of Peblis*, ed. T.D. Robb (Edinburgh, 1920), viii.
78 Barbour, *The Bruce*, ed. Jamieson (Edinburgh, 1820), viii.
79 Thus R. Miles sent his best respects to "Mrs Pinkerton" around 1789 (Nat. Lib. Scot. MS 1709, f. 192). The letter is undated, but it is addressed to JP at Knightsbridge and contained observations on the 2nd edition of the *Essay on Medals* (1789).
80 Adam de Cardonnel sent "best compliments from Mrs. D.C. and myself to you and family" on 27 Sept. 1790 (Nat. Lib. Scot. MS 1709, f. 38).
81 Thorkelín to JP, 29 Aug. 1791 (JP, *Literary Correspondence*, Vol. 1, 268).
82 JP to Thorkelín, 24 Oct. 1791 (Edin. Univ. Lib. MS La. III. 379/666).
83 Ibid., 26 Nov. 1791 (ibid., MS La. III. 379/664).
84 Thorkelín to JP, 18 Jan. 1792 (JP, *Literary Correspondence*, Vol. 1, 280); similar refs. occur in other letters from Thorkelín, e.g., to "Mrs. Pinkerton and your lovely daughters" on 10 Aug. 1792 (LWL MSS Vol. 82.1).

85   Notice must be taken of what must be a coincidence. On 12 July 1790 John Pinkerton and Rachel Hudson were married in St George's Church, Hanover Square (J.H. Chapman and G.J. Armytage, eds., *The Register Book of Marriages belonging to the Parish of St. George, Hanover Square*, 4 vols. [London, 1886–97], Vol. 2, 45). I examined this register in Somerset House, London, in the 1960s (reference lost) and the signature is not JP's, which is highly distinctive. London had other Pinkertons besides JP. See also David Macpherson's comments below. He says JP "dismissed" the woman he'd been living in "concubinage" with and "passing for his wife," to marry another. A concubine is "A woman who cohabits with a man without being his wife; a kept mistress" (*OED*).

86   JP to Douce, 27 Jan. 1823 (Bodl. Lib. MS Douce d. 37, f. 49).

87   JP to Thorkelin, 3 May 1792 (Edin. Univ. Lib. MS La. III. 379/670).

88   JP to Thorkelin, 22 Sept. 1792 (ibid., MS La. III. 379/671).

89   JP to Lord Buchan, 29 Jan. 1793 (JP, *Literary Correspondence*, Vol. 1, 320).

90   JP, *Ancient Scotish Poems*, Vol. 2, 383.

91   Craigie, "Macpherson on Pinkerton," 441.

92   "I have put David Macpherson upon writing an alphabetical description of Scotland to serve as a book of reference" (JP to Thorkelin, 24 Oct. 1791, Edin. Univ. Lib. MS La. III. 379/666).

93   JP to Napier, 31 Oct. 1818 (B.L. Add. MS 34612, f. 234). Sarah Kemble (d. 1807) was the wife of theatre manager Roger Kemble and mother of Sarah Siddons. The date of Sarah Kemble's death is a clue to the identity of the woman JP "shook off." Sarah Kemble could not have given advice about the women JP had similar trouble with later, i.e., in 1809–13 (see ch. 6). The likelihood, therefore, is that what he said to Napier applied to the woman in Kentish Town.

94   JP wrote a chapter on "Drunkenness" in *Recollections of Paris*, Vol. 2, 338–61, conceding "This sermon ... is, as usual, preached by a sinner" (358).

95   Ibid., 172.

96   Ritson to Paton, 21 July 1795 (*Letters to Paton*, 32).

97   J.S. Harford, *Life of Thomas Burgess* (London, 1840), 1–2.

98   As JP told Lord Buchan later, 21 Apr. 1794 (JP, *Literary Correspondence*, Vol. 1, 350).

99   On 27 Nov. 1798 Thomas Burgess sent JP a draft for £300, noting that before then he had been unable "to fulfill a promise, which I was as anxious to discharge, as you could be to receive" (LWL MSS Vol. 82.2). JP later complained that Burgess had led him to marry his sister "by splendid promises never performed" (JP to Macvey Napier, 31 Oct. 1818, B.L. Add. MS 34612, f. 234).

100 *Hampshire Parish Registers*, eds. W.P.W. Phillimore et al., 16 vols. (London: 1899–1914), Vol. 6, 66. JP's wife is here called Henry Maria, which is repeated elsewhere, e.g., *Horace Walpole's Correspondence*, Vol. 12, 10 n.; but her name was Henrietta Maria, as it is stated in a Chancery document dated 18 June 1810, National Archives, London, C13/113. Information about the ceremony from the parish register was provided by Rev. A.L. Bryan, late Rural Dean at Odiham.

101 Harford, *Life of Thomas Burgess*, 2.

102 For a brief acct. of Hampstead at the time, incl. the stage, see Anna Letitia Le Breton, *Memoir of Mrs. Barbauld* (London, 1874), 62 ff.

103 Detail from an inventory made by JP c. 1805 (Nat Lib. Scot. MS 1709, ff. 155–6).

104 JP told Douce, 27 Dec. 1817, that the piano had double pedals and cost 20 gns. (Bodl. Lib. MS Douce d. 37, f. 20); he wrote a poem "To Mrs. P. on her exquisite performance of Scotish music" (Nat. Lib. Scot. MS 1711, f. 166).

105 JP wrote of her "malignity" to Douce, 27 Jan. 1823 (Bodl. Lib. Douce MS d. 37, f. 49).

106 JP to Lord Buchan, 21 Apr. 1794 (JP, *Literary Correspondence*, Vol. 1, 350).

107 As in n. 106.

108 *Gentleman's Magazine*, Vol. 58, pt. i (Feb., 1788), 125–7; monthly to year's end, Vol. 58, pt. ii (Supplement, 1788), 1149–51. Identified as JP's in letter to Gibbon 23 July 1793 (JP, *Literary Correspondence*, Vol. 1, 330) and elsewhere. Full list in Bibliography, II (2).

109 *Gentleman's Magazine*, Vol. 58, pt. ii (Supplement, 1788), 1150.

110 *Critical Review*, 2nd ser., Vol. 1 (Apr. 1791), 428. Rev. of Thorkelín's *Sketch of the Character of the Prince of Denmark*. See above, n. 43.

111 *Critical Review*, 2nd ser., Vol. 5 (Appendix, 1792), 557. Rev. of Edward Ledwich's *Antiquities of Ireland*. These remarks are similar to those in the "Philistor" series, *Gentleman's Magazine*, Vol. 58, pt. ii (July 1788), 592, and in the 1794 reissue of the *Enquiry*, Vol. 1, 8 (Advertisement). The review of Ledwich extends over two issues of the *Critical*. In the first, Vol. 5 (Aug., 1792), the reviewer says (p. 396) he is sorry to see "Mr. Pinkerton classed among the authors who support the northern colonization of Ireland, while that writer is, perhaps, the first to argue that the Gothic colonies in Ireland proceeded from Belgic Gaul." In the second (Appendix, 1792), 559, he quotes Ledwich's praise for Pinkerton's *Dissertation* and *Enquiry*.

112 Gibbon, *History of the Decline and Fall of the Roman Empire*, Vol. 4, 98 n.

113 "Has Robert Heron (I forget his real name) published his lives of the Scottish saints to which I am a subscriber," Gibbon wrote to Peter Elmsley,

28 Mar. 1789. See Gibbon's *Letters*, ed. J.E. Norton, 3 vols. (London, 1956), Vol. 3, 145, also JP, *Vitae Antiquae Sanctorum*, xiv. For comments on JP's *Dissertation, Enquiry,* and other works, see Gibbon's *Miscellaneous Works,* ed. Lord Sheffield, 2 vols. (London, 1796), Vol. 2, 714–15.

114 JP to Gibbon, 23 July 1793 (JP, *Literary Correspondence*, Vol. 1, 329).

115 On 21 Apr. 1794 (ibid., 351).

116 Gibbon to JP, 25 July 1793 (ibid., 331–3).

117 JP to Gibbon, 28 Oct. 1793 (B.L. Add. MS 34886, f. 416). A copy of JP's Address is in Nat. Lib. Scot. MS 1711, ff. 154–9. Most of it is in Gibbon's *Miscellaneous Works*, ed. Sheffield, 5 vols. (London, 1814), Vol. 3, 582–90.

118 JP to Thomas Astle, 6 Dec. 1793 (MS James Osborn Collection, Yale Univ. Lib.)

119 Astle to JP, 10 Dec. 1793 (JP, *Literary Correspondence*, Vol. 1, p. 337). JP had been introduced to Astle as "a man of strict honour" by Walpole in 1790 (*Horace Walpole's Correspondence*, Vol. 42, 279–80).

120 Boswell, *Private Papers,* ed. G. Scott and F. Pottle, 18 vols. (New York, 1928–34), Vol. 18, 234.

121 Butler to JP, 23 Dec. 1793 (JP, *Literary Correspondence*, Vol. 1, 341).

122 Banks to JP, 8 Dec. 1793 (ibid., 336–7); JP to Douce, 9 Dec., asking for "assistance and support in this great national undertaking" (Bodl. Lib. MS Douce d. 37, f. 65).

123 Gibbon, *Miscellaneous Works* (1814), Vol. 2, 494 n.

124 Its title in Gibbon, *Miscellaneous Works* (1796), Vol. 2, 707–17, the text followed here, is "An Address &c." An edit. of it is in *The English Essays of Edward Gibbon*, ed. P.B. Craddock (Oxford, 1972), 534–45.

125 Gibbon, *Miscellaneous Works* (1796), Vol. 2, 714–15.

126 JP to John Nichols, 20 Jan. 1794 (JP, *Literary Correspondence*, Vol. 1, 344).

127 G.M. Young, *Gibbon* (London, 1948), 175. Young says JP approached Gibbon with the scheme, but this contradicts Gibbon's own statement, quoted above. JP told Lord Buchan 5 Apr. 1794 that the plan had not originated with him: "It was solely Mr. Gibbon's" (JP, *Literary Correspondence*, Vol. 1, 347).

128 A one page sketch of "Scriptores Rerum Picticarum et Scoticarum" is Nat. Lib. Scot. MS 1711, f. 163.

129 In Gibbon's *Miscellaneous Works* (1796), Vol. 2, 714–16.

130 Walpole to JP, 25 Jan. 1795, *Horace Walpole's Correspondence*, Vol. 16, 321. For interest in portraits, see *Gentleman's Magazine*, Vol. 62, pt. ii (Dec. 1792), 1101; 63, pt. i (Apr. 1793), 322–4; 63, pt. i (May 1793), 393, 394–7, *passim*. Thomas Birch's *Heads of Illustrious Persons of Great Britain* (1743 and later eds.) had helped stir it up. See JP's comments on Birch's

"splendid work" in letterpress on Mary Queen of Scotland, *Iconographia Scotica* (London, 1794–7).

131 JP to Lord Buchan, 19 Oct. 1795 (JP, *Literary Correspondence*, Vol. 1, 393).

132 James L. Caw, *Scottish Portraits*, 2 vols. (Edinburgh, 1903), Vol. 1, xx. For his leaving Andrew of Saltoun's periwig unfinished while catching "the Soul," 113–14.

133 See his comment, JP, *Literary Correspondence*, Vol. 1, 271 n.

134 Herbert took over employment of engravers after Silvester Harding overcharged him (JP to Buchan, 27 Dec. 1794, ibid., 370).

135 Walpole to JP, 25 Jan. 1795 (*Horace Walpole's Correspondence*, Vol. 16, 324).

136 This is inserted near the end of *Iconographia Scotica*, amidst a group of six portraits, neither of which is commented on by JP. Perhaps this reflects strife between him and Herbert. Three of them are of "Mary," otherwise unidentified. "Mary Queen of Scotland" is depicted in an earlier portrait, "From a Painting in Kensington Palace." The offending one was "From a Painting in Lord Buchan's Possession." It shows a devout, saintly Mary, writing.

137 As he admitted in *Gentleman's Magazine*, Vol. 68, pt. i (Apr. 1798), 302.

138 JP to Malcolm Laing, 7 Jan. 1800 (JP, *Literary Correspondence*, Vol. 2, 122).

139 *Monthly Magazine*, Vol. V (Feb. 1798), 82, where he revealed he'd demanded that his name be removed from the title-page. Herbert was an "*unaccountable*," he told Lord Buchan on 6 July 1795 (JP, *Literary Correspondence*, Vol. 1, 386).

140 JP to Lord Buchan, 28 Feb. 1796 (ibid., 399).

141 JP, *Modern Geography* (1802), Vol. 2, 577.

142 Note on the first image of Mary Queen of Scots.

143 JP to Lord Buchan, Jan. 10, 1795 (JP, *Literary Correspondence*, Vol. 1, 373).

144 JP, *Enquiry* (1794), Vol. 1, Advertisement, 1. The text was a reissue of the leftover 1789 copies, with a new Advertisement, where in addition to answering some of his critics he makes the "acknowledgement" that "the attacks on the Celts, and Celtic writers, are too repeatedly urged" in the earlier text (9). Certain offending passages are also deleted and replaced – which meant reprinting sheets. Cf., for instance, Vol. 1, 121–4, in the two texts, where his initial scoffing at the notion that the Picts were merely "a branch of the Gaelic race, who went from Ireland to the west of Scotland, where they were known in all ages, as at present, for a set of Celtic savages, incapable of any progress in society" is deleted, and the section wherein it appears rewritten. But to omit the many short stabs at Celts would have required an entirely reworked edition. JP could not "omit to express his regret that he has not a further opportunity of

removing blemishes of this kind, from this, and some of his preceding publications" (1). Yet he admitted that his attack on the Celts arose "partly from a wish that their want of civilization might be branded with due disgrace, not as a lasting reproach, but as a stimulus to future improvement" (10).

145 JP to Banks, 28 Jan. 1800 (Natural History Museum, Botanical Library, Dawson Turner copies of Banks's corres., Vol. 12, f. 25).

146 See Godwin to JP, 13 Oct. 1799 (JP, *Literary Correspondence*, Vol. 2, 105–6).

147 JP to Lord Buchan, 21 Apr. 1794 (ibid., Vol. 1, 134–5).

148 *Critical Review*, 2nd ser., Vol. 12 (Oct. 1794), 134–5. The "typographic error" pointed to is proof of JP's authorship; the general point he makes is made elsewhere; e.g., cf. the passage quoted with JP's defence of the *Enquiry* in *Gentleman's Magazine*, Vol. 59, pt. ii (Nov. 1789), 979–82, part of which is quoted in ch. 2 above. Note his point there about the injustice of taking "two or three warm pages from a thousand cool ones, to give an estimate of a book."

149 Ritson to Paton, 5 Mar. 1794 (*Letters to Paton*, 12).

150 Ritson, *Scotish Song*, 2 vols. (London, 1794), Vol. 1, xv n.

151 *Critical Review*, 2nd ser., Vol. 13 (Jan. 1795), 53. Ritson recognized JP's hand in the review. JP was, he told Paton, 19 May 1795, *"the gentleman* to whom I and my little publication are so much obliged" (*Letters to Paton*, 20; the review is attributed to JP in this book, and printed on 37–49).

152 The certificate (drawn up 20 Feb. 1795) is printed in Thomas Byerley and Joseph Robertson, *The Percy Anecdotes* (London, 1821–3), Vol. 1, 43. See Ritson to William Laing, 1 Dec. 1796, in Nichols, *Illustrations*, Vol. 3, 779.

153 Ritson to Paton, 19 May 1795 (*Letters to Paton*, 20).

154 Hamilton to JP, 15 July 1802 (JP, *Literary Correspondence*, Vol. 2, 226–9); 30 Sept. 1802 (LWL MSS Vol. 82.3).

155 JP to Lord Buchan, 21 Nov. 1791 (JP, *Literary Correspondence*, Vol. 1, 272).

156 JP to Lord Buchan, 23 Sept. 1793 (ibid., 334).

157 JP to Douce, 9 Dec. 1793 (Bodl. Lib. MS Douce d. 37, f. 65).

158 JP to Lord Buchan, 5 Apr. 1794 (JP, *Literary Correspondence*, Vol. 1, 346).

159 See his letter to Lord Buchan, 20 Feb. 1792 (ibid., 296–8).

160 JP to Lord Buchan, 23 Sept. 1793 (ibid., 334).

161 JP to John Bradfute, 26 Sept. 1793 (PO'F coll.).

162 Walpole read part "with great delight" and made "few trifling corrections" (Walpole to JP, 11 Apr. 1794, *Horace Walpole's Correspondence*, Vol. 16, 319); JP read some of the MS to Percy, evidently in 1792, and he "expressed great satisfaction at the historical painting and animation"

(JP to Lord Buchan, 23 Sept. 1793, *Literary Correspondence*, Vol. 1, 334–5); he wrote Beattie in 1794, asking for "his thoughts on the best characteristics of historic style" (Margaret Forbes, *Beattie and his Friends*, 285); he wrote Astle, 5 Mar. 1794, asking for an introduction to Bishop Douglas ("a good judge of style") and whether Astle himself would care to peruse the MS. (MS James Osborn Coll., Yale Univ. Lib.)

163   The immediate effects of the decision were that he compiled a list of Gibbonisms for his use and wrote a note on "the pomp and magnificence of Gibbon's diction" (Nat. Lib. Scot. MS 1711, ff. 189, 187).

164   He still considered treating the earlier periods as late as 5 Apr. 1794 (JP to Lord Buchan, *Literary Correspondence*, Vol. 1, 346), but these were treated in Vol. 1 of Robert Heron's 6 vol. *History of Scotland* which appeared soon afterwards. JP was not displeased with that vol., which relied heavily on the *Enquiry*. See ch. 4.

165   Gibbon said in his "Address" on the monkish historians that "a skilful judge has assured me, after a perusal of the manuscript [of the *History of Scotland*], that it contains more new and authentic information than could be fairly expected from a writer of the eighteenth century (*Miscellaneous Works* [1796], Vol. 2, 715).

166   JP to Dilly, 23 July 1795 (*Report on the Laing Manuscripts preserved in the University of Edinburgh*, 2 vols. [London, 1914–25], Vol. 2, 578–9).

167   Dilly to JP, 25 July 1795 (LWL MSS Vol. 82.2).

168   Disc. in Dilly to JP, 18 Sept. 1795, 17 Jan. 1796 (ibid.); 11 Jan. 1796 (JP, *Literary Correspondence*, Vol. 2, 395–6). The offer is in the letter of 11 Jan.

169   An "Accompt" for the period 1 Jan. 1788 to 14 Oct. 1789 (LWL MSS Vol. 82.4, [appendix]) shows income from rent and dividends in Edinburgh at £65/7/10 against expenses of £92/3/9½. The expenses included two payments to Mary Heron of £16/13/4.

170   "Bond of Annuity" (LWL MSS Vol. 82.4, [appendix]). The annuity was purchased from the city of Edinburgh on 6 Sept. 1794; half-yearly payments to Mary Heron to begin 6 Mar. 1795 for the half-year preceding; the amount to be £16/13/4.

171   "And for 15 years 1782–97 it cost me at least £40 a year in correspondence copies of MSS &c. &c. of Scotish poetry and history of which £600 I received £20" (JP to Macvey Napier, 28 Jan. 1818, B.L. Add. MS 34612, f. 172).

172   JP to Archibald Constable, 6 Oct. 1796 (Nat. Lib. Scot. MS 2524, f. 24).

173   His wife was always "sick and expensive," he told Douce 14 Feb. 1817 (Bodl. Lib. Douce d. 37, f. 15). He repeated this 23 Oct. 1823: "For my wife's feeble constitution and five or six children who died at an early age led to many extra expences which almost consumed my little share

in the public funds (about £700)" (ibid., f. 53). In JP's inventory of contents of the cottage in Hampstead, made c. 1805, his wife's clothing is valued at £150, his at £50 (Nat. Lib. Scot. MS 1709, f. 155).

174 Two daughters survived, Mary Margery and Harriet Elizabeth.

175 See Walpole to JP on 11 Feb. 1788 (*Horace Walpole's Correspondence*, Vol. 16, 294–5), from which it is evident JP planned to apply for a vacant position if Joseph Planta resigned (he didn't). On 20 May 1790 JP asked Thorkelín if he intended to apply for the position of assistant librarian, left vacant by the death of Charles Woide; "If you do not … I shall use my little interest to procure the place" (Edin. Univ. Lib. MS La. III. 379/667). The vacancy wasn't filled. See *Horace Walpole's Correspondence*, Vol. 16, 311–12.

176 JP to Banks, 29 Jan. 1795 (B.L. Add. MS 31299, f. 18).

177 JP later noted (in referring to "some European countries") "the indecent avarice of the clergy" in seizing upon "places and pensions" (*Recollections of Paris*, Vol. 1, 211–12).

178 Walpole to JP, 5 Feb. 1795 (*Horace Walpole's Correspondence*, Vol. 16, 327).

179 On 30 Jan. 1794, the title he had in mind ended with "the death of James V" (JP, *Literary Correspondence*, Vol. 1, 345).

180 Ibid., 395–6.

181 JP to Lord Buchan, 30 May 1796 (ibid., 413).

182 See Dilly to JP 26 Jan. 1797 and Gillies to JP 16 Feb. 1797 (ibid., 438, 447). Gillies wrote the review in the *Monthly*, 2nd ser., Vol 23 (May 1797), 1–10; the *Critical* review extended over two numbers, May and July 1797 (2nd ser., Vol. 20, 1–8, 288–97).

183 JP, *The History of Scotland.*, 2 vols. (London, 1797), Vol. 1, [v]. Announced in *St. James's Chronicle*, Feb. 16/18, 1797.

184 By Silvester Harding, engraved by William Gardiner; see Sher, *The Enlightenment & the Book*, 184–5. Beattie said it was "a striking likeness" (Forbes, *Beattie and his Friends*, 302).

185 JP, *History of Scotland*, Vol. 1, viii–ix.

186 Nat. Lib. Scot. MS 1709, f. 70.

187 See the Bibliography, I (20), for details of reviews. In Oskar Wellens, "John Pinkerton: *Critical* Reviewer," *Notes and Queries*, Vol. 27, 5 (Oct. 1980), 419, the assertion is made that JP reviewed his *History of Scotland* in the *Critical*. JP was not writing for the journal in May 1797 (see above). The review shows no marks of his hand.

188 *British Critic*, Vol. 11 (Apr. 1798), 345–58; author identified in E.L. De Montluzin, "Attributions of Authorship in the *British Critic*," *Studies in Bibliography*, Vol. 51 (1998), 250. Whitaker wrote *The Genuine History of the Britons Asserted*, noted in ch. 2.

189 An announcement that JP's *History* was published "this day" was in
    St. James's Chronicle, Feb. 20/22, 1798. This was a common device to
    promote slow-selling books.
190 William Anderson, *Answer to an Attack* (Edinburgh, 1797), 30–2.
191 Ibid., 36.
192 JP and Constable had quarrelled; JP roundly abused the publisher in a
    letter on 6 Oct. 1796 (Nat. Lib. Scot. MS 2524, ff. 23–4). In it he noted that
    Anderson "whom you recommended" had overcharged him. Anderson
    in his pamphlet (40) says "His bookseller, his engraver, and many other
    persons in Edinburgh" had "received marks" of JP's "liberality." See
    Thomas Constable, *Archibald Constable and his Literary Correspondents*,
    3 vols. (Edinburgh, 1873), Vol. 1, 22. On 28 Jan. 1818 JP told Macvey
    Napier that Constable had brought an action against him "for a few
    books ... before he sent me the accot. (B.L. Add. MS 34612, f. 172).
193 JP, *History of Scotland*, Vol. 1, 507 n.
194 See his ridicule of one of JP's emendations of Shakespeare in *Letters of
    Literature* in Shakespeare, *Plays*, eds. Johnson & Steevens, rev. Isaac Reed,
    15 vols. (London, 1793), Vol. 10, 500–1 n.
195 Steevens to Percy, 12 Apr., 1 May 1797 (Nichols, *Illustrations*, Vol. 7, 19–21).
196 *European Magazine*, Vol. 31 (May 1797), 328.
197 The feeling was not unanimous; see Malcolm Laing's opinion of
    Anderson, in JP, *Literary Correspondence*, Vol. 2, 24–5.
198 JP to Lord Buchan, 15 Dec. 1787 (ibid., Vol. 1, 174).
199 JP, *History of Scotland*, Vol. 1, 60–1.
200 Ibid., 85.
201 Ibid., x.
202 Ibid., 94.
203 Ibid., 194.
204 Ibid., 252.
205 Ibid., Vol. 2, 300.
206 Ibid., Vol. 2, 105, Vol. 1, 335.
207 Ibid., Vol. 2, p. 312, Vol. 1, 178.
208 Ibid., Vol. 2, 119, 194.
209 Ibid., 132.
210 Ibid., 364–5.
211 Ibid., 108–9.

4. Reviewer and Geographer, 1798–1802

1 JP to Douce, 14 Feb. 1817 (Bodl. Lib. MS Douce d. 37, f. 15).
2 JP to Malcolm Laing, 22 Jan. 1798 (JP, *Literary Correspondence*, Vol. 2, 20).

3 Joseph Cooper Walker to JP, 6 May 1799 (ibid., 57–8).

4 See JP, *History of Scotland*, Vol. 1, xi–xii; also JP to Malcolm Laing, 28 Jan. 1797, where he says his immediate plan was "to resume, in 2 vols. 4to, the early part, from Agricola to 1371" (JP, *Literary Correspondence*, Vol. 1, 440).

5 *Critical Review*, 2nd ser., Vol. 35 (May 1802), 98–9. Review of John Leyden's edition of *The Complaynt of Scotland* (see below).

6 £100; Edward Harding to JP, 22 Nov. 1796 (JP, *Literary Correspondence*, Vol. 2, 55 n.).

7 JP to Malcolm Laing, 6 Apr. 1800 (ibid., 141).

8 JP to Laing, 12 Aug. 1799 (ibid., 84).

9 Morison & Son to JP, 18 Nov.,1796, on Johnson's death (ibid., Vol. 1, 423–5).

10 See William Hazlitt, *The Life of Thomas Holcroft*, ed. Elbridge Colby, 2 vols. (London, 1925), Vol. 2, 151, 204–5.

11 A notice that JP had a volume of poems in the press is in the *British Critic*, Vol. 11 (Mar. 1798), 344. It was printed at JP's expense; see his letter, 7 Feb. 1800, in Nichols, *Illustrations*, Vol. 5, 677. "Public Happiness," in *Other Juvenile Poems*, 32–42, was written in 1781. JP reprinted "The Castle of Argan" in the vol., 22–5.

12 Laing to JP, 9 Feb. 1797 (JP, *Literary Correspondence*, Vol. 1, 446).

13 JP attributed to Walpole remarks about "Lord B." ("I am plagued with his correspondence, which is full of stuff"); see JP, *Walpoliana*, vol. 1, p. 96. Buchan was insulted; see *Horace Walpole's Correspondence*, Vol. 15, 192–3 n. Buchan's last letter to JP was on 29 July 1799; he said his "great seclusion" from society made it "much more difficult than formerly to promote your literary designs" (LWL MSS Vol. 82.2) – a polite dismissal. JP was soon telling Laing how Buchan had "deluded" him; JP to Laing, 6 Apr. 1800 (JP, *Literary Correspondence*, Vol. 2, 142). In a note written on a letter from JP to him, 1 July 1798, Buchan charged JP with "gross ingratitude to his literary benefactor" (LWL Misc. MSS). But JP had dedicated *Scotish Poems* to him.

14 JP to Douce, 11 Mar. 1798 (Bodl. Lib. MS Douce d. 37, f. 3).

15 Note by Douce (ibid., f. 4).

16 JP, *Literary Correspondence*, Vol. 2, 61 n.

17 Hazlitt, *Life of Thomas Holcroft*, Vol. 2, 140.

18 JP, *Literary Correspondence*, Vol. 2, 158–9.

19 Holcroft to JP, 11 May 1799 (Nat. Lib. Scot. MS 1709, f. 81).

20 John Fraser to JP, 30 July 1798 (LWL MSS Vol. 82.2). She was c. 78. JP on 26 June 1796 said she was "now about 76" (Anderson, *Answer to an Attack*, 20).

21 Mary Heron to JP, 3 Mar. 1790 (LWL MSS Vol. 82.1).
22 Laing to JP, 29 Apr. 1799 (LWL MSS Vol. 82.2; from a paragraph omitted in letter in JP, *Literary Correspondence*).
23 "Registered Commission," 6 May 1800 (LWL MSS Vol. 82.4, [appendix]).
24 Godwin to JP, 13 Oct. 1799 (JP, *Literary Correspondence*, Vol. 2, 106, misdated Oct. 10).
25 Hazlitt, *Life of Thomas Holcroft*, Vol. 2, 153.
26 Ibid., 195.
27 Info. from Godwin's diary used in this paragraph was provided by Prof. Lewis Patton.
28 These from *The Treasury of Wit*, Vol. 2, 80–1 (Dauphin's illness); 100–1 (Duke of Cumberland); 103 (Only three crowns); 137 (Locke on universities); 141 (Quin and Charles I); 146 (Will and Tom); 198 (Irish baronet), were all repeated, with slight variations, in *Walpoliana*.
29 JP, *Walpoliana*, Vol. 1, vi.
30 A.T. Hazen, *A Bibliography of Horace Walpole* (Folkestone & London, 1973), 149; Phillips' delighted response to receiving the "lively & elegant biographical sketch" is in a letter to JP, 25 Oct. 1799 (LWL MSS Vol. 82.2).
31 *Walpoliana*, Vol. 1, xlvi. The Latin means: believe the expert.
32 JP to Lord Buchan, 19 Dec. 1791, 10 Jan. 1795 (JP, *Literary Correspondence*, Vol. 1, 277, 374); the Latin *homo umbratilis* means man of shadows, i.e., retiring, private by nature. Literato is Spanish: man of letters.
33 *Walpoliana*, Vol. 1, xxxiii.
34 Ibid., xvi.
35 Ibid., xxi, 74–6. Fontenelle's *Entretiens sur la Pluralité des Mondes* (1686) posits life on the moon, planets, and other planetary systems. JP quotes Walpole: "It certainly requires more credulity to believe that there is no God, than to believe that there is. This fair creation, those magnificent heavens, the fruit of matter and chance! O impossible!"
36 *Walpoliana*, Vol. 1, xxiv–xxv.
37 Ibid., xxviii.
38 Ibid., xxxiii.
39 Ketton-Cremer quoted a part of it (from *Walpoliana*, Vol. 1, xl–l) in *Horace Walpole*, 297–9. It also forms the basis of Austin Dobson's sketch, "A Day at Strawberry Hill," in *Eighteenth Century Vignettes* (London, 1906), 151–60.
40 *British Critic*, Vol. 15 (Feb. 1800), 208.
41 *European Magazine*, Vol. 36 (Dec. 1799), 395.
42 MS note in Douce copy of the second edition of *Walpoliana* (London, 1800) in the Bodl. Lib., shelf-marked Douce W. 103. Douce told JP on 24 Mar. 1800 (LWL MSS Vol. 82.2) that he already had the *Monthly Magazine* version of *Walpoliana* and couldn't afford the book.

43 Malone to Percy, 6 Dec. 1800, in *Correspondence of Thomas Percy & Edmond Malone*, ed. Arthur Tillotson (Louisiana, 1944), 86.

44 "Upon the *Anti-Christian Doctrines* of the Monthly Magazine," in *The Porcupine*, 25 Dec. 1800. Intro. by Malone.

45 *Monthly Magazine*, Vol. 6 (July 1798), 19; (Sept. 1798), 196; Vol. 7 (Feb. 1799), 112. Another such attack sent to the magazine was printed in the appendix to *Walpoliana*, Vol. 2, 178–80.

46 *Monthly Review*, 2nd ser., Vol. 32 (June 1800), 179.

47 I.e., one who made models in clay or wax to be copied by the sculptor (from *OED*, modelling).

48 The scheme was outlined in the *Monthly Magazine,* Vol. 7 (Jan. 1799), 55.

49 See Laing to JP, 6 Oct. 1799 (JP, *Literary Correspondence*, Vol. 2, 102).

50 He would "enter into no further expense for the present," JP to Laing, 28 June 1800 (ibid., 168); yet correspondence on portraits went on for some considerable period after, with Laing and many others. He put off another volume for "three or four years" (ibid., 141). His attention was turning to geography and geology.

51 Vol. 1 of Andrews' work in *Critical Review*, 2nd ser., Vol. 26 (June 1799), 176–83. Among other signs of JP, the review contains a puff of *Iconographia Scotica* (p. 181). For the Macpherson review, see n. 60. I also think JP wrote the review of his friend Andrew Stuart's *Genealogical History of the Stewarts* (1798), ibid., Vol. 26 (May 1799), 1–10.

52 He told J.B. Nichols, 31 Mar. 1800, that he was paid "eight guineas per sheet for all I send to other magazines" (meaning other than the *Gentleman's Mag.*), in JP, *Literary Correspondence*, Vol. 2, 392, misdated 1810; corrected by LWL MSS Vol. 82.2, JP to Nichols, 31 Mar. 1800, copy.

53 See O. Wellens, "The 'Critical Review': New Light on its Last Phase," *Revue belge de philologie et d'histoire*, Vol. 56 (1978), 681, citing a letter of Robert Southey's in May 1799; also Roper, *Reviewing before the Edinburgh*, 20–2.

54 Information in this paragraph mostly from letters by Macpherson in a biographical sketch in Wyntoun, *Orygynale Cronykil*, ed. David Laing, 3 vols. (Edinburgh, 1872–79), Vol. 3, xxxvii–xlix.

55 Wyntoun, *Orygynale Cronykil*, ed. Macpherson, 2 vols. (London, 1795), Vol. 2, 441.

56 Ibid., Vol. 1, xiv.

57 Ibid., Vol. 2, 495, 502.

58 Ibid., 364.

59 Ibid., 484.

60 *Critical Review*, 2nd ser., Vol. 26 (May 1799), 90–3. The puffing marks it as JP's, and also the fact that he approved of Macpherson's procedure

in focusing on the Scottish elements in Wyntoun's work and omitting voluminous extraneous matter. This was just what he'd recommended in *Ancient Scotish Poems*. The scoffing at retention of textual minutiae is typical of JP. Note too his use of the spelling Winton, characteristic of him, and of the last phrase about the ostrich, which he was fond of, as in writing of Chalmers in his *Enquiry* (1814), Vol. 1, xxi ("The ostrich hides his head and thinks no one sees him").

61 Craigie, "Macpherson on Pinkerton," 439.
62 *Critical Review*, 2nd ser., Vol. 31 (Jan. 1801), 73–9. The opinion of Ramsay as a poet is essentially the same as that in JP's *Ancient Scotish Poems*, Vol. 1, cxxxii–cxxxv. Chalmers is termed the *"knight of the leaden mace* (p. 73), quoting T.J. Mathias's words in *Pursuits of Literature*. JP used the same ammunition against Chalmers in his *Enquiry* (1814), Vol. 1, xx. The review puffs "the Maitland collection of ancient poems," i.e., *Ancient Scotish Poems* (77). Note the repeated spelling "Scotish," approved of by both JP and Ritson.
63 JP, *Modern Geography* (1802), Vol. 1, ix.
64 See Robert Mayhew, "William Guthrie's *Geographical Grammar*, the Scottish Enlightenment and the Politics of British Geography," *Scottish Geographical Journal*, Vol. 115, 1 (1999), 19–34. Becky Sharp was still reading Guthrie while the nineteenth century "was in its teens" (Thackeray, *Vanity Fair*, ch. 3).
65 Formally, Association for Promoting the Discovery of the Interior Parts of Africa, founded in 1788.
66 For Banks' achievements and influence, see John Gascoigne, "Joseph Banks, mapping and the geographies of natural knowledge," in Miles Ogborn and Charles Withers, eds., *Georgian Geographies: Essays on space, place and landscape in the eighteenth century* (Manchester, 2004), 151–73.
67 See Browne to JP, 19 July 1799 (JP, *Literary Correspondence*, Vol. 2, 72); also Hazlitt, *Life of Thomas Holcroft*, Vol. 2, 242, 249.
68 Browne to JP, 23 June 1799 (JP, *Literary Correspondence*, Vol. 2, 63–4).
69 Banks to JP, 26 June 1799 (ibid., 69).
70 JP to Nicol, 26 June 1799 (MS Yale Univ. Library).
71 Nicol to JP, 28 June 1799 (ibid.); Browne to JP, 27 Jan. 1800 (JP, *Literary Correspondence*, Vol. 2, 125–7), is an attempt to untangle the business.
72 The relative importance of the two African trips may be indicated by the space given to each in J.N.L. Baker, *A History of Geographical Discovery and Exploration* (New York, 1967), 304–5 (Park), 309 (Browne, three sentences).
73 *Critical Review*, 2nd ser., Vol. 26 (July 1799), 249–59, completed Vol. 27 (Oct. 1799), 199–207. I do not attribute these to JP.

74 Southey to William Taylor, 22 Oct. 1799; in J.W. Robberds, *Memoir of the Life and Writings of the late William Taylor*, 2 vols. (London, 1843), Vol. 1, 301.

75 *Critical Review*, 2nd ser., Vol. 26 (Aug. 1799), 366, 368, 375–7; full rev. 361–79, Vol. 27 (Nov., 1799), 286–98. JP admitted he wrote this rev. in *Modern Geography* (1807), Vol. 3, 837 n. He told Banks he "furnished some materials" for it. Browne knew he wrote it (see JP, *Literary Correspondence*, Vol. 2, 125–7), as did Southey.

76 JP to Cadell and Davies, 7 Oct. 1799 (MS Historical Society of Pennsylvania).

77 Cadell and Davies, answering JP, 9 Oct. 1799, asked him to provide "a Sketch of your Plan" (MS Historical Society of Pennsylvania). He agreed to do so on 14 Oct. (Edin. Univ. MS La. II. 597 [13]), and a fortnight or so later sent them his "Idea of a new system of Geography" (B.L. Add. MS 34886, ff. 16–23).

78 A.A. Wilcock, "'The English Strabo': the geographical publications of John Pinkerton," *Transactions of the Institute of British Geographers*, no. 61 (Mar. 1974), 38.

79 Dated 10 Mar. 1800; among the "Cadell Purchase" papers, 11 (a), in the archives of Longman and Co., London. (Now in Reading Univ. Lib.) But a letter from Cadell & Davies to JP, 11 Mar. 1800 (LWL MSS Vol. 82.2) suggests the publishers were then still deliberating.

80 Murray to Archibald Constable, 1 June 1807 (Constable, *Archibald Constable*, Vol. 1, 373).

81 As early as 4 Apr. 1800 JP complained to Cadell and Davies of Arrowsmith's delays and "self-importance" (MS Pierpont Morgan Library). The publishers refused to get rid of him. JP attacked Arr. in *Modern Geography* (1807), Vol. 1, xli–xlii, xlii n.; see ch. 6 below.

82 JP to Banks, 18 Jan. 1800 (Natural History Museum, Botanical Library, Dawson Turner copies of Banks' corres., Vol. 12, f. 11).

83 Banks to JP, 19 Jan. 1800 (ibid., 12, f. 13).

84 JP to Banks, 28 Jan. 1800 (ibid., 12, f. 24).

85 Rennell, *The Geographical System of Herodotus* (London, 1800), 725–6 n. (the *Critical* note); 210 n. (the *Dissertation*). In an edition of Rennell's *Memoir of a Map of Hindoostan* (London 1792), 97, JP is credited with "great judgment and discrimination." JP later often had praise for Rennell; see, e.g., *Modern Geography* (1802), Vol. 2, 234, 235; also ch. 6 below.

86 JP to Banks, 24 Mar. 1800 (as in n. 82, Vol. 12, ff. 52–3).

87 Banks to JP, 26 Mar. 1800 (ibid., f. 54).

88 *Critical Review*, 2nd ser., Vol. 29 (May 1800), 27, 29; (July 1800), 252. The review is part of the JP-Rennell spat, outlined in the text. "We detest

despotism and monopoly in any branch of science," says the reviewer, a familiar JP refrain. Cf. his censure of Rennell in *Modern Geography* (1807), Vol. 3, 837 n., to the effect that "monopolies of commerce may be allowed, but monopolies of science!" Monopoly was a big word with him.

89 He may have written them in their entirety. In a letter to JP in Paris, 30 Sept. 1802 (LWL MSS 82.3) Samuel Hamilton noted that an apology had been inserted the Appendix in the current issue of the *Critical* for the omission of the article on maps; he and "Dr. P." – unidentified – "thought that preferable to the insertion of meagre accounts." He hoped JP would "furnish some very strong articles" for the next Appendix. This suggests JP had been in charge of assembling, or had been writing himself, the prior "Review of Maps and Charts."

90 *Critical Review*, 2nd ser., Vol. 27 (October 1799), 129–30. The MS, in JP's handwriting, is Nat. Lib. Scot. MS 1711, f. 150.

91 *Critical Review*, 2nd ser., Vol. 12 (Sept. 1794), 68. The comments on Heron's first vol. are identical with those in the 1794 reissue of the *Enquiry*, Vol. 1, 3 (Advertisement).

92 Heron, *New General History of Scotland*, 5 vols. (Perth, 1794–9), Vol. 4, vi.

93 *Critical Review*, 2nd ser., Vol. 27 (Nov. 1799), 315. Full rev. 313–21; Vol. 26 (Aug. 1799), 379–86. Godwin evidently recognized that JP wrote the first instalment of the review. The "letters from Scotland" remark also points to JP. The reviewer's criticisms of Heron's account of the Stuarts coincide with the views and facts in JP's *History of Scotland*.

94 Or he might have been tipped off about the vitriolic nature of the second half, due out in November.

95 Godwin to JP, 13 Oct. 1799 (JP, *Literary Correspondence*, Vol. 2, 106–7; the quotes from JP are from Godwin's letter).

96 *Critical Review*, 2nd ser., Vol. 28 (Jan. 1800), 48–57. JP asked Laing on 22 Aug. 1799 for information about Campbell's book, and Laing replied on 11 Sept. that it was "a ridiculous quarto" (JP, *Literary Correspondence*, Vol. 2, 87, 94). On 7 Jan. 1800 JP told Laing that Campbell's slurs against his *Iconographia Scotica* and edition of the *Bruce* were without basis. Herbert, he wrote, was responsible for the poor engravings in the first, and the MS of the *Bruce* had been copied by an amanuensis. He asked Laing to write an article in "one of your magazines," pointing out these misrepresentations (ibid., 122–3). Before Laing could do so, JP was defended on both points in the *Critical* (see the rev., 53–4). The reviewer also puffed *Ancient Scotish Poems* and said that Laing ("a learned gentleman in Scotland") would soon publish an exposure of the Ossian

fraud (50). Laing then realized JP was a *Critical* reviewer (JP, *Literary Correspondence*, Vol. 2, 131).

97 A. Campbell, *An Introduction to the History of Poetry in Scotland*, 2 pts. (Edinburgh, 1798–9), pt. 1, 189.

98 Perhaps even earlier; Gibbon says in his "Address" of 1793 that JP "retains no antipathy to a Celtic savage" (*Miscellaneous Works* [1796], Vol. 2, 714).

99 *Gentleman's Magazine*, Vol. 58, pt. i (June 1788), 500–1, pt. ii (Aug. 1788), 689 (parts of his "Philistor" series; see ch. 3, n. 108); answered by "Owain o Feirion" (probably either William Owen, known also as William Owen Pughe; or Edward Williams, i.e., Iolo Morganwg) in 58, pt. ii (July 1788), 606–8, (Sept. 1788), 821, and (Oct. 1788), 865–7. The same opponent published "Authentic Documents of Ancient British History" in 59, pt. i (Jan. 1789), 30–2, (Apr. 1789), 335–6, pt. ii (July 1789), 603–5, (Dec. 1789), 1077–8, 60, pt. i (Mar. 1790), 214, and pt. ii (Nov. 1790), 989–90. Another item called forth by JP's remarks was "King Arthur, a Poem," communicated by A. Crocker, and printed in 58, pt. ii (Sept. 1788), 820–1, (Nov. 1788), 1012, and (Supplement 1788), 1172–5.

100 *Gentleman's Magazine*, Vol. 58, pt. ii (Aug. 1788), 689 (in the "Philistor" series).

101 JP, *The Bruce*, Vol. 1, xii. This was the manuscript of *Beowulf.*

102 Ibid., xiii.

103 *Critical Review*, 2nd ser., Vol. 9 (Oct. 1793), 169. The consistency of views between this and other statements by JP on Welsh poetry is sufficient to determine authorship; Iolo Morganwg spotted JP as the author (Constantine, *The Truth Against the World*, 130–1).

104 Gibbon said in his "Address" that "For the sake of the Saxon Chronicle, the editor [JP] will probably improve his knowledge of our mother tongue" (Gibbon, *Miscellaneous Works* [1796], Vol. 2, 716).

105 Still viewed with considerable skepticism by academic historians; see Christopher A. Snyder, *The Britons* (Malden, Mass., 2003), 93–4.

106 Turner, *The History of the Anglo-Saxons* (London, 1799), 20 n.

107 *Critical Review*, 2nd ser., Vol. 28 (Jan. 1800), 18–19. Turner, as I read his sentence, identified the reviewer of Jan. 1800 as JP in his *Vindication* (London, 1803), iii. So did Southey; see Robberds, *Memoir of Taylor*, Vol. 1, 514, 516.

108 Caradoc of Llancarfan, twelfth-century Welsh author; he wrote a life of the sixth-century historian Gildas. JP's letters from Thomas Burgess, 21 Sept. 1794, 20 Apr. 1799 (LWL MSS Vol. 82.2) deal with Caradoc. JP had asked Burgess to try to find information about him.

109 *Critical Review*, 2nd ser., Vol. 28 (Jan. 1800), 24.

110 Ibid., Vol. 33 (Oct. 1801), 122; on the Triads, see J.L. Lloyd, *A History of Wales*, 2 vols. (London, 1954), Vol. 1, 123. William Skene rejected all three series of Triads as "entirely spurious" (*Celtic Scotland*, 3 vols. [Edinburgh, 1886–90], Vol. 1, 23; 3, 100 n.). The third series was a forgery by Iolo Morganwg.

111 P.c., from a MS note, National Library of Wales (reference lost).

112 Southey to Wm. Taylor, 8 Apr. 1804 (Robberds, *Memoir of Taylor*, Vol. 1, 498).

113 Turner, *Vindication*, 255–67; on the alleged mistranslation of just one word, evidently the single reference of Gerald to rhyme, ibid., 261–2.

114 R.H. Hodgkin, *A History of the Anglo-Saxons*, 2 vols. (Oxford, 1935), Vol. 1, 195–6.

115 See, e.g., the use of Taliesin, Llywarch, and Aneirin in Peter Berresford Ellis, *Celt and Saxon: The Struggle for Britain AD 410–937* (London, 1993), 66–7, 74–7, 109.

116 Nora K. Chadwick, *Celtic Britain* (London, 1963), 103; see Snyder, *the Britons*, 217–19. It is also used in Scottish historical writing, e.g., in Houston and Knox, *The New Penguin History of Scotland*, 35–6.

117 JP to Cadell and Davies, 7 Oct. 1799 (MS Historical Society of Pennsylvania).

118 What is now termed earth science or geology; see ch. 6 for JP's work in this field.

119 *Sketch of a New Arrangement of Mineralogy* (London, 1800). See Arthur Aikin to JP, 7 Sept. 1800 (JP, *Literary Correspondence*, Vol. 2, 188–90).

120 In Laing, *History of Scotland*, 2 vols. (London, 1800), Vol. 1, 527–44.

121 John Aikin, consulted on "the giddiness & other uneasy feelings in your head," advised JP to be sparing "in the use of all vinous & fermented liquors" as well as "the total disuse of spirits" (Aikin to JP, 17 Mar. 1800, LWL MSS Vol. 82.2). See also Aikin to JP, 8 Apr. 1800, and JP to Laing, 8 July 1800 (JP, *Literary Correspondence*, Vol. 2, 148, 176).

122 Details in a letter from his tenant, A. Neil, to JP, 26 May 1800 (LWL MSS Vol. 82.3). JP said later that two houses owned by him in Somers Town had burnt. "Loss by fire of two of my houses in Somerstown bought to increase my little income" is one item on a list of losses in a letter to Douce, 15 Feb. 1817 (Bodl. Lib. MS Douce d. 37, f. 15).

123 JP to Laing, 8 July 1800 (JP, *Literary Correspondence*, Vol. 2, 176). From JP's bio. in *Public Characters of 1800–1801*, 22, it is apparent this "original infirmity of nerve" dated from childhood.

124 John Aikin later produced *Geographical Delineations: or, a Compendious View of the Natural and Political State of all Parts of the Globe* (London, 1806), which evoked a charge of plagiarism from JP.

125 Thomas Burgess's phrase, letter to JP, 12 Dec. 1793 (LWL MSS Vol. 82.2).
126 JP to Cadell and Davies, 5 June 1801 (Nat. Lib. Scot. MS 948, No. 7).
127 JP to Cadell and Davies, 28 Jan. 1802 (MS Henry E. Huntington Lib.).
128 JP, *Modern Geography* (1802), Vol. 1, xi.
129 JP, *Recollections of Paris*, Vol. 1, 53.
130 JP, *Modern Geography* (1802), Vol. 1, 2; quotes below from the Preface, iii ff.
131 See, e.g., William Taylor's definition of the genre: "the geographer
     must renew for each generation his perishable toil. From every new
     travel, from every new voyage, he draws something to interpolate in
     his system: war ploughs up the land-marks he had mapped ... plague
     thins the population he had enumerated ... earthquake mars the city he
     had described ... every annual register, every newspaper even, urges
     the alteration of pages – how can he hope for more than a metonic cycle
     of celebrity?" (*Annual Review*, Vol. 1 [1802], 437.) A metonic cycle is
     c. 19 years. Author identified in Robberds, *Memoir of Taylor*, Vol. 2, 42.
132 JP undertook "the necessary, but inglorious drudgery of laborious
     compilation," said the *Edinburgh Review*, Vol. 3 (Oct. 1803), 68.
133 "Who is to say what is and what is not geography, or who is or who is
     not a geographer?" (J.N.L. Baker, *The History of Geography* [New York,
     1963], 101); on JP's use of Rennell, ibid., 149–2.
134 Douce to JP, 5 Nov. 1802 (JP, *Literary Correspondence*, Vol. 2, 233–4).
135 JP, *Modern Geography* (1802), Vol. 1, 3 n.; the named source being Didier
     Robert de Vaugondy's *Essai sur l'histoire de Géographie* (Paris, 1755).
136 O.F.G. Sitwell, "John Pinkerton: An Armchair Geographer of the Early
     Nineteenth Century," *The Geographical Journal*, Vol. 138, 4 (Dec. 1972), 474.
137 JP, *Modern Geography* (1802), Vol. 1, 2.
138 JP, *Modern Geography* (1807), Vol. 1, 3 n.
139 *Edinburgh Review*, Vol. 3 (October 1803), 72–3.
140 Sitwell, "John Pinkerton: An Armchair Geographer," 477.
141 For a relevant disc., see R.E. Dickinson and O.J.R. Howarth, in *The
     Making of Geography* (Westport, Conn., 1976), 120–3.
142 JP, *Modern Geography* (1802), 10, 103.
143 Ibid., Vol. 2, 537.
144 See Rives' comment in Tacitus, *Agricola; Germania*, xxxvii; also David
     Hume's "Of the Populousness of Ancient Nations," *Selected Essays*,
     eds. Stephen Copley and Andrew Edgar (Oxford, 1993), 226. Such
     commentary is found elsewhere in British literature; see mockery of it in
     Sterne's *Tristram Shandy*, Vol. 1, ch. 21; Vol. 2, ch. 19; Vol. 3, ch. 20.
145 Dickinson and Howarth, *The Making of Geography*, 123.
146 Letter XXVII, 179–89.

147 JP, *Modern Geography* (1802), Vol. 1, x.
148 Ibid., 6. Europe was not always "first treated" by British geographers, and JP here may be thinking of classical authors, e.g., Strabo, Pliny, and Ptolemy; he mentions Ptolemy just after this passage. To treat Asia first was sometimes seen as in keeping with the Biblical account of creation. John Payne, in his *Universal Geography*, 2 vols. (Dublin, 1794), Vol. 1, 11, admitted the superiority of Europe in "arts, arms, civilization, polished manners, and extensive commerce," but put Asia first because "Divine revelation leads us to suppose that Man when first created was placed" there. Guthrie's *Geographical Grammar* had Asia before Europe too. As late as 1819 Christopher Kelly, in *A New and Complete System of Universal Geography* (Dublin, 1819), Vol. 1, 5, put Asia ahead of Europe because there "the eternal Jehovah first smiled propitiously on the works of his own creation, and emphatically pronounced them "very good." Alexander Adam, however, in *Summary of Geography and History* (London, 1802), 131, places Europe ahead owing to "the genius of its inhabitants." John Aikin in *Geographical Delineations* (1806) also deals with Europe first, then Asia.
149 *Modern Geography* (1802), Vol. 1, 15–16.
150 Ibid., Vol. 2, 15.
151 Ibid., Vol. 1, 104–5.
152 Sitwell, "John Pinkerton: An Armchair Geographer," 475.
153 *Modern Geography* (1802), Vol. 1, 1; he seemed aware of the danger of slipping into topographical description; see, e.g., a comment in Vol. 2, 558.
154 Ibid., 465.
155 Griffith Taylor, ed., *Geography in the Twentieth Century* (New York, 1953), 35.
156 Robert Mayhew, *Enlightenment Geography: The Political Languages of British Geography, 1650–1850* (Houndmills, UK, 2000), 185, 187.
157 JP, *Modern Geography* (1802), Vol. 2, 513.
158 JP, *Recollections of Paris*, Vol. 2, 247.
159 JP, *Modern Geography* (1802), Vol. 2, 570.
160 JP, *Modern Geography*, 2 vols. (Philadelphia, 1804), Vol. 2, 435 n. See Barton's "Note concerning the article America," Vol. 2, 611–12.
161 Ibid., p. 442 n.; see *Modern Geography* (1802), Vol. 2, 575.
162 JP, *Modern Geography* (1804), Vol. 2, 446 n.; *Modern Geography* (1802), 581 n.; see also 513 n.
163 JP, *Modern Geography* (1804), Vol. 2, 611.
164 *Modern Geography* (1802), Vol. 2, 122.

165 Ibid., 558.

166 Ibid., 325.

167 Two years after Rennell's death; see Mayhew, *Enlightenment Geography*, 198–9.

168 JP, *Literary Correspondence*, Vol. 2, 226. He'd expressed a wish to visit Paris prior to this. See a letter to Lord Buchan, 16 Mar. 1789 (ibid., Vol. 1, 214).

169 Robert Townson to JP, 20 May 1800 (ibid., Vol. 2, 156–7).

170 JP to Napier, 31 Oct. 1818 (B.L. Add. MS 34612, f. 234).

171 JP to Douce, 27 Jan. 1823 (Bodl. Lib. MS Douce d. 37, f. 49).

172 A holiday is mentioned in Walker's letters to JP 27 June and 8 Nov. 1797 (JP, *Literary Correspondence*, Vol. 2, 9, 12).

173 Henrietta Pinkerton to JP, 11 May 1797 (LWL MSS Vol. 82.2).

174 JP, *Modern Geography* (1802), Vol. 2, 506.

175 Information from the agreement with Cadell and Davies for the geography and receipts for money paid. Longman and Co., Cadell papers, 11 (a), (b), (c), and (d).

176 Aikin to JP, 26 Aug., 13 Nov. 1802 (JP, *Literary Correspondence*, Vol. 2, 229–30, 236–7). Aikin by November had almost completed what he had been authorized to do – see ch. 6 – but very little of the vol. had been printed.

177 Samuel Hamilton's letter to JP, 30 Sept. 1802 (LWL MSS Vol. 82.3) says he had hopes of JP "furnishing some very strong articles" on maps for the next Appendix of the journal; but the appendices for 1802 and 1803 show no overt sign of his hand, nor is there such a sign anywhere following the destruction of the review's premises in Feb. 1803.

178 *Critical Review*, 2nd ser., Vol. 35 (May 1802), 95–9 (Leyden), (June 1802), 130–42 (Stewart). Leyden recognized JP's hand in both reviews and wrote a defence of his own work in the *Scots Magazine*, Vol. 64 (July 1802), 568–73.

179 He asked Laing on 15 May 1802 to express to Dalyell "my pleasure in his publication" (JP, *Literary Correspondence*, Vol. 2, 226).

180 *Scots Magazine*, Vol. 64 (July 1802), 568.

181 See Robberds, *Memoir of Taylor*, Vol. 2, 25, 68. The review passed into many hands after the destruction of its premises by fire in 1803; its last year of operation was 1817.

5. Paris Interlude, 1802–1805

1 JP, *Modern Geography* (1807), Vol. 1, xliv.

2 See Cadell & Davies to JP in Nov. 1802 (JP, *Literary Correspondence*, Vol. 2,

237 n.) and Longman & Co. (which had a share in the work) to JP on 14 Nov. 1804 (ibid., p. 314). Cadell & Davies printed 250 extra copies on large paper at £6 a copy; all hadn't been sold by 1805. The rumoured second edition is referred to in the *Monthly Review*, 2nd ser., Vol. 55 (Mar. 1808), 289 n.

3 In the agreement on the abridgment (Longman & Co., Cadell Papers, 11 [e]), Cadell & Davies reserved the right to print as many copies as they wished. They printed 5,000 copies of the second edition. Perhaps the first had a similar run. "The abridgement made by Pinkerton of his valuable book found a ready admittance into almost every school," wrote a critic in the *Gentleman's Magazine*, Vol. 79, pt. i (Feb. 1809), 145.

4 *Annual Review*, Vol. 1 (1802), 448.

5 *Critical Review*, 2nd ser., Vol. 39 (Sept. 1803), 69. Not written by JP. But it was a puff. A letter to JP, 20 Mar. 1802, from a *Critical* reviewer in Exeter (with signature cut off) gives this assurance: "You may depend on my best attention to yr. Work, I will take care to have the reviewing of it" (LWL MSS Vol. 82.3).

6 *Anti-Jacobin Review*, Vol. 18 (May 1804), 13.

7 *British Critic*, Vol. 21 (June 1803), 581–8; Vol. 22 (Aug. 1803), 148–55; *Edinburgh Review*, Vol. 3 (Oct. 1803), 67–81.

8 Charles Southey, ed., *The Life and Correspondence of Robert Southey*, 6 vols. (London, 1849–50), Vol. 2, 247.

9 JP, *Recollections of Paris*, Vol. 2, 262; but later said in the 1817 edition of the *Geography*, Vol. 1, 760 [780], that the work was "written throughout in an English sense and spirit, though liberal … to all nations."

10 See the treatment of JP's *Atlas* in ch. 6 for related discussion.

11 JP, *Modern Geography* (1802), Vol. 1, 255.

12 Bodl. Lib. MS Montagu d. 9, f. 148.

13 JP to Douce, 29 Oct. 1802 (Bodl. Lib. MS Douce d. 37, f. 63).

14 Phrases from *Monthly Review*, 2nd ser., Vol. 52 (Feb., 1807), 173, 181.

15 JP, *Recollections of Paris*, Vol. 1, 38, 354, 357, 509, 37.

16 Taken from the Vatican to Paris after Napoleon's conquest of Italy; returned after his defeat.

17 *Recollections of Paris*, Vol. 1, 10, 58, 77.

18 Ibid., Vol. 2, 97, Vol. 1, 310, 35.

19 Ibid., Vol. 2, 93–4, Vol. 1, 186–7, 264–5.

20 Ibid., Vol. 2, p. 102, Vol. 1, 32, 299, 301, 337.

21 Ibid., 479, 266, 404–5, Vol. 2, 351.

22 " … in England society was very small and select, whereas in France it was much larger and more diverse … salon society in Paris admitted

of a far more heterogeneous assemblage of people" (Robin Eagles, *Francophilia in English Society, 1748–1815* [Houndmills, UK, 2000], 128).

23 Hume in "My Own Life" wrote of his "real satisfaction in living at Paris, from the great number of sensible, knowing, and polite company with which that city abounds above all places in the universe" (*History of England*, Vol. 1, xi).

24 JP, *Recollections of Paris*, Vol. 1, 30.

25 *Edinburgh Review*, Vol. 8 (July 1806), 414.

26 *Critical Review*, 3rd ser., Vol. 9 (Nov. 1806), 287, 289.

27 *Monthly Review*, 2nd ser., Vol. 52 (Feb. 1807), 173.

28 *Universal Magazine*, 2nd ser., Vol. 8 (Aug. 1807), 137.

29 *Scots Magazine*, Vol. 68 (Sept. 1806), 685–90.

30 *Quarterly Review*, Vol. 3 (Feb. 1810), 197.

31 *Recollections of Paris*, Vol. 1, 302–3, Vol. 2, 209 n.

32 Jeremy Black, *Natural and Necessary Enemies: Anglo-French relations in the eighteenth century* (Athens, Ga., 1986), 179.

33 In *Religio Universalis et Naturalis* (Paris, 1818), 13, he wrote that the existence of the human race is ruled by "fame, siti, somno, amore animale" – hunger, thirst, sleep, and carnal love. See chapter 7 for disc. of this book.

34 *Recollections of Paris*, Vol. 1, 24.

35 Ibid., 64.

36 Ibid., 192.

37 Ibid., Vol. 2, 27.

38 Ibid., Vol. 1, 277–9.

39 Ibid., 132.

40 See his comment, ibid., 127.

41 Mawe to JP, 16 Mar. 1802 (JP, *Literary Correspondence*, Vol. 2, 208).

42 On this translation, see Wilcock, "'The English Strabo'," 40.

43 Longman's in a letter to JP, 14 Nov. 1804, said the French work had been "translated, as you informed us, under your own direction" (LWL MSS 82.3). JP denied he had any part in it, in his reply, 8 Dec. 1804 (*Literary Correspondence*, Vol. 2, 316–17).

44 Nat. Lib. Scot. MS 1709, f. 214. The minister may have been Joseph Fouché. See *Recollections of Paris*, Vol. 1, 473.

45 JP, *Modern Geography* (1802), Vol. 1, 255.

46 See JP, *Géographie Moderne*, 6 vols. (Paris, 1804), Vol. 1, 59. The translator, C.A. Walckenaer, expanded JP's 4 pp. on France's political geography to 16 pp., mostly through footnotes, subtly adjusting the perspective in the text.

47 Nicolas Louis François de Neufchâteau, *Tableau des Vues* (Paris, 1804), 3–4.

48 Ibid., 29, 62.

49 JP, *Recollections of Paris*, Vol. 1, 473.

50 JP, *Modern Geography* (1807), Vol. 1, 278–9 n.; he said in *Recollections of Paris*, Vol. 2, 258, that he was "menaced with the Temple."

51 In *Recollections of Paris*, Vol. 2, 241–66, a surprisingly mild riposte.

52 JP to Longman & Co., 8 Dec. 1804: "... if I asked leave to go, I should probably be sent to Verdun" (JP, *Literary Correspondence*, Vol. 2, 317).

53 See Mentelle's review of the French abridgment of JP's geography in *Bibliothèque Francaise* (1 July 1805), 34–41, (15 July 1805), 27–34. JP retaliated in *Modern Geography* (1807), Vol. 1, xxxv n. Malte-Brun denounced the abridgment in *Journal de l'Empire* (10 June 1806), attacking JP and those defending him ("Il y a quelque chose de pis qu'un Anglais, c'est un Anglomane"). Malte-Brun's pilferings from JP's geography were exposed in Jean-Gabriel Dentu's *Moyen de parvenir en Littérature, ou Mémoire à consulter sur une Question de Propriété littéraire* (Paris, 1811). Dentu tried to prove his case in court, an affair that caused a stir, but what Malte-Brun had done was not considered a forgery. See JP's comments in *Modern Geography* (1817), Vol. 1, 761[781] n. Malte-Brun attacked JP often, and he retaliated in various pp. of his geography, e.g., *Modern Geography* (1807), Vol. 1, xxxv–xxxvi. See a favourable review of *Géographie Moderne* in *Magasin Encyclopédique*, Vol. V (1804), 416–8. R.E. Dickinson, in *The Makers of Modern Geography* (New York, 1969), 18–20, mentions JP's name only among those writing early "compilations" of geography, but gives two pages to Malte-Brun's "great work," published in 1810–29, under the title "The First Universal Geography: Malte-Brun." See also Wilcock's similar treatment of Malte-Brun, "'The English Strabo'," 41–3.

54 It was rumoured that JP had been hired by the British government to write the dissertation. For JP's suspicions about the author of the rumours, see *Modern Geography* (1807), Vol. 1, xxxvi n., xlv n. Elie Jondot, who admired JP's geography, wrote condescendingly of the *Recherches sur l'Origine des Scythes ou Goths* in *Journal des Debats* (24 May 1804).

55 JP to David Bailie Warden, 17 May 1809 (MS Maryland Historical Society).

56 JP to Cadell & Davies, 29 Nov. 1804 (Nat. Lib. Scot. MS 1709, f. 178). He was trying to extract money from the publishers (see below); the longer he could say his detention was, the more likely they would be to pay. So he likely thought. He repeated the figure of "a year and a half" to Longman's in December (as in n. 52 above).

57 As in n. 52, 316; JP to Cadell & Davies, 29 Nov. 1809 (MS Yale Univ. Library).

58 JP to Spottiswoode, Jan. 1805 (MS Yale Univ. Library).

59 *Recollections of Paris*, Vol. 1, 79.
60 Ibid., 473–4.
61 JP to David B. Warden, 17 May 1809 (MS Maryland Historical Society).
62 *Recollections of Paris*, Vol. 1, vii.
63 See JP, *Literary Correspondence*, Vol. 2, 320–1 n.
64 JP to Warden, undated, likely early 1805 (MS Maryland Historical Society).
65 The high taxes in England induced many to flee to France; see Eagles, *Franciophilia in English Society*, 110, 113.
66 JP to Warden, 13 May 1805 (MS Maryland Historical Society).
67 JP to Warden, 12 Feb. 1806 (ibid.).
68 *Recollections of Paris*, Vol. 2, 408–9, 422–3, 444, 479, 498–9. Latin: let it be perpetual.
69 On 16 Aug. 1805 Cadell & Davies paid JP £25; the signed receipt is in Longman & Co., Cadell Papers, 11 (f).
70 The charge at customs was £74/10, he told Cadell & Davies on 5 Feb. 1806. In the same letter he wrote: "I have just received a letter from my banker at Paris, M. Perregaux, to say that he has drawn on Mr. Coutts [JP's London banker] for £93/15 – the balance of my accompt when I was detained in France, and for which I am greatly indebted to his kindness as I had no other resource" (MS Yale Univ. Lib.) The letter he refers to, dated 3 Jan. 1806, is in LWL. MSS Vol. 82.3.

6. The Dishonoured Veteran, 1806–1814

1 Laing and JP's factor, Gibson, had been active in securing the annuity; see a veiled ref. to it in JP to Laing, 15 May 1802 (JP, *Literary Correspondence*, Vol. 2, 226); on 22 Mar. 1803, Laing told JP that Gibson was in London but had given him "his word to make your affairs wound up as soon as he returns" (LWL MSS Vol. 82.3). Much difficulty occurred in establishing JP's title to the properties. On 14 Aug. 1805 Gibson told JP that he had "repeatedly" written to him in Paris to say that his annuity had been arrested to pay a debt (ibid.) JP told Henrietta of the annuity a week later (see below). The annuity was subject to 10 per cent tax at source (£7/10) and small handling fees.
2 See Geo. Chalmers to Archibald Constable on 27 Oct. 1803: "But there seems to be a Pinkerton mania in Scotland" (Constable, *Archibald Constable*, Vol.1, 411).
3 *Public Characters of 1800–1801* (1801), 20–31; *Universal Magazine*, 2nd ser., Vol. 2 (Nov. 1804), 393–7; *European Magazine*, Vol. 51 (June 1807), 411–13.
4 *Critical Review*, 3rd ser., Vol. 2 (May 1804), 17.

5 *Edinburgh Review*, Vol. 2 (July 1803), 376.

6 JP, *Recollections of Paris*, Vol. 1, 109.

7 "In the Romantic period women writers became more numerous and
women readers more influential, and so caused increasingly acute anxieties
to their male contemporaries" (Jacqueline Pearson, *Women's Reading in
Britain 1750–1835: A Dangerous Occupation* [Cambridge, 1999], 33).

8 Nat. Lib. Scot. MS 1711, f. 164.

9 JP, *Recollections of Paris*, Vol. 1, 122.

10 Ibid., 110.

11 JP, *Modern Geography* (1817), Vol. 1, 762 [782].

12 JP, *Enquiry* (1794), Advertisement, 10; *Modern Geography* (1802), Vol. 2, 471,
474; related comment, 489, 491, 513; he had been influenced by Walpole,
perhaps, who regarded Rousseau's ideas about primitive life as "puerile."

13 Published in 1754; JP takes his texts, translating himself, from the
Amsterdam *Oeuvres diverses* of Rousseau (1766).

14 JP, *Recollections of Paris*, Vol. 1, 165.

15 Cited by JP, ibid., 218. Rousseau understood private property as "a tragic
fall from a state of primordial innocence from which we cannot go back,"
says Simon Swift in *Romanticism, Literature and Philosophy* (London, 2006),
28–9. See his reading of the *Discours sur l'inégalité*, 31–2.

16 JP, *Recollections of* Paris, Vol. 1, viii, 132–3, 219–20, 225, 286–7, 440–1.

17 See, e.g., Shelley's response to the *Discours sur l'inégalité*, cited in
Edward Duffy, *Rousseau in England* (Berkeley, 1979), 92. Hazlitt's "On
the Character of Rousseau" in *The Round Table*, 2 vols. (Edinburgh, 1817),
45–55, shows the impact of Rousseau on him, and, by inference, on
Wordsworth. For Rousseau's influence on Wordsworth, see also Thomas
McFarland, *Romanticism and the Heritage of Rousseau* (Oxford, 1995),
66–74. Coleridge was "never more than a very lukewarm admirer" (64).

18 Longman & Co., Cadell Papers, II (a) and (e).

19 JP to Cadell & Davies, 13 Feb. 1802 (MS Boston Public Lib.).

20 Cadell & Davies to JP, 27 Apr. 1803 (JP, *Literary Correspondence*, Vol. 2,
251–2).

21 See Butler to JP, 21 Mar. 1802 (ibid., 214).

22 JP agreed to proceed to arbitration in a letter to Cadell & Davies on 15
Apr. 1802 (B.L. Add. MS 34886, f. 14). See also JP, *Literary Correspondence*,
Vol. 2, 214 n.

23 Prior to JP's leaving London, Aikin told him his father, John Aikin, and
Mrs. Barbauld had "suggested a few alterations," and JP authorized him
"to consult with my father upon the propriety of such alterations & the
general conduct of the work" (A. Aikin to JP, 19 Apr. 1803, LWL MSS
Vol. 82.3).

24 As in n. 20 above, 252–3. Aikin explained to JP how they had learned what his role was in his letter on 19 Apr. 1803 (as in n. 23). John Aikin blurted it out.

25 As in n. 20 above, 253.

26 The arbitration bond dated 17 Mar. 1804, with Butler and Andrew Strahan named as arbitrators, is a MS at Yale Univ. Library.

27 Longman's to JP, 14 Nov. 1804, JP's reply (draft), 8 Dec. 1804 (LWL MSS Vol. 82.3). JP's remark on Aikin omitted in letter in *Literary Correspondence*.

28 Longman & Co., Cadell Papers, II (g) and (h).

29 Yet knew that "By the love of money the character becomes degraded, the generous feelings obliterated, the very mind paralysed" (*Modern Geography* [1802], Vol. 2, 564).

30 Gillies, *Memoirs*, Vol. 2, 124.

31 Longman & Co., Cadell Papers, II (g).

32 JP to Cadell & Davies, 17 Dec. 1805 (MS Yale Univ. Lib.)

33 Longman & Co., Cadell Papers, II (n.), dated 6 Dec. 1806.

34 Thomas Shearer and Arthur Tillotson, "Percy's Relations with Cadell and Davies," *The Library*, 4th ser., Vol. 15 (Sept. 1934), 236.

35 J.R. Barker, "Cadell and Davies and the Liverpool Booksellers," ibid., 5th ser., Vol. 14 (Dec. 1959), 274–80.

36 See Browne to JP, 5 Jan., 18 Feb., 25 Feb., 1799 (JP, *Literary Correspondence*, Vol. 2, 39–40, 43, 45). Browne was initially promised £1,000 for the manuscript, but was persuaded to accept £600 and to get the remaining £400 once the 1,250 copies were sold. He was miffed.

37 See Butler to JP, 21 Mar. 1802 (ibid., 214).

38 Dean, *James Hutton*, 62.

39 Cadell & Davies to JP, 20 May 1803 (LWL MSS Vol. 82.3).

40 Hamilton to JP, 15 July, 30 Sept. 1802 (ibid.).

41 As she noted in an inventory of goods made in 1810 (Nat. Lib. Scot. MS 1709, f. 157).

42 Henrietta's annotation on this letter: "I do not know what the annuity in Scotland is." Robert Ingram had done some work for JP, and successfully arrested the annuity to get paid. An irate letter from him to JP, dated 4 Mar. 1803, is in LWL MSS Vol. 82.3. He mentions JP's wife and daughters in the letter as having "misrepresented" "some matters" to him. James Gibson's letter to JP, 14 Aug. 1805 (ibid.), tells of the success of Ingram's action for payment.

43 Nat. Lib. Scot. MS 1709, f. 104. Henrietta's annotation, f. 105.

44 I.e., second editions of *Modern Geography* (1807) and the abridgment (1806). For the latter he was paid on publication £133.

45 *Recollections of Paris* (1806).

46 Nat. Lib. Scot. MS 1709, f. 106.

47 JP to Walker, 2 Oct. 1806 (JP, *Literary Correspondence*, Vol. 2, 345).

48 That I have seen.

49 The agreement is Nat. Lib. Scot. MS 1711, f. 129; he got £300.

50 *Critical Review*, 3rd ser., Vol. 9 (Sept. 1806), 29.

51 Coxe, *Vindication of the Celts*, 4–5.

52 Referring to JP's *Letters of Literature*, where both writers are termed imitators. JP notes too that Horace, "like a Sabine puppy, was impudent enough to prescribe an absolute rule of poetry from his own skull" (230).

53 *Anti-Jacobin Review*, Vol. 25 (Nov. 1806), 256–8. The fiercely orthodox Whitaker and Polwhele were writing for the journal at this time. In the rev. of *Modern Geography* in the *Anti-Jacobin*, Vol. 18 (June 1804), 169, JP's remark (Vol. 2, 325) on "the great antiquity of the earth" was said to be "perfectly harmless" and involved "no consequences inconsistent with the doctrine of Moses." The reviewer had noted a willingness of JP's to "carry the antiquity of the human race rather higher than Scripture chronology will warrant," but said "this was really not his intention" (168). Also JP's "due respect" for "revelation" (164) was pointed out.

54 *Modern Geography* (1807), Vol. 1, xx, xxii, xxiv.

55 *Edinburgh Review*, Vol. 10 (Apr. 1807), 157, 169; full rev. 154–71.

56 Ibid., 171.

57 See Allen to JP, 24 Oct. 1810 (JP, *Literary Correspondence*, Vol. 2, 360–4); the reviewer was William Stevenson.

58 JP, *A Modern Atlas* (London, 1815), [i].

59 JP, *Modern Geography* (1817), Vol. 1, 762 [782].

60 Samuel Johnson, "An Ode to Friendship," *Poems*, ed. E.L. McAdam, Jr (New Haven and London, 1964), 71.

61 *Quarterly Review*, Vol. 25 (Apr., 1821), 176 n. See also Vol. 3 (Feb. 1810), 197 (quoting approvingly Malte-Brun's censure of JP); Vol. 6 (Dec. 1811), 508 (JP a "heedless plagiarist"); Vol. 7 (June 1812), 348 (JP's "absurdity"); Vol. 23 (July 1820), 326 (JP's "ignorance" and "want of feeling").

62 JP, *Modern Geography* (1807), Vol. 1, cxxxii.

63 In 2012 copies of the *Atlas* were on sale by map dealers for about £8,000.

64 JP to Warden, 29 Apr. 1807 (MS Maryland Historical Society).

65 JP to Longman & Co., 5 Apr. 1810 (JP, *Literary Correspondence*, Vol. 2, 394).

66 JP to Longman & Co. (rough draft), written soon after 2 Mar. 1810 (Nat. Lib. Scot. MS 1709, f. 167).

67 JP, *Modern Geography* (1817), Vol. 1, 760 [780] n.

68 Young's letter responding to JP's reopening of the correspondence is dated 5 Jan. 1807 (LWL MSS Vol. 82.3).

69 See JP to Laing, 6 Dec. 1805, Laing to JP, 12 Mar. 1806 (*Literary Correspondence*, Vol. 2, 327, 334–6. No second edition of the *History* appeared. JP had led Laing into error on the Welsh bards. See Laing's *History of Scotland*, Vol. 2, 436 n., and Turner's chastisement, *Vindication*, 265 n.

70 Walker to JP, 26 Aug. 1806, JP to Walker, 14 Sept. 1806 (JP, *Literary Correspondence*, Vol. 2, 341, 344).

71 JP to Walker, 15 Jan. 1807, with Walker's ans., 27 Jan. (ibid., 353–5). But Walker had praised him as "the profound historian of Scotland" and said *Letters of Literature* displayed "great ingenuity of criticism, and deep and original thinking" in his *Historical Memoir on Italian Tragedy* (London, 1799), vii, 221 n.

72 Banks to JP, 5 June 1807 (JP, *Literary Correspondence*, Vol. 2, 359–60).

73 JP, *A General Collection of the Best and most Interesting Voyages and Travels in all Parts of the World*, 17 vols. (London, 1808–14), Vol. 1, 621–734. This is Uno Von Troil's *Letters on Iceland*, which is an account of the expedition Banks organized.

74 Dawson Turner, in his statement about this matter dated 4 July 1835 (LWL MSS Vol. 82.1, prefixed) says she "lived with" JP "for 4 years from the age of 16 to 20" and was "driven from him" in Sept. 1809; a letter from the lawyer George Tennant to JP, 22 Jan. 1813 (ibid., Vol. 82.4), says she told him she was "not now more than 21." If she was 20 in 1809 she could not be 21 in 1813. Tennant may have misheard what she said in a tearful interview. Did she say she was not more than 21 when she and JP split? If so, she could have been 17 when the relationship began. Hard facts on this matter may be lost through misunderstanding or attempted recrimination.

75 Info. about the daughter from Tennant's letter, as in n. 74; John Tims to JP, 11 Sept 1809 (LWL MSS Vol. 82.4) says the daughter was then "of the age of 8 months or thereabouts" and that Hester was "again with child by you." Dawson Turner (as in n. 74) says this second child was his son.

76 Tennant to JP, as in n. 74.

77 See JP to Longman & Co., 5 Apr. 1810 (JP, *Literary Correspondence*, Vol. 2, 394–5).

78 Copy of agreement, Nat. Lib. Scot. MS 1711, f. 43.

79 For biographical information on Neele and Herbert, see Josephine French et al., eds. *Tooley's Dictionary of Mapmakers*, 4 vols. (Tring, Herts., UK, 1999–2004), Vol. 2, 301, Vol. 3, 311–12.

80 JP to Cadell & Davies, 10 June 1808 (MS Yale Univ. Library). He said later that he thought "draftsmen and engravers ... might occasionally be employed under my own eye" (JP to Longman's, 5 Apr. 1810, JP, *Literary Correspondence*, Vol. 2, 394).

81 The "first impressions" from subscribers had been requested in the advertisement for the work (as in n. 62).

82 As he told Cadell & Davies 19. Mar 1809 (Edin. Univ. Lib. MS La. II. 597/6).

83 Joseph Planta and George Shaw, M.D., were employees of the B.M., Shaw a naturalist who helped in the second edition of JP's geography. See *Modern Geography* (1807), Vol. 1, xxvii.

84 JP to Douce, 11 Nov. 1808, with draft of Douce's reply, and memorandum (Bodl. Lib. MS Douce d. 37, ff. 10–12).

85 JP, *Letters of Literature*, 365.

86 JP, *The Treasury of Wit*, Vol. 2, 5.

87 *Quarterly Review*, Vol. 1 (1809), 112–31.

88 Reprinted in Roberts' *Letters and Miscellaneous Papers* (London, 1814), 336–55. Roberts died in 1810.

89 JP, *Modern Geography* (1807), Vol. 1, xxvi.

90 JP to Cadell & Davies, 19 Mar. 1809 (Edin. Univ. Lib. MS La. II. 597/6).

91 He'd moved there (now Bloomsbury Way) by 19 Aug. 1809 (W. Richardson to JP, that date, that address, containing details of the furnishings in the house, Nat. Lib. Scot. MS 1709, f. 125). He was still at Queen Anne St. on 17 May 1809 (address in a letter from Warden on that date, MS Maryland Historical Society).

92 JP to Cadell & Davies, 2 Oct. 1809 (MS Yale Univ. Library).

93 As in n. 75 above; a letter of Tims to a Mr. Kinrich, another official, on 14 Sept. 1814 (LWL MSS Vol. 82.4), concerns the arrangements being discussed with JP for support of Hester and the children.

94 As in n. 74 above.

95 JP to Cadell & Davies, 14 Oct. 1809 (MS Yale Univ. Library).

96 Henrietta to JP, n.d., likely early 1810 (Nat. Lib. Scot. MS 1709, f. 159).

97 JP to Douce, n.d., likely early 1810 (Bodl. Lib. MS Douce d. 37, f. 64).

98 Henrietta to John B. Burgess, n.d. (Nat. Lib. Scot. MS 1709, f. 157). The note says that she had left "sundries packed up in Packing cases" on 29 Mar. 1810; JP had "ordered" her to take them with her, but then, on her leaving, forbade her to remove them until her brother called on him.

A Chancery document dated 18 June 1810 (National Archives, London, C13/113) relating to a dispute over property in Odiham describes her as "the wife but now living seperate [*sic*] and apart from John Pinkerton."

99 I assume this was the part of his property that formed the collection known as "The Pinkerton Papers," given to the Nat. Lib. Scot. in 1934 by Mrs B.M.H. Riddel, JP's great-grandaughter. See the Bibliography, VIII (10).

100 Nat. Lib. Scot. MS 1709, f. 157, contains a list of what she had recently delivered to Pinkerton, and which she hoped could now be taken back and sold.

101 The few documents relating to the separation make it difficult to piece the whole story together.

102 JP to Cadell & Davies, 17 Feb. 1810 (MS Yale Univ. Library).

103 JP to Longman's, 24 Feb. 1810 (Edin. Univ. Lib. MS La. II. 597/7).

104 Cadell & Davies to Longman's, 2 Mar. 1810, in JP, *Literary Correspondence*, Vol. 2, 387–90; a rough draft of JP's letter to Longman's in response is Nat. Lib. Scot. MS 1709, ff. 167–70. A fragment of a letter of JP to Cadell & Davies, 6 Mar. 1810, is in LWL Misc. MSS. It appears to be an effort to make peace, though it recommends strongly replacing Neele with Lowry.

105 JP to Longman's, 5 Apr. 1810 (JP, *Literary Correspondence*, Vol. 2, 395).

106 Cadell & Davies to Longman's, 2 Mar. 1810, noted that the draughtsman was "obliged to read carefully large volumes of Travels, &c. that are put into his hands" by JP (as in n. 104, 389).

107 JP to Cadell & Davies, 1 Mar. 1811; reply (rough draft) 2 Mar. (Bodl. Lib. MS Montagu d. 15, f. 104).

108 The figure he gives in letter to Douce, 14 Feb. 1817 (Bodl. Lib. MS Douce d. 37, f. 15).

109 JP to Warden, 14 Dec. 1811 (MS Maryland Historical Society).

110 The names of the last two are here given from the maps; in the preliminary list, they are called Polar. Northern and Polar. Southern).

111 JP, *A Modern Atlas*, vii (with list of maps).

112 The two remaining at the end being of the World on Mercator's Projection: Western Part and Eastern Part.

113 Mayhew, *Enlightenment Geography*, 12.

114 Porter, *The Making of Geology*, 237.

115 This may vary, depending on the copy consulted. The *Atlas* was not printed as most books are, but patched together from separately published maps.

116 JP, *A Modern Atlas*, v.

117 E.g., in a "Catalogue of Maps, and of Books of Voyages and Travels," *Modern Geography* (1802), Vol. 2, 781–91.

118 On the use of such explanatory matter on maps, see J.H. Andrews, *Maps in Those Days: Cartographic Methods before 1850* (Dublin, 2009), 409–12.

119 JP, *A Modern Atlas*, v.

120 Ibid., iv; see his advertisement for the *Atlas* (as in n. 62). The claim is doubtful. See G.R. Crone, *Maps and Their Makers* (London, 1968), 125–7; P.D.A. Harvey, *The History of Topographical Maps* (London, 1980), 181–3, for the complex history of relief in mapping. Contour lines were not yet in general use to depict topographic features.

121 See Andrews, *Maps in Those Days*, 289.

122 JP, *A Modern Atlas*, v.

123 "In cartography, the use of colour is often metaphorical," J.S. Keates, *Understanding Maps* (Harlow, UK, 1996), 220.

124 Cadell & Davies to Longman's, 2 Mar. 1810 (JP, *Literary Correspondence*, Vol. 2, 388.

125 JP, *A Modern Atlas*, v.

126 Ibid., vii.

127 JP doubted it existed; see *Modern Geography* (1802), Vol. 2, 523. Alexander Dalrymple in 1770 had posited a great southern continent extending from 30° south to the Pole, with probably over fifty million people.

128 See his comments, *Modern Geography* (1802), Vol. 2, 532–3.

129 In his geography he called it "Notasia, or New Holland" (ibid., 449); he preferred the former and tried to affix it (433–4), but it didn't take hold. He persisted in *Petralogy*, Vol. 1, 296 n. The name Australia was given to the island by Matthew Flinders, the circumnavigator of 1801–3; see Gascoigne, "Joseph Banks," 155. "Notasia or New Holland" was still in use in 1824, in the *Supplement to the Encyclopædia Britannica* published that year on an Arrowsmith map from 1816 (Vol. 2, facing 2), as well as in the account of Australasia by John Barrow, 1–18.

130 On these mountain ranges see Charles Bricker's comments, in R.V. Tooley et al., *A History of Cartography: 2500 Years of Maps and Mapmakers* (London, 1969), 180; Simon Garfield, *On the Map* (New York, 2013), 204–9.

131 J.B. Harley, *The New Nature of Maps: Essays in the History of Cartography* (Baltimore and London, 2001), 76.

132 "… the small-scale map of a continent is a summary interpretation, founded on a judgement of what is held to be important" (Keates, *Understanding Maps*, 170).

133 "… all maps embody a theory (perhaps scientific) about the nature of the world," says A.M.C. Godlewska, in D.N. Livingstone and C.W.J. Withers, eds., *Geography and Enlightenment* (Chicago, 1999), 250.

134   Faheino Mawe (North Island); Tavai Poenammoo (South Island); see
        Modern Geography (1802), Vol. 2, 487, where he gives a much longer one
        for the North Island, and comments that the names "equal the Russian in
        length" and "might well be contracted."
135   E.g., maps 50 and 51 feature the "South Pacific Ocean." The Grand
        Ocean was "falsely called the Pacific," he said (*Petralogy*, Vol. 1, 497). For
        Notasia, see n. 129 above.
136   Jean Baptiste Bourguignon d'Anville (1697–1782); but JP had criticized
        him sharply in the *Dissertation on the Scythians or Goths*, 9–10.
137   See a note on "Hydrography" in his "Catalogue of Maps," *Modern
        Geography* (1802) Vol. 2, 789–90. He makes differing comments on the
        importance of charts elsewhere.
138   See his account of the "Great Bank," i.e., Grand Bank (ibid., 540; also 553).
139   E.g., in Tooley et al., *A History of Cartography*, 141–4, 179–80; Crone, *Maps
        and Their Makers*, 137–40.
140   *Modern Geography* (1807), Vol. 1, "Memoir," xxxiii.
141   Ibid., xli–iii.
142   Prospectus for JP's *Voyages and Travels*; this occupies 13–16 of a pamphlet
        of prospectuses issued by Cadell & Davies in 1807. The pamphlet is
        No. 16 in Vol. 2 of a collection of pamphlets made by James Maidment
        (B.L. 816. 1. 47).
143   Ibid.
144   JP, *Voyages and Travels*, Vol. 11, 1.
145   This Preface was written late and inserted in reissues of Vol. 1. The
        "general table of contents" was similarly inserted towards the end of the
        project.
146   But not this edition.
147   It is a sad irony that the work of this secret and bitter enemy was given
        so much space.
148   JP cites this work in describing ironstone in *Petralogy*, Vol. 1, 95–6.
149   "Translators" were employed; see JP to Longman's, 5 Apr. 1810 (JP,
        Literary Correspondence, Vol. 2, 394).
150   I mean Johnson mangled it; see Johnson, *A Voyage to Abyssinia*, ed. Joel
        J. Gold (New Haven and London, 1985), l–lv. JP cut off the "Sequel" and
        15 dissertations "on various Subjects" relating to Abyssinia that follow
        Lobo's main text in Johnson's version.
151   JP, *Voyages and Travels*, Vol. 1, vi.
152   Ibid., Vol. 14, 878 n. The passage in question, a gruesome one from
        Churchill's *Collection*, tells what happens in a certain aboriginal tribe of
        Brazil when a girl of marriageable age is not courted by anyone. "A very
        indelicate passage is omitted," is JP's note.

153 See below for further discussion of Vols. 16 and 17.

154 Thomas Rees in London to JP in Edinburgh, 4 Dec. 1812, explains that the "copy" he'd been instructed by JP to insert in Volume 16 "must some of it be considerably curtailed" (LWL MSS Vol. 82.4).

155 JP, *Voyages and Travels*, Vol. 1, v.

156 As when, say, a note refers the reader to p. 60 when the text occupies pp. 90–9.

157 See a letter complaining of errors, *Universal Magazine*, 2 ser., Vol. 15 (Mar. 1811), 208–9.

158 *European Magazine*, Vol. 67 (June, 1815), 528.

159 G.R. Crone and R.A. Skelton, "English Collections of Voyages and Travels 1625–1846," in Edward Lynam, ed., *Richard Hakluyt & His Successors* (London, 1946), 135; on JP, 134–7.

160 See Weber to JP, 7 Apr. 1813 (JP, *Literary Correspondence*, Vol. 1, 407), explaining and regretting deficiencies in the Catalogue. On internal evidence, JP was part-author.

161 The 6-volume American edition supplies them; see the Bibliography, I (27b).

162 But the latter without JP's name on the title-page.

163 JP, *Petralogy*, Vol. 1, iv.

164 Except for "veinstones," defined in an appendix to Vol. 2, 563–90. These "have often been confounded with rocks" (589).

165 Defining it as "the doctrine of the constitution of this globe, which rather belongs to natural philosophy" (ibid., Vol. 1, xliii). Not everyone used the word in this way. See Dennis R. Dean, "The Word 'Geology'," *Annals of Science*, Vol. 36 (1979), 43.

166 *Petralogy*, Vol. 1, xxxii.

167 The *Edinburgh Review*, Vol. 23 (Apr. 1814), 79, said about 550 of the 1,200 pages that comprise the body of the book (omitting appendages) consisted of quoted passages.

168 *Petralogy*, Vol. 1, xlvi.

169 Ibid., 55–6.

170 Porter, *The Making of Geology*, 171.

171 *Petralogy*, Vol. 1, 102.

172 Ibid., vii.

173 Ibid., Vol. 2, 39.

174 Ibid., 315.

175 Ibid., 313; Vol. 1, 31.

176 The extent to which Hutton was indeed the founder of the science of geology is discussed in Porter's *The Making of Geology*, 184–215, where the Plutonist/Neptunist debate is cast in quite a different light.

177 See his letter to JP, commenting on *Modern Geography*, 10 Oct. 1802 (JP, *Literary Correspondence*, Vol. 2, 231–3); also 9 May 1809 (on granite, 386–7).

178 Dennis R. Ryan, *James Hutton and the History of Geology* (Ithaca, 1992), 164.

179 JP, *Petralogy*, Vol. 2, 162.

180 Leigh and Sotheby sale catalogues, dated 6 Apr. 1812 and 7 Jan. 1813, B.L. S. C.S. 75 (9), 78 (7).

181 Sarah Couper, entry on JP in the *Oxford Dictionary of National Biography*, online; from parish records, St Mary Magdalen, Old Fish Street, London.

182 JP to Macvey Napier, 31 Oct. 1818 (B.L. Add. MS 34612, f. 234).

183 Samuel Taylor (for the Burgess family) to JP, 27 Dec. 1813, Tennant & Harrison (for JP) to JP, 15 Dec. 1814, 7 Jan. 1815 (LWL MSS Vol. 82.4), deal with the separation agreement.

184 Thomas Rees and John Britton, *Reminiscences of Literary London from 1779 to 1853* (London, 1896), 60. See n. 208 below.

185 *Critical Review*, 2nd ser., Vol. 34 (Jan. 1802), 58–68. JP, however, found much to praise. The similarity of opinions about Stonehenge and its connection to ancient Belgae (to JP a Germanic nation) in this review to those in the *Enquiry*, and other signs, show JP's hand.

186 Nat. Lib. Scot. MS 1709, f. 168. Rough draft, written soon after 2 Mar. 1810.

187 Young to JP, 28 Sept. 1812 (LWL MSS Vol. 82.4; cited in JP, *Literary Correspondence*, Vol. 2, 403 n.).

188 See his comments on the "interest required" to get a child into the Hospital, *Recollections of Paris*, Vol. 2, 172–3. But a new rule made in 1801 stipulated that the main purpose of the Hospital was to support illegitimate children (Ruth McClure, *Coram's Children: The London Foundling Hospital in the Eighteenth Century* [New Haven and London, 1981], 250–1); this, depending on what he was prepared to disclose about the child's parentage, might have been an impediment.

189 Information about the child is given in letters from John Smith to JP, 17 Dec. 1812 and 29 Nov. 1814; LWL MSS, Vol. 82.4. (Transcripts sent to PO'F by W.S. Lewis.)

190 Robert Ferguson's letter to JP, 22 Nov. 1812 (Nat. Lib. Scot. MS 2670, ff. 174–5) indicates that JP was then in Edinburgh and had not been there over a week.

191 J.G. Lockhart to J.W. Croker, 10 Aug. 1840; in Alan Lang Strout, "Some Unpublished Letters of John Gibson Lockhart to John Wilson Croker," *Notes & Queries*, Vol. 187 (23 Sept. 1944), 135.

192 See, e.g., "*Remarks on* Pinkerton's *Etymology of the* Picts," by "Milo," in *Scots Magazine*, Vol. 74 (Sept. 1812), 654–5.

193 Scott to JP, 24 Apr. 1802 (JP, *Literary Correspondence*, Vol. 2, 218–9); it was sent at Laing's suggestion (ibid., 207); JP to Scott, 16 Apr. 1802, acknowledges the "polite present" (Nat. Lib. Scot. MS 3874, f. 138).

194 Scott to Benjamin Haydon, 27 Dec. 1820; Scott, *Letters*, ed. H.J.C. Grierson, 12 vols. (London, 1932–7), Vol. 6, 320.

195 Siddons to JP, 1 Dec. 1812 (JP, *Literary Correspondence*, Vol. 2, 403–4).

196 Scott to JP, 15 Jan. 1815 (Scott, *Letters*, Vol. 12, 414–15).

197 Scott to Sarah Smith, 5 Apr. 1813 (ibid., Vol. 3, 249).

198 A note from Kemble to JP, dated 17 Mar. 1812, is Nat. Lib. Scot. MS 1709, f. 132.

199 Siddons to JP, 22 Feb. 1813 (ibid., f. 144).

200 Gillies, *Memoirs*, Vol. 2, 123.

201 Quoted lines from a MS of the play, Nat. Lib. Scot. MS 1712, ff. 9, 25, 28.

202 *Scots Magazine*, Vol. 75 (Mar. 1813), 164. Other details about the production are from this account.

203 Nat. Lib. Scot. MS 1709, ff. 146–7.

204 Letter on 21 Mar 1813 (Scott, *Letters*, Vol. 3, 237–8).

205 Vol. 75 (Mar. 1813), 164.

206 Lockhart to Croker, 10 Aug. 1840; in Strout, "Some Unpublished Letters," 135.

207 Gillies, *Memoirs*, Vol. 2, 132; see also Gillies, *Recollections of Sir Walter Scott*, 188–90.

208 It didn't sell until 1814; for 490 gns., half of what JP paid for it. Rees also sent news, as in his letters of 4 Dec. 1812 and 5 Mar. 1813 (LWL MSS Vol. 82.4), of the progress of the sale of the house on Tavistock Place, which – despite John Britton's statement that it was rented – JP may have owned as well. James Saltwell also sent JP word about the disposal of the houses.

209 As he noted, in a letter to JP, 4 Dec. 1812 (LWL MSS Vol. 82.4). See n. 154.

210 Rees to JP, 21 May 1813 (as in n. 209), notes that Neele had called on him with two drawings, of Asia and Africa. Neele wanted to know if they should be sent to JP, "as they only contained in a general way" what JP had inspected "in detail." Rees "thought it unnecessary to send them down – advised Mr Neale to put them in hand – but if you wish to see them they may yet be sent." He says further that Hebert had made "some progress" on a map of Abyssinia. Rees wanted to know if JP had any objection to its insertion.

211 A letter of James Saltwell (evidently a relative of Betsey's) to JP, 8 Jan. 1813, quotes JP as saying "Betsey will write in a few days" and was indisposed (as in n. 209); but on 8 Feb 1813 he writes to JP again, expressing concern over "your disagreements," and hoping JP will "by no means put any impediment in the way" of Betsey's "leaving Scotland as you promised when she left Milton." (It is not clear which community named

Milton is referred to.) If JP breaks his promise, "both brother and aunt are determined to come after her." He notes that JP has said in his letter "so much against her." What was "against her" is not apparent. But the rift between JP and Betsey was permanent. Subsequent letters from Saltwell to JP, 30 Dec. 1813, 11 Jan., 1 Mar. 1814 (ibid.) express anger, and convey threats of legal action over payment of support for "your wife & child." Such threats may explain, in part, JP's decision to head back to Paris.

212 Whence he wrote to William Blackwood on Apr. 20 (Nat. Lib. Scot. MS 4001, f. 196).

213 See *Letters of George Dempster to Sir Adam Fergusson 1756–1813*, ed. James Fergusson (London, 1934), 340. JP was in Dunnichen in July 1813, not, as stated in this source, July 1812.

214 Charles Rogers, *Leaves from my Autobiography* (London, 1876), 23.

215 JP to Blackwood, 19 & 24 July 1813 (Nat. Lib. Scot. MS 4001, ff. 198, 200).

216 JP, *Enquiry* (1814), Vol. 1, Advertisement, [iii]–vi.

217 Celadonite, a mineral.

218 A rock: pudding-stone.

219 Gillies, *Memoirs*, Vol. 2, 123, 124, 126–7.

220 Robertson to JP, 22 Dec. 1813 (JP, *Literary Correspondence*, Vol. 2, 420–2).

221 James Grant, *Cassell's Old and New Edinburgh,* 3 vols. (London, 1884–7), Vol. 3, 58.

222 See George Dempster to JP, 30 Nov. 1813 (JP, *Literary Correspondence,* Vol. 2, 415–16; Lord Frederick Campbell to Dempster, 15 Feb. 1814 (ibid., 424–5); Dempster to JP, 18 Feb. 1814 (ibid., 424 n.).

223 *Edinburgh Review*, Vol. 23 (Apr. 1814), 63–79.

224 JP to William Roscoe, Aug. 1, 1814 (MS Liverpool Public Libraries).

225 For the significance of Roscoe, see the essay on him in Washington Irving's *Sketch Book*; also M.S. Phillips, *Society and Sentiment: Genres of Historical Writing in Britain, 1740–1820* (Princeton, 2000), 272–3.

226 Grenville to JP, 3 Aug., 11 Oct. 1814 (JP, *Literary Correspondence*, Vol. 2, 436–40, 454–5).

227 Thomas Grenville to Lord Grenville, 11 Oct. 1814 (*The Manuscripts of J.B. Fortesque, Esq., preserved at Dromore,* 10 vols. [London, 1892–1927], Vol. 10, 391).

228 Thomas Longman to JP, 24 Oct. 1814 (JP, *Literary Correspondence*, Vol. 2, 455–6).

229 Ibid., 456–7.

230 JP to Roscoe, 20 Nov. 1814 (MS Liverpool Public Libraries).

231 John Smith to JP, 29 Nov. 1814. "Your child is at No. 4 Lion Street Newington, Surrey about a mile and a half from the end of London

Bridge. He is under the care of Mrs. Parsons the same woman at whose house you first left him at Bexley, but who has since taken a house in the situation I describe. I believe the child is healthy, though he was formerly much otherwise. Should you determine to take him to France there is no necessity for you to trouble yourself to settle with Mrs. Parsons. I will willingly discharge the last quarters acct. I shall ever consider it as unfortunate that you took offence at my not getting the boy into the Foundling Hospital. On my part none was intended, and with the high opinion I have ever entertained of your talents and character I shall reflect with pleasure that it has been my lot to have been in any degree useful to you. My sincere wishes for a cessation of those distresses of which you complain" (as in n. 189).

232 JP in a letter to Douce in Nov. 1816 said "a ridiculous chancery suit brought by my blessed wife" had forced him to leave England (Bodl. Lib. MS Douce d. 37, f. 13), and mentions the suit again in a letter to Douce on 14 Feb. 1817 (ibid., f. 15). I have failed to find evidence of such a suit at this stage of his life.

7. A Banished Man, 1815–1826

1 A phrase from Douce's acct. of JP in 1820; see n. 31 below.

2 He'd received £200 from Robert Ferguson; Ferguson to JP, 8 Dec. 1814 (LWL MSS Vol. 82.4). Ferguson said he could pay the money back only when he found it convenient to do so, i.e., it was a handout. He had given JP £100 in 1805.

3 Constable asked JP on 12 Jan. 1813 for his help in completing the *Supplement to the fourth, fifth, and sixth editions of the Encyclopædia Britannica*, of which Napier was the editor (LWL MSS Vol. 82.4). Napier on 30 Mar. 1814 asked JP to write nine entries on geographical topics, taken from the list of JP's suggestions, at 15 gns. per printed sheet (ibid.). JP wrote none of them, and no other entry. He later charged that the proprietors of the *Encyclopædia* stole "nearly the whole" of his *Essay on Medals* (*Modern Geog.* [1817], Vol. 1, 761 [781]). JP's relations with Longman's had not deteriorated as had those with Cadell & Davies; Thomas Rees, employed variously by the firm, was a friend.

4 Undated letter, written, however, in 1815 (MS Maryland Historical Society).

5 *London Magazine*, Vol. 2 (4 Jan. 1823), 12. Identified by JP as his in a letter to Douce, 23 Oct. 1823 (Bodl. Lib. MS Douce d. 37, f. 53).

6 JP to Warden, 14 May 1815 (MS Maryland Historical Society).

7 For Longman & Co.'s trade in old books, see Rees and Britton, *Reminiscences*, 47–8.

8 Constable, *Archibald Constable*, Vol. 2, 115. JP told Douce, 14 Feb. 1817, he'd lost money by their bankruptcy (Bodl. Lib. MS Douce, d. 37, f. 15).

9 Ibid., f. 13.

10 JP to Douce, 14 Feb. 1817 (ibid., ff. 14–15).

11 Ibid., 9 July 1817, ibid., f. 18.

12 The first in *Scots Magazine,* 2nd ser., Vol. 1 (Nov. 1817), 310–12. For a list, see the Bibliography, II (6). They are signed "J.P." and are identified as JP's in letters to Napier.

13 JP, *Modern Geography* (1817), Vol. 1, 762 [782]. A note indicating payment of £100 is in Longman & Co., Cadell Papers, 11 (q). JP's "Supplement" to Vol. 1 occupies pp. 759–67 [779–87]; an "Advertisement to the third edition," i–vii, is dated Sept., 1811 – presumably one borrowed by the publishers from advertising for the third edit. of the abridgment (see Bibliography, I [25i]), though the running title calls it Advertisement to "this edition."

14 JP to Napier, 28 Jan. 1818 (B.L. Add. MS 34612, f. 173).

15 Normally ether; JP used the spelling derived from Aristotle (ἀιθήρ); for comment on the history of this non-substance, see C.S. Lewis, *The Discarded Image* (Cambridge, 1964), 3–5, 32–3.

16 *Religio Universalis et Naturalis*, 29–30; æther, 12, 14; Ladda M – Translator for Rossion Inc.

17 Ibid., 26–7.

18 Ibid., 47.

19 Ibid., 17.

20 Dr Robert Willan.

21 JP to Douce, 17 Aug. 1818 (Bodl. Lib. MS Douce d. 37, f. 24).

22 JP to Warden, 25 May 1806 (MS Maryland Historical Society).

23 *Recollections of Paris*, Vol. 2, p. 86; his list of "finest" French wines, 204–5.

24 JP to Napier, 31 Oct. 1818 (B.L. Add. MS 34612, f. 34).

25 JP to Douce, 30 Mar. 1819 (Bodl. Lib. MS Douce d. 37, f. 28).

26 JP to Douce, 23 Oct. 1819 (ibid., f. 30).

27 JP to Douce, 11 Dec. 1820 (ibid., f. 34).

28 JP to Joseph Van Praet, 12 [Feb.?] 1823 (Bib. Nat. French MS 12773, f. 128).

29 Douce to JP, 10 Apr. 1819 (Bodl. Lib. MS Douce d. 2, f. 11).

30 Ibid., f. 13.

31 Douce to JP, June 1820 (ibid., f. 20). In a letter to Thorkelín, 30 Apr. 1819, Douce, responding to T.'s inquiry, says he "was at Paris the year before last [i.e., 1817] where I saw your old acquaintance Mr. Pinkerton who has resided there for some years past, & I suppose may be considered as a banished man … He is living in mean

lodgings, and exists partly by gleaning scarce books on the Parisian stalls, and supplying some of the London dealers in that article" (Edin. Univ. Lib. MS La. III. 379, No. 269, f. 482). The letter is referred to in Wood's Introduction to *Correspondence of Thomas Percy & John Pinkerton*, xxi. I doubt this visit took place. Douce gives no other particulars about JP's location, not even his address, says nothing that he couldn't know or guess from JP's letters, and passes over his alleged visit quickly; he was perhaps trying to discourage any attempt by T. to correspond with or visit JP.

32 On 18 Dec. 1821; sale catalogue, B.L. S. C.S. 127 (7).
33 JP to Douce, 6 Mar. 1822 (Bodl. Lib. MS Douce d. 37, f. 42).
34 JP to Douce, 27 Jan. 1823 (ibid., f. 49).
35 Douce to JP, 16 Feb. 1823 (ibid., MS Douce d. 2, f. 40).
36 Royal Literary Fund archives, London, case No. 486. The grant was made 12 Mar. 1823. One sentence in the report reads: "Mr. Pinkerton is further stated, on the authority of Francis Douce Esq., to be now living in Extreme poverty in Paris."
37 JP to Douce, 8 Apr. 1823 (Bodl. Lib. MS Douce d. 37, ff. 50–1).
38 Douce to JP, 30 July 1823 (ibid., MS Douce d. 2, ff. 44–5; answering JP's note, 24 July, MS Douce d. 37, f. 52).
39 Ibid., MS Douce d. 37, f. 56. This is a copy of the letter, sent to Douce for his approval.
40 JP to Douce, 19 June 1824, ibid., f. 58.
41 Ibid., f. 59. The July 5 letter is a rough draft.
42 Somerset House, London. Wills for 1826, "Swabey," f. 284. See the Bibliography, VIII (15).
43 JP to Scott, 25 June 1825 (Nat. Lib. Scot. MS 3900, f. 284).
44 That I have seen.
45 Bodl. Lib. MS Douce d. 37, f. 60.
46 Acte de Décès, 168/211 16, Archives de la Seine et de la Ville de Paris, Paris.
47 A file of marked graves is in the Archives de la Seine et de la Ville de Paris. Pinkerton's name is not mentioned in it.
48 "Mrs. P. died about three years ago" (Douce to JP, 16 Feb. 1823, Bodl. Lib. MS Douce d. 2, f. 40).
49 Harford, *Life of Thomas Burgess*, p. 480.
50 M.M. Pinkerton to Douce, 17 Mar. 1826 (Bodl. Lib. MS Douce d. 37, ff. 61–2).
51 Ibid., f. 62. An undated rough draft.
52 Harford, *Life of Thomas Burgess*, p. 480; *Gentleman's Magazine*, Vol. 101, pt. ii (Aug. 1831), 187.

53 *Gentleman's Magazine*, Vol. 103, pt. i (May 1833), 462; Harford, *Life of Thomas Burgess*, 480.

54 Mrs. Riddell gave the Pinkerton Papers to the Nat. Lib. Scot. in 1934.

55 Information from memo by Turner, 4 July 1835 (LWL MSS Vol. 82.1, prefixed).

56 Thomas Curley, *Samuel Johnson, the Ossian Fraud, and the Celtic Revival in Great Britain and Ireland*, 122. Curley dismisses JP (8) as one of the "outright forgers," unlike (presumably) the fabricators of Ossian and parliamentary debates, though on a glance these seem just as much "outright" hoaxes as JP's. Nick Groom, a major apologist for forgers, says of JP's tampering with *Hardyknute* that he "completed [it] with more mischief than sense" (*The Making of Percy's Reliques* [Oxford, 1999], 63 n.), without explaining why JP's meddling with his source was mischievous while Macpherson's with his wasn't.

57 Groom, *The Forger's Shadow*, 101.

58 Iolo Morganwg's term for him in 1794; see chapter 2.

59 *The Penny Cyclopædia* (London, 1840), Vol. 18, 167.

60 Trevor-Roper, *The Invention of Scotland*, 145.

61 See Trevor-Roper's treatment of JP, ibid., 145–50, and the response of William Ferguson, *The Identity of the Scottish Nation*, 250–7; also David McCrone, "Tomorrow's Ancestors: Nationalism, Identity and History," in Cowan and Finlay, eds., *Scottish History*, 253–71.

62 Colley, *Britons: Forging the Nation*, 125.

63 JP, *An Essay on Medals* (1784), 173.

64 JP to Lord Buchan, 30 July 1794 (JP, *Literary Correspondence*, Vol. 1, 354).

65 *Recollections of Paris*, Vol. 2, 20.

66 When writing of the perils awaiting "An Englishman at Paris," he says the French "pretend that *our* robbery of the Hindoos authorizes them to rob *us*" (ibid., Vol. 1, 127); or, when writing of English vs. French art, "*We* lions are not painters" (174, my italics); or, when he notes the unexpected hospitality of a French woman to him and his companions – "to mere strangers, and those strangers English" (261); or, in commenting on being a dinner guest, "an Englishman is … surprised when he finds not a bit of carpet in his bedroom" (Vol. 2, 97); or, in discussing "the French custom of sticking [sirloin] with little morsels of lard, [which] is to an English palate truly nauseous" (196); "an English patriot" (257).

67 In 1794 it was suggested he might be a candidate for Keeper (librarian) of the Advocates' Library, an office Hume had once occupied. He was flattered but demurred, and said if the vacancy reoccurred he might

consider it, provided year-round residence were not required. See JP, *Literary Correspondence*, Vol. 1, 348.

68 JP, *Enquiry* (1814), Vol. 1, Advertisement, vi.
69 JP, *Enquiry* (1794), Vol. 1, Preface, x.
70 JP, *The Treasury of Wit*, Vol. 1, 43.
71 JP to Lord Buchan, 10 Nov. 1789 (JP, *Literary Correspondence*, Vol. 1, 237).
72 JP to Buchan, 19 July 1790 (ibid., 256).
73 JP, *Modern Geography* (1802), Vol. 1, 145–209.

Bibliography

I. Pinkerton's books, with reviews

No texts beyond 1900 are included. The lists of reviews and of editions and derivative texts outside Britain may be incomplete. I may not have located all such editions and texts within Britain for items 25 and 28. Some of JP's works were published initially in numbers and parts, e.g., those of maps and voyages for his *Atlas* and travel collection; such numbers and parts, variously titled, sometimes appear in library catalogues, as do, occasionally, prospectuses, which for some titles were issued separately, and short extracts from his writing. No numbers, parts, prospectuses, or short extracts are listed here or elsewhere in this bibliography. Most titles in his books appeared in large or small block capitals; I have not duplicated that feature, and have capitalized words in such titles where needed or where I felt was appropriate. I have italicized titles, which again differs from the normal practice in JP's books. I have shortened many titles of 2nd, 3rd, and other editions, and merely noted those of some derivative texts. I have converted roman to arabic numerals in dates of publication. JP on occasion adorned title-pages with quotes, normally from classical sources. I have not included any of those. Many of the texts listed here and elsewhere in the bibliography are available online.

1. *Craigmillar Castle. An Elegy.* Edinburgh: 1776.
   ["Printed for the Author, and sold by all the Booksellers."]
2. *Rimes.* London: Charles Dilly, 1781.
   Critical Rev., 51 (Mar. 1781), 216–19; *Monthly Rev.*, 65 (July 1781), 13–17.
2a. [2nd ed.] *Rimes by Mr. Pinkerton.* London: Charles Dilly, 1782.
   European Mag., 2 (July 1782), 45–50; *Gentleman's Mag.*, 52 (Mar. 1782), 131.
   [Advertisement in 2 repeated, except for closing para., and extended.]

2b. [Reissue.] *Rimes by Mr. Pinkerton*. London: Charles Dilly, 1786.
   [A reissue of unsold copies of the 2nd edit. Extension of Advertisement
   in 2a. cancelled; list of "Corrections, 1786" on A4v.]
 3. *Scottish Tragic Ballads*. London: J. Nichols, 1781.
   Critical Rev., 52 (Sept. 1781), 205–8; *Gentleman's Mag.*, 51 (June 1781),
   279–80; *Monthly Rev.*, 66 (Apr. 1782), 292–4.
   [A page follows the title-page, extending the title, as follows: *Hardyknute, an
   heroic ballad, now first published complete; with the other more approved Scottish
   ballads, and some not hitherto made public, in the tragic stile. To which are prefixed
   two dissertations, I. On the oral tradition of poetry. II. On the tragic ballad.* JP
   identified as the "ingenious editor" in *Gentleman's Mag.*, 52 (Mar. 1782), 131 n.]
3a. [2nd ed.] *Select Scotish Ballads*. 2 vols. London: J. Nichols, 1783.
   Critical Rev., 56 (Aug. 1783), 129–33; *English Rev.*, 2 (July 1783), 68–9;
   Gentleman's Mag., 53, pt. ii (Aug. 1783), 690–1; *Monthly Rev.*, 71 (Sept.
   1784), 226–7; *New Rev.*, 3 (May 1783), 421.
   [The second title-page (as in 3) reappears in Vol. 1; JP identifies himself,
   Vol. 1, x.]
 4. *Two Dithyrambic Odes. I. On Enthusiasm. II. To Laughter. By the author of
   Rimes*. London: C. Dilly, 1782.
   Gentleman's Mag., 52 (Mar. 1782), 131; *Monthly Rev.*, 68 (Apr. 1783), 355.
 5. *Tales in Verse. By Mr. Pinkerton*. London: James Dodsley, 1782.
   Critical Rev., 53 (May 1782), 343–7; *Gentleman's Mag.*, 52 (May 1782), 243;
   Monthly Rev., 67 (Aug. 1782), 109–12.
 6. *An Essay on Medals*. London: James Dodsley, 1784.
   Critical Rev., 58 (Sept. 1784), 199–205; *English Rev.*, 5 (May 1785), 325–9;
   Gentleman's Mag., 54, pt. ii (July 1784), 521–3; *Monthly Rev.*, 71 (Sept.
   1784), 201–4; *Scots Mag.*, 46 (Sept. 1784), 477–9.
6a. [2nd ed.] *An Essay on Medals: or, An Introduction to the Knowledge of
   Ancient and Modern Coins and Medals; especially those of Greece, Rome, and
   Britain. By John Pinkerton. A new edition, corrected, greatly enlarged, and
   illustrated with plates.* 2 vols. London: J. Edwards and J. Johnson, 1789.
   Analytical Rev., 4 (Aug. 1789), 401–4; *Gentleman's Mag.*, 59, pt. ii (Sept.
   1789), 837; *Monthly Rev.*, 81 (Aug. 1789), 139–40.
6b. [3rd ed.] [Title as in 6a] *The third edition, with corrections and additions.*
   2 vols. London: Cadell and Davies; and Longman, Hurst, Rees, and
   Orme, 1808.
   Quarterly Rev., 1 (1809), 112–31.
6c. [Fr. trans.] *Dissertations sur la Rareté, les différentes Grandeurs et la
   Contrefaction des Médailles Antiques, avec des Tables du Dégré de Rareté des
   Médailles des Anciens Peuples, Villes, Rois et des Empereurs Romains. Le tout*

traduit de l'Anglois et augmenté des indices nécessaires par Jean Godefroi Lipsius. Avec une géographie numismatique des anciens peuples, villes et rois. Dresden: les frères Walther, 1795.

6d. [Ger. trans.] *Johan Pinkerton's Abhandlung von der Seltenheit, den verschiedenen Grössen, und der Nachahmung alter Münzen. Ebendesselben tabellarische Üebersicht von dem Grade der Seltenheit der Münzen alter Völker, Städte, Könige, und Rom Kaiser.* Trans. Johann G. Lipsius. Dresden: In der Waltherischen Hofbuchhandlung, 1795.

7. *Letters of Literature. By Robert Heron, Esq.* London: G.G. and J. Robinson, 1785.
   Critical Rev., 60 (Dec. 1785), 405–13, 61 (Jan. 1786), 18–26; *Edinburgh Mag.*, 2 (Sept. 1785), 160–2; *English Rev.*, 7 (Jan. 1786), 33–7; *European Mag.*, 8 (Aug. 1785), 106–10, 8 (Sept. 1785), 195–200, 8 (Oct. 1785), 290–3, 8 (Nov. 1785), 376–9; *Gentleman's Mag.*, 55, pt. ii (July 1785), 544–6, 55, pt. ii (Sept. 1785), 717–20; *Monthly Rev.*, 74 (Mar. 1786), 175–82; *New Rev.*, 8 (Jul, 1785), 34–50. [The only book by JP with this nom de plume.]

8. *Ancient Scotish Poems, never before in print. But now published from the MS. collections of Sir Richard Maitland, of Lethington, Knight, Lord Privy Seal of Scotland, and a Senator of the College of Justice. Comprising pieces written from about 1420 till 1586. With large notes, and a glossary. Prefixed are An essay on the origin of Scotish poetry. A list of all the Scotish poets, with brief remarks. And an appendix is added, containing, among other articles, an account of the contents of the Maitland and Bannatyne MSS.* 2 vols. London: Charles Dilly, and for William Creech, Edinburgh, [1785].
   Critical Rev., 61 (Mar. 1786), 169–76, 61 (Apr. 1786), 268–73; *Gentleman's Mag.*, 56, pt. i (Feb. 1786), 147–50; *Monthly Rev.*, 76 (Feb. 1787), 121–9; *New Rev.*, 9 (Jan. 1786), 52–7; *Scots Mag.*, 49 (Feb. 1787), 76–9. [Date given on title-page as MDCCLXXXVI; JP's name at end of Preface, Vol. 1, xix.]

9. *The Treasury of Wit; being a methodical selection of about twelve hundred, the best, apophthegms and jests; from books in several languages. In two volumes. Volume I. Containing Greek. Roman. Eastern. Spanish. Italian. German. Newly translated. [Volume II. Containing French and English: many of the latter before unpublished. With a discourse on wit and humour.] By H. Bennet, M.A.* London: Charles Dilly and Thomas Evans, 1786.
   Monthly Rev., 76 (May 1787), 445–6. [On the authorship of this work see ch. 2, note 179, ch. 4, note 28.]

10. *A Dissertation on the Origin and Progress of the Scythians or Goths. Being an Introduction to the Ancient and Modern History of Europe. By John Pinkerton.* London: George Nicol, 1787.

Critical Rev., 64 (Sept. 1787), 167–75; *English Rev.*, 10 (Aug. 1787), 131–7; *European Mag.*, 12 (Nov. 1787), 382–4; *Monthly Rev.*, 77 (Oct. 1787), 318–19. [Reissued as an appendix to 11, 11a, and 11b below.]

10a. [Fr. trans.] *Recherches sur l'Origine et les divers Établissemens des Scythes ou Goths, Servant d'Introduction à l'Histoire ancienne et moderne de l'Europe; accompagnées De plusiers Eclairessemens sur la Géographie ancienne de cette partie du monde; ouvrage traduit sur l'original Anglois De J. Pinkerton. Avec des augmentations et des corrections faites par l'Auteur, et une carte du monde connu des anciens.* Paris: Imprimerie de la République, 1804.
Journal des Débats (24 May 1804).

11. *An Enquiry into the History of Scotland preceding the Reign of Malcom III. or the year 1056. Including the authentic history of that period. In two volumes. By John Pinkerton.* London: George Nicol, and for John Bell in Edinburgh, 1789.
Analytical Rev., 5 (Sept. 1789), 1–13; *Critical Rev.*, 69 (Apr. 1790), 361–71, 70 (July 1790), 11–22, 70 (Oct, 1790), 364–74; *Monthly Rev.*, 2nd ser., 1 (Apr. 1790), 387–404.

11a. [Reissue] [Title as in 11] 2 vols. London: B. & J. White and I. Herbert, 1794. [With an Advertisement and certain new sheets.]

11b. [2nd ed. of 11] *An Enquiry into the History of Scotland preceding the Reign of Malcolm III …* [as in 11] *A new edition, with corrections and additions.* 2 vols. Edinburgh: Bell & Bradfute, William Laing, Doig & Stirling, William Blackwood, and Oliphant, Waugh, & Innes, 1814.
[With a new "Advertisement to this edition."]

12. *Vitae Antiquae Sanctorum qui habitaverunt in ea parte Britanniae nunc vocata Scotia vel in ejus insulis. Quasdam editit es MSS. quasdam collegit Johannes Pinkerton qui et variantes lectiones et notas pauchulas adjecit.* London: John Nichols, 1789.
Gentleman's Mag., 59, pt. ii (July 1789), 635–6.

12a. [New ed.] *Pinkerton's Lives of the Scottish Saints.* Ed. W.M. Metcalfe. 2 vols. Paisley: Alexander Gardner, 1889.
["Revised and enlarged" by Metcalfe; English intro. by him; Latin text.]

12b. [Trans.] *Ancient Lives of Scottish Saints.* Trans. W.M. Metcalfe. Paisley, Alexander Gardner, 1895.
[Trans. of "the principal Lives" in 12a.]

13. *The Bruce; or, The History of Robert I. King of Scotland. Written in Scotish verse By John Barbour. The first genuine edition, published from a MS. dated 1489; with notes and a glossary By J. Pinkerton.* 3 vols. London; G. Nicol, 1790.
Analytical Rev., 9 (Appendix, 1791), 512–15; *Critical Rev.*, 70 (Aug. 1790), 164–70; *English Rev.*, 17 (Apr. 1791), 254–7; *Monthly Rev.*, 2nd ser., 15 (Dec. 1794), 431–7.

14. *The Medallic History of England to the Revolution. With forty plates.* London: Edwards & Sons, and Faulder, 1790.
    Analytical Rev., 7 (June 1790), 131–4; *Critical Rev.,* 2nd ser., 1 (Feb. 1791), 131–8; *Monthly Rev.,* 2nd ser., 3 (Nov. 1790), 288–90.

14a. [2nd ed.] *The Medallic History of England. Illustrated by forty plates.* London: E. Harding and J. Scott, 1802.

15. *A New Tale of a Tub, written for the delight and instruction of every British subject in particular, and all the world in general.* London: J. Ridgway, 1790.
    Analytical Rev., 9 (Mar. 1791), 339–40; *Critical Rev.,* 2nd ser., 1 (Apr. 1791), 464; *Gentleman's Mag.,* 61, pt. i (Apr. 1791), 305–6.
    [On the title-page of the B.L. copy is this annotation: "by John Pinkerton, Esq." It is not in JP's handwriting, but the work is nevertheless assigned, with a query, to JP in the catalogue of printed books. On a front fly-leaf of the Bodl. Lib. copy is this note: "F. Douce from the author." It is in pencil and blurred, but I am confident it is in JP's handwriting. Another proof is the close resemblance between the ideas in this book and JP's ideas elsewhere, including those in the *Critical Review*'s "Review of Public Affairs," which he was writing, in part or whole, soon after the publication of *A New Tale of a Tub.* We find in *A New Tale* sympathy towards polytheism, ridicule of Celts, abuse of George Steevens, and various other marks of his pen. The publisher James Ridgway was a friend of JP's in 1790. JP arranged with him to publish Thorkelín's *Sketch of the Character of the Prince of Denmark* in 1791 and remained on good terms with him for two or three years afterwards.]

16. *The Spirit of all Religions.* ?Amsterdam: 1790.
    Analytical Rev., 9 (Mar. 1791), 339–40; *Critical Rev.,* 2nd ser., 1 (Apr. 1791), 464; *Gentleman's Mag.,* 61, pt i (Apr. 1791), 361.
    [The faded word "Pinkerton" is written in pencil on the title-page of the Bib. Nat. copy, and on the strength of that annotation the work is attributed to JP in the catalogue of printed books. But JP did not make the annotation. Collateral testimony is supplied by one of his statements in a letter to his children on 13 Jan. 1809: "I want some Latin books and believe she [Henrietta Pinkerton] may send all in that language as they can be of no use to her or you. She will also find among the pamphlets a very small one called 'A new system of Religion' or 'The Spirit of all Religions' which I want" (Nat. Lib. Scot. MS 1709, f. 120). JP doesn't say he wrote the obscure pamphlet, but his words show he was not only familiar with its existence but knew it had been published under two titles. When the work was published in 1818 in an expanded Latin edition in Paris, where he was living, he wrote to Macvey Napier on 28 Jan. of that year and praised it extravagantly: "The work is full of rational piety and universal

benevolence and one professed intention is to recommend devotion even to proud philosophers who despise the vulgar superstitions by presenting an argumentative system worthy of the sublime magnificence of the material universe as disclosed by modern discoveries. It is also remarked that the more sublime any system of religion is, it is the more likely to be true; and certainly this is the most sublime system that has ever been presented to the world and forms an epoch in the progress of the human mind" (B.L. Add. MS 34612, f. 173). JP was personally acquainted with Antoine Renouard, who published the new edition. In the letter to Napier he is clearly describing his own book. We can add that in *Modern Geography* (1802) he sometimes expresses knowledge of or sympathy for polytheism (the system espoused in the pamphlet). He notes, e.g., that the Chinese "are so far from being atheists that they are in the opposite extreme of polytheism, believing even in petty demons who delight in minute acts of evil, or good" (Vol. 2, 90). Such demons figure in the system advanced in his pamphlet. At another point he says: "… as the Schamanians admit one chief infernal deity and his subalterns, authors of evil, so they believe in one supreme, uncreated beneficent being, who commits the management of the universe to inferior deities, who delegate portions of it to subaltern spirits. With more philosophy they might suppose that evil cannot exist except in matter and that an evil spirit is a contradiction in terms" (Vol. 2, 48). The last phrase is quite close to phrasing in the pamphlet – "An evil deity is a contradiction in terms" (34, Eng. edit.) – and the entire quote is consonant with the system the pamphlet advances. See also his comments on the native religion of the Society Isles, *Mod. Geog.* (1802), Vol. 2, 507. One of the letters in JP's proposed second series of his *Letters of Literature* was to be called "New Theory Origin of Evil (Polytheism)" (Nat. Lib. Scot. MS 1719, f. 19).]

16a. [Reissue.] *A New System of Religion.* ?Amsterdam: 1790.
    Critical Rev., 70 (Aug. 1790), 210; *English Rev.*, 18 (Nov. 1791), 389–90.

16b. [New ed.] *Religio Universalis et Naturalis; Disquisitio Philosophica, et exemplis omnium nationum et rationibus variis consolidata.* Paris: Antoine Renouard, 1818.

17. *Scotish Poems, reprinted from scarce editions. The Tales of the Priests of Peblis. The Palice of Honour. Squire Meldrum. Eight Interludes, by David Lindsay. Philotus, a Comedy. Gawan and Gologras, a Metrical Romance. Ballads, first printed at Edinburgh, 1508. With three pieces before unpublished. Collected by John Pinkerton, F.S.A. Perth, Honorary member of the Royal Society of Icelandic Literature at Copenhagen, and of the Royal Society of Sciences at Drontheim. In three volumes.* London: John Nichols, 1792.

Critical Rev., 2nd ser., 9 (Sept. 1793), 77–85; *Gentleman's Mag.*, 63, pt. i (May 1793), 446–7; *Monthly Rev.*, 2nd ser., (June 1793), 172–6.

18. *Iconographia Scotica or Portraits of Illustrious Persons of Scotland Engraved from the most Authentic Paintings &c with short Biographical Notices. By John Pinkerton, F.S.A. Perth.* London: I. Herbert, Barrett, 1794–7.
    Analytical Rev., 24 (Nov. 1796), 518–19, 25 (Mar. 1797), 296–7, 25 (May 1797), 487; *British Critic*, 11 (Apr. 1798), 433–4; *Gentlemen's Mag.*, 65, pt. ii (Supp. 1795), 1100–1, 66, pt. ii (Oct. 1796), 858–9.
    [Published in numbers; full text, 1797.]

19. *Guiccciardini's Account of the Ancient Flemish School of Painting. Translated from his description of the Netherlands, published in Italian at Antwerp, 1567. With a Preface, by the translator.* London; I. Herbert, 1795.
    Critical Rev., 2nd ser., 16 (Feb. 1796), 238–9; *Gentleman's Mag.*, 66, pt. i (Jan. 1796), 53.
    [Pamphlet, 30 pp. in thin main text, mainly a list of painters with brief biographies; 10-page Preface; skimpy piece of hackwork.]

20. *The History of Scotland from the Accession of the House of Stuart to that of Mary. With appendixes of original papers. By John Pinkerton. In two volumes.* London: C. Dilly, 1797.
    Anti-Jacobin Rev., 2 (Feb. 1799), 113–22, 2 (Mar. 1799), 262–7, 3 (June 1799), 113–20), 3 (July 1799), 246–59; *British Critic*, 11 (Apr. 1798), 345–58; *Critical Rev.*, 2nd ser., 20 (May 1797), 1–8, 20 (July 1797), 288–97; *Monthly Rev.*, 2nd ser., 23 (May 1797), 1–10.

21. *The Scotish Gallery; or, Portraits of Eminent Persons of Scotland: many of them after pictures by the celebrated Jameson, at Taymouth, and other places. With brief accounts of the characters represented, and an introduction on the rise and progress of painting in Scotland. By John Pinkerton.* London: E. Harding, 1797–9.
    Monthly Mag., 9 (Supp. 1800), 631; *Monthly Rev.*, 2nd ser., 31 (Jan. 1800), 19–24; *Scots Mag.*, 62 (Feb. 1800), 104–5.
    [Published in numbers; full text 1799.]

22. *Other Juvenile Poems, by the author of Rimes.* [London: printed by John Nichols for JP, 1798.]

23. *Walpoliana.* 2 vols. London: R. Phillips, [1799].
    British Critic, 15 (Feb. 1800), 208; *Critical Rev.*, 2nd ser., 28 (Feb. 1800), 212–19; *European Mag.*, 36 (Dec. 1799), 395–6; *Monthly Mirror*, 8 (Nov. 1799), 284; *Monthly Rev.*, 2nd ser., 32 (June 1800), 178–84.
    [JP's name not on title-page or elsewhere. For editions 1800–30, see Hazen, *A Bibliography of Horace Walpole*, 145–7.]

24. *Sketch of a New Arrangement of Mineralogy.* [Printed by Samuel Hamilton for JP.] 1800.

24a. [Trans.] *Esquisse d'une Nouvelle Classification de Minéralogie; suivie de quelques remarques sur la nomenclature des roches. Par M. Jean Pinkerton.* Trans. H.J. Jansen. Paris: H.J. Jansen, 1803.
Journal de Physique, 56 (June 1803), 476.
[…"quelques changemens et quelques additions" by JP.]

25. *Modern Geography. A Description of the Empires, Kingdoms, States, and Colonies; with the Oceans, Seas, and Isles; in all parts of the World: including the most recent discoveries, and political alterations. Digested on a new plan. By John Pinkerton. The astronomical introduction by the Rev. S. Vince, A.M. F.R.S. and Plumian Professor of Astronomy, and Experimental Philosophy, in the University of Cambridge. With numerous maps, drawn under the direction, and with the latest improvements, of Arrowsmith, and engraved by Lowry. To the whole are added a Catalogue of the best Maps, and Books of Travels and Voyages, in all Languages: And an ample Index.* 2 vols. London: Cadell and Davies; Longman, Rees, 1802.
Annual Rev., 1 (1802), 437–48; *Anti-Jacobin Rev.,* 17 (Apr. 1804), 377–86, 18 (May 1804), 13–20; 18 (July 1804), 266–77; *British Critic,* 21 (June 1803), 581–8, 22 (Aug. 1803), 148–55; *Critical Rev.,* 2nd ser., 36 (Sept. 1802), 1–10, 37 (Mar. 1803), 249–65, 39 (Sept. 1803), 58–69; *Edinburgh Rev.,* 3 (Oct. 1803), 67–81; *Monthly Mag.,* 14 (Supp. 1803), 586.

25a. [2nd ed.] … *With numerous maps, revised by the author. To the whole are added* [as in 25] *A new edition, greatly enlarged.* 3 vols. London: Cadell and Davies; Longman, Hurst, Rees, and Orme, 1807.
Edinburgh Rev., 10 (Apr. 1807), 154–71; *Monthly Rev.,* 2nd ser., 55 (Mar. 1808), 287–300.

25b. [3rd ed.] … *The astronomical introduction by M. La Croix, Member of the Institute of France; translated by John Pond Esq. Astronomer-Royal. With numerous maps, revised by the author, and engraved by Mr. Lowry.* [as in 25] *A new edition, with additions and corrections to the year 1817.* 2 vols. London: Cadell and Davies; Longman, Hurst, Rees, Orme, and Brown, 1817.

25c. [Amer. ed.] … *With numerous maps … engraved by the first American artists.* 2 vols. [Vol. 2 adds, following Vince's credentials, *The article America corrected and considerably enlarged, by Dr. Barton, of Philadelphia.*] Philadelphia: John Conrad & Co., and M. & J. Conrad, Baltimore, et al., 1804.

25d. [Fr. trans.] *Géographie Moderne, rédigée sur un nouveau plan, Ou description historique, politique, civile et naturelle des Empires, Royaumes, Etats et leurs Colonies; avec celle des Mers et des îles de toutes les parties du monde … Par J. Pinkerton. Traduite de l'anglais, avec des notes et augmentations considérables, Par C.A. Walckenaer. Précédée d'une Introduction a la Géographie mathématique et critique, Par S.F. Lacroix … Accompagnée d'un Atlas … de 42 Cartes, dressées par Arrowsmith … Revues et corrigées par J.N. Buache … Avec un Catalogue des meilleures Cartes et Livres de voyages, et un Index très-ample.* 6 vols. Paris: Dentu, 1804.

Magasin Encyclopédique, 5 (1804), 416–18.

[This text is extensively annotated and altered. The promised 42 maps form a separate seventh vol., though not so numbered. That vol. retains the title-page as above, with the word *Atlas* highlighted and on its own line after *très-ample.* It properly belongs in Aaron Arrowsmith's bibliography rather than JP's; but see JP's comments on Arrowsmith's maps in ch. 6 above.]

25e. [?Another Fr. trans.] Paris, 1811.

[A "very strange and unwarrantable French translation of one or two volumes, (the rest having happily never appeared,) … published at Paris in 1811," JP, *Modern Geography* (1817), Vol. 1, 760 (780). Not seen by PO'F.]

25f. [?It. trans.] Trans. Louis M. Galanti. 8 vols. Rome: 1805–10.

[Referred to by JP in *Modern Geography* (1817), Vol. 1, 765 (785). As in 25e.]

25g. [Eng. abr.] … *Carefully abridged from the larger work, in two volumes, quarto.* London: Cadell and Davies; Longman, Rees, 1803.

Annual Rev., 2 (1803), 371.

[Adapted somewhat by A. Aikin; see ch. 6 above, esp. n. 23.]

25h. [Eng. abr. 2nd ed.] … *The second edition, revised by the author.* London: Cadell and Davies; Longman, Hurst, Rees, and Orme. 1806.

25i. [Eng. abr. 3rd ed.] … *The third edition, revised and enlarged by the author.* London: Cadell and Davies; Longman, Hurst, Rees, Orme, and Brown, 1811.

[Sometimes mistaken, as in Baker, *Hist. of Geog.,* 149, for the 3rd ed. of 25.]

25j. [Another Eng. abr.] *An Introduction to Mr. Pinkerton's Abridgment of his Modern Geography, for the use of schools.* By John Williams. London: Longman, Hurst, Rees, and Orme, 1808.

Annual Rev., 7 (1809), 61.

25k. [An ed. of 25j.] *An Introduction to Mr. Pinkerton's Abridgement …* London: Longman, Hurst, Rees, Orme, and Brown; Cadell and Davies, 1820.

25l. [Amer. abr.] *Pinkerton's Geography, epitomised, for the use of schools.* By David Doyle. Philadelphia: Samuel F. Bradford, 1805.

25m. [Fr. abr.] *Abrégé de la Géographie Moderne, rédigée sur un nouveau plan.* Abr. Jean N. Buache. Paris: Dentu, 1805.

Bibliothèque Française (1 July 1805), 34–41, (15 July 1805), 27–34; *Journal de l'Empire* (10 June 1806).

25n. [2nd ed. of 25m.] Paris: Dentu, 1806.

25o. [Another Fr. abr.] Abr. C.A. Walckenaer. 2 vols. Paris: Dentu, 1811.

25p. [Another Fr. abr.] Abr. C.A. Walckenaer and Jean B. Eyriès. 2 vols. Paris: Dentu, 1827.

26. *Recollections of Paris, in the years 1802–3–4–5. By J. Pinkerton. In two volumes.* London: Longman, Hurst, Rees, and Orme; Cadell and Davies, 1806. *Annual Rev.,* 5 (1806), 30–41; *Critical Rev.,* 3rd ser., 9 (Nov. 1806), 286–95; *Edinburgh Rev.,* 8 (July 1806), 413–21; *Monthly Rev.,* 2nd ser., 52 (Feb. 1807), 173–82; *Scots Mag.,* 68 (Sept. 1806), 685–90; *Universal Mag.,* 2nd ser., 8 (Aug. 1807), 135–8, 8 (Sept. 1807), 237–9.

26a. [Ger. trans.] *Ansichten der Hauptstadt des Französischen Kayserreichs vom Jahre 1806 an. Von Pinkerton, Mercier und C.F. Kramer. Erster Band. Mit Kupfern.* 2 vols. Amsterdam: Kunst-und Industrie-Comptoir, 1807. [Much of JP's work is in this comp.]

27. *A General Collection of the Best and Most Interesting Voyages and Travels in all Parts of the World; many of which are now first translated into English. Digested on a new plan. By John Pinkerton, author of Modern Geography, &c. &c. Illustrated with plates.* 17 vols. London: Longman, Hurst, Rees, and Orme; Cadell and Davies, 1808–14.
European Mag., 67 (June 1815), 524–8.
[Published in numbers; and in groups of vols. in consecutive yrs.: Vols. 1–2 (1808), 3–6 (1809), 7–10 (1811), 11–13 (1812), 14 (1813), 15–17 (1814).]

27a. [Reissue.] *A General Collection of the Best and Most Interesting Voyages and Travels in various Parts of America* [as in 27]. 4 vols. London: [as in 27], 1819. [Vols. 11, 12, 14, 15 of 27.]

27b. [Amer. incomplete ed.] *A General Collection of the Best and Most Interesting Voyages and Travels, in all Parts of the World* [as in 27]. *Illustrated and adorned with numerous engravings.* 6 vols. Philadelphia: Kimber and Conrad, William Falconer, New York, et al., 1810–12.

27c. [Selection] *Early Australian Voyages. Pelsart. Tasman. Dampier. By John Pinkerton.* Ed. Henry Morley. London: Cassell & Co., 1886.

27d. [Selection] *Voyages and Travels of Marco Polo.* Ed. Henry Morley. New York: Cassell & Co., 1886.

28. *A Modern Atlas, from the latest and best authorities, exhibiting the various divisions of the world, with its chief empires, kingdoms, and states, in sixty maps, carefully reduced from the largest and most authentic sources. Directed and superintended by John Pinkerton, author of Modern Geography, &c.* London: Cadell and Davies; Longman, Hurst, Rees, Orme, and Brown, 1808–15. [Published in numbers; full text 1815.]

28a. [Amer. ed.] Philadelphia: T. Dobson and Son, 1818.

28b. [Abr.] *Pinkerton's School Atlas.* London: Cadell and Davies; Longman, Hurst, Rees, and Orme, 1809.

28c. [Ed. of 28b.] London: [as in 28], 1817.

28d. [Ed. of 28b.] London: [as in 28b], 1821.

28e. [Amer. abr.] *An Atlas, for the use of schools, Selected from Pinkerton's larger Work* ... Boston: Thomas & Andrews, 1816.

 29. *Petralogy. A Treatise on Rocks, by J. Pinkerton.* 2 vols. London: White, Cochrane, and Co., 1811.
Edinburgh Rev., 23 (Apr. 1814), 63–79; *Monthly Mag.,* 33 (Supp. 1812), 664–74; *Monthly Rev.,* 2nd ser., 72 (Sept. 1813), 1–18.

29a. [Another impression.] ... *by Pinkerton.* London: [as in 29], [1811].
[The publication date on the title-page is erroneously given as 1800. Citations (as in 29) are from works as late as 1811. Text identical with 29 except for some rearrangement of opening sections of Vol. 1 and 2 pp. of advertisements by the publisher at the end of Vol. 2.]

II.   Pinkerton's contributions to magazines

1. *Critical Review,* 2nd ser., 1 (Apr. 1791), 425–30 (Thorkelín, *Sketch*); 430–3 (Blind Harry, *Wallace*); 433–5 (Tertullian, *Address*); 2 (July 1791), 300–8 & (App., 1791), 550–6 (Lodge, *Illustrations of British History*); 4 (Jan. 1792), 55–8 (Ritson, *Pieces*); 5 (May 1792), 17–25 (Macleod, *Casus Principis*); (Aug. 1792), 394–401 & (App. 1792), 552–61 (Ledwich, *Antiquities of Ireland*); (Aug. 1792), 402–10 & (App. 1792), 561–70 (Soc. of Antiq. of Scotland, *Transactions*); 6 (Oct. 1792), 129–41 (Webb, *Analysis*); (Nov. 1792), 283–93 (Ritson, *Ancient Songs*); 9 (Oct. 1793), 169–70 (Llywarch Hen, *Heroic Elegies*); (Nov. 1793), 267–72 (Polwhele, *Historical Views*); 290–9 (Thomas Robertson, *Hist. of Mary*); 10 (Feb. 1794), 152–6 (David Robertson, *Tour through the Isle of Man*); 12 (Sept. 1794) 67–8 & 26 (Aug. 1799), 379–86 & 27 (Nov. 1799), 313–21 (Heron, *Hist. of Scotland*); 12 (Oct. 1794), 134–9 (Buchanan, *Defence of the Scots*); 13 (Jan. 1795), 49–58 (Ritson, *Scotish Song*); 26 (May 1799), 90–3 (Wyntoun, *Orygynale Cronykil*); (June 1799), 176–83 (Andrews, *Hist. of Great Britain*); (Appendix, 1799), 524–8 (Pougens, *Essai*); 26 (Aug. 1799), 361–79 & 27 (Nov. 1799), 286–98 (Browne, *Travels*); 27 (Oct. 1799), 121–8 & (Dec. 1799), 389–400 (Tooke, *View of the Russian Empire*); 28 (Jan. 1800), 12–24 & 33 (Oct. 1801), 121–34 (Turner, *Hist. of the Anglo-Saxons*); 28 (Jan. 1800), 48–57 (Alex. Campbell, *Intro. to the Hist. of Poetry in Scotland*); 29 (May, 1800), 27–41 & (July, 1800), 249–58 (Rennell, *Geographical System*); 30 (Dec., 1800), 361–77 (King, *Munimenta Antiqua*); 31 (Jan. 1801), 73–9 (Allan Ramsay, *Poems*); (Mar. 1801), 251–60 (Laing, *Hist. of Scotland*); 276–88 (Leyden, *Historical and Philosophical Sketch*); 34 (Jan. 1802), 58–68 (Britton, *Beauties of Wiltshire*); 35 (May 1802), 76–8 (Playfair, *Statistical Brieviary*); 95–9 (*Complaynt of Scotland*); (June 1802), 130–42 (Dugald Stewart, *Acct. ... of William Robertson*).
[See chs. 3 and 4 for discussion of JP's connection with the *Critical Review.* I list here reviews of books, and of those only ones (from a list of c. 130

that I think are his) that, from external and/or internal evidence, I am
confident were written by him. All are from the second series of the review.
JP wrote pieces other than reviews for the journal. External evidence, as
indicated in ch. 3, definitely shows him writing both the "Review of Public
Affairs" and "Retrospect of Foreign Literature" in 1791. But I hesitate to
aver that he wrote all such pieces, or all of any one of them, in the period
1791–5. In his second shift with the journal I suspect he wrote much if not
all of the "Review of Maps and Charts" that appeared in the appendices
of vols. 31–34 (1801–2); yet again I am reluctant to say these, or any one of
them, were solely his work. One non-review item that can be ascribed to
him with certainty is the chart named "Linnaean Table of the Nations and
Languages in Europe and Asia," 27 (Oct. 1799), 129–30, the MS of which, in
JP's hand, is Nat. Lib. Scot. MS 1711, f. 150.]

2. *Gentleman's Magazine*, 55 pt. i (Mar. 1785), 164–5 ("J. Black"); 55 pt. ii (Nov.
   1785), 887–8 ("Eusebes"); (Dec. 1785), 959 ("Philarchion"); 56 pt. i (Feb.
   1786), 95 (*"The Author of Letters of Literature"*); (May 1786), 390 ("Hint"); 56
   pt. ii (Sept. 1786), 773–4 (rev. of Thorkelín, *Diplomatarium Arna-Magnaeum*);
   (Nov. 1786), 942–4 & (Dec. 1786), 1021–3 & 57 pt. i (Feb. 1787), 120–1
   ("Vindex"); (Mar. 1787), 242–4 (summary and rev. of Perabo's *Valsei*); (May
   1787), 397–9 ("Vindex"); 58 pt. i (Feb. 1788), 125–7 & (Mar. 1788), 196–8 &
   (Apr., 1788), 284–6) & (May 1788), 404–5 & (June 1788), 499–501 & 58 pt. ii
   (July 1788), 591–2 & (Aug. 1788), 689 & (Sept. 1788), 777–8 & (Oct. 1788),
   877–8 & (Nov. 1788), 967–9 & (Dec. 1788), 1056–8 & (Supp. 1788), 1149–51
   ("Philistor"); 59 pt. ii (July 1789), 583 (signed by JP); (Sept. 1789), 801–2
   (signed by JP); (Oct. 1789), 909–10 ("Zenodotus"); (Nov. 1789), 979–82 (JP's
   open letter to *Analytical* reviewers); (Nov. 1789), 984 (signed by JP); 65 pt. i
   (Jan. 1795), 40 & 65 pt. ii (Nov. 1795), 902 ("Fabius Pictor").
3. *London Museum*, 2 (4 Jan. 1823), 11–12 ("Letter of a Veteran. Death of
   Josephine – Anecdotes of Buonaparte").
4. *Monthly Magazine*, 5 (Feb. 1798), 82 (signed by JP); (Mar. 1798), 197–9 &
   (Apr. 1798), 278–80 & (May 1798), 356–9 & (June 1798), 436–40 & 6 (July
   1798), 36–8 & (Aug. 1798), 115–18 & (Oct. 1798), 276–9 & (Nov. 1798), 356–8
   & (Dec. 1798), 442–4 & 7 (Jan.,1799), 39–41 & (Apr. 1799), 216–18 & (May
   1799), 300–1 (*Walpoliana*).
5. *Oriental Collections*, 1 (Jan.–Mar., 1797), 91 (signed by JP).
6. *Scots Magazine*, 2nd ser., 1 (Nov. 1817), 310–2 & (Dec. 1817), 412–13 & 3
   (July 1818), 12–15 & (Nov. 1818), 407–8 & (Dec. 1818), 497–8 & 5 (July 1819),
   20–4 & (Nov. 1819), 455–6 & (Dec. 1819), 527–30 ("Anecdotes, Historical,
   Literary, and Miscellaneous," by "J.P.")

III.   Pinkerton's fugitive publications not in magazines

For exclusions, see the introductory comment to section I above.

1. Trans. of the fragment "Nordymra sive Historia Rerum in Northumbria a Danis Norvegisque Gestarum, Seculis IX. X. et XI." into English. In Grímur Thorkelín, *Fragments of English and Irish History in the ninth and tenth century* (London: Nichols, 1788), 2–59.
2. "An Historical Dissertation on the Gowrie Conspiracy." In Malcolm Laing, *The History of Scotland, from the Union of the Crowns on the Accession of James VI. to the Throne of England, to the Union of the Kingdoms in the Reign of Queen Anne*, 2 vols. (London: Cadell and Davies; Manners and Miller, Edinburgh, 1800), Vol. 1, 527–44.
3. *"Extract from an Address to the Eminent, the Learned, and the Lovers of the early Literature and History of England, concerning an intended Publication to be intituled 'Rerum Anglicarum Scriptores;' or a Collection of the Original Historians of England, chronologically arranged; collated with the Manuscripts, illustrated with Notes, Chronological Tables, Maps, Complete Indexes, &c."* In Edward Gibbon, *Miscellaneous Works*, ed. J.B. Holroyd (Lord Sheffield), 5 vols. (London: John Murray, 1814), Vol. 3, 582–90.
4. "Hymn to Liberty." In JP, *Literary Correspondence* (1830), Vol. 1, 206–8. This song, written by JP for the Glasgow Revolution's Club celebration of the centenary of the Glorious Revolution, was "printed on a sheet for the use of drinkers and singers."
5. *"Proposal for the publication of the Ancient Historians of England, arranged in a regular chronological series, from the earliest accounts till the year 1500."* In JP, *Literary Correspondence*, Vol. 2, 436–40 n. A sentence in it indicates that 100 copies, "not intended for the public eye," may have been printed in 1814.

IV.   Books attacking Pinkerton and his works

Listed chronologically; includes only books solely devoted to attacking JP or with lengthy assaults on him.

1. William Pettman, *A Letter to Robert Heron, Esq. Containing a Few Brief Remarks on his Letters of Literature by one of the Barbarous Blockheads of the Lowest Mob, Who is a true Friend to Religion and a sincere Lover of Mankind.* London: G. and T. Wilkie, 1786.

2. William Webb, *An Analysis of the History and Antiquities of Ireland, prior to the fifth Century. To which is subjoined, a Review of the general History of the Celtic Nations.* Dublin: W. Jones, 1791.
   [Attacks, 173–230, JP's *Dissertation on the Scythians or Goths.*]

3. John Lanne Buchanan, *A Defence of the Scots Highlanders, in general; and some Learned Characters in particular: with a new and satisfactory Account of the Picts, Scots, Fingal, Ossian, and his Poems: As also, of the Macs, Clans, Bodotria. And Several other Particulars respecting the High Antiquities of Scotland.* London: J. Egerton, W. Stewart, and W. Richardson, 1794.
   [An attack on JP's *Dissertation on the Scythians or Goths* and *Enquiry.*]

4. James Tytler, *A Dissertation on the Origin and Antiquity of the Scottish Nation.* London: for the booksellers, 1795.
   [An attack on JP's *Enquiry.*]

5. William Anderson, *Answer to an Attack, made by John Pinkerton, Esq. of Hampstead, In his History of Scotland, lately Published, upon Mr. William Anderson, Writer in Edinburgh. Containing an Account of the Records of Scotland, and many strange letters by Mr. Pinkerton, accompanied with suitable comments necessarily arising from the Subject.* Edinburgh: Manners and Miller, 1797.

6. Robert Macfarlan, *George Buchanan's Dialogue concerning the Rights of the Crown of Scotland translated into English; with two Dissertations prefixed; one Archeological inquiring into the pretended Identity of the Getes and Scythians, of the Getes and Goths, and of the Goths and Scots; and the other Historical vindicating the Character of Buchanan as an Historian, and containing some Specimens of his Poetry in English Verse.* London: Cadell & Davies; W. Creech, Edinburgh, 1799.
   [The attack on JP's *Dissertation on the Scythians or Goths* and *Enquiry* is 1–52.]

7. [William Coxe], *A Vindication of the Celts, from Ancient Authorities; with Observations on Mr. Pinkerton's Hypothesis Concerning the Origin of the European Nations, in his Modern Geography, and Dissertation on the Scythians, or Goths.* London: E. Williams, 1803.

8. Sharon Turner, *A Vindication of the Genuineness of the Ancient British Poems of Aneurin, Taliesin, Llwarch Hen, and Merdhin, with Specimens of the Poems.* London: E. Williams, 1803.
   [An attempt to refute JP's arguments on the Welsh bards in the *Critical Review.*]

9. Nicolas Louis François de Neûfchateau, *Tableau des Vues que se propose la Politique Anglaise dans toutes les Parties du Monde.* Paris: Baudouin, 1804.
   [An attack on JP's *Modern Geography.*]

10. [?Joseph Binnie], *A few Remarks in reply to Mr. Pinkerton, upon The Scarcity of Men of Genius at Naples, in recent times, as asserted in his "Modern Geography." By a Neapolitan.*

V.   Biographical notices of Pinkerton and studies of his works

1. *Public Characters of 1800–1801* (London, 1801), 20–31.
2. *Universal Mag.*, 2nd ser., 2 (Nov., 1804), 393–7.
3. *European Mag.*, 51 (June, 1807), 411–13.
4. *Monthly Mag.*, 2nd ser., 1 (May 1826), 546–7.
5. *Gentleman's Mag.*, 96, pt. i (May, 1826), 469–72.
6. *Annual Register ... of the Year 1826* (London, 1827), 232–5.
7. *Edinburgh Annual Register for 1826* (Edinburgh, 1828), 19, pt. ii, 141–5.
8. Nichols, *Illustrations*, Vol. 5, 665–73.
9. *The Literary Correspondence of John Pinkerton*, ed. Dawson Turner, 2 vols. (London: Henry Colbourn and Richard Bentley, 1830), 1, v–x.
10. *The Penny Cyclopædia* (London, 1840), Vol. 18, 166–7.
11. S. Austin Allibone, *Critical Dictionary of English Literature* (Philadelphia and London: 1859–71), Vol. 2, 598–9.
12. Robert Chambers, *A Biographical Dictionary of Eminent Scotsmen*, ed. Thomas Thomson (London: 1868–70), Vol. 3, 245–8.
13. T.F. Henderson in *Dictionary of National Biography*, Vol. 45 (1896), 316–18.
14. W.A. Craigie, "Macpherson on Pinkerton: Literary Amenities of the Eighteenth Century," *PMLA*, 42 (1927), 433–42.
15. O.F.G. Sitwell, "John Pinkerton: An Armchair Geographer of the Early Nineteenth Century," *The Geographical Journal*, 138 (1972), 470–9.
16. A.A. Wilcock, "'The English Strabo' the geographical publications of John Pinkerton," *Transactions of the Institute of British Geographers*, no. 61 (Mar. 1974), 35–45.
17. P. O'Flaherty, "John Pinkerton (1758–1826): Champion of the Makars," *Studies in Scottish Literature*, 13 (1978), 159–95.
18. Wellens, O., "John Pinkerton: *Critical* Reviewer," *Notes and Queries*, 27, 5 (1980), 419–20.
19. H.H. Wood, "Introduction," *The Correspondence of Thomas Percy & John Pinkerton* (New Haven and London, Yale Univ. Press, 1985), vii–xxxv.
20. Sarah Couper in *Oxford Dictionary of National Biography* (Oxford Univ. Press, online ed., 2008).
21. Jean Culp Flanigan in *Eighteenth-Century British Historians*, ed. Ellen J. Jenkins; *Dictionary of Literary Biography*, Vol. 306 (Detroit: Gale, online, 2010), 275–80.
22. Colleen Franklin in *Eighteenth-Century British Literary Scholars and Critics*, ed. Frans De Bruyn; *Dictionary of Literary Biography*, Vol. 356 (as in 21, online, 2010), 239–47.

VI.   Letters by Pinkerton in print

Full titles of sources are elsewhere in the Bibliography.

1. William Anderson, *Answer to an Attack, made by John Pinkerton, Esq.* (1797). Letters to Anderson, 12 June 1796 – 28 Feb. 1797.
2. Gibbon, *Miscellaneous Works* (1814), Vol. 3, 578–81. Letter to J.B. Holroyd, 24 Oct. 1814.
3. Nichols, *Illustrations*, Vol. 5 (1828), 673–77. Letters to John Nichols, 28 Nov. 1782, 3 Oct. 1783, 2 Apr. 1784, 20 Jan. 1794, 28 Jan. 1794, 7 Feb. 1800; to William Herbert, 22 Feb. 1790; to J.B. Nichols, 31 Mar. 1800.
4. JP, *Literary Correspondence* (1830). 74 of JP's letters, 1782–1814, out of a total of c. 387 in the book.
5. Sir John Sinclair, *Correspondence* (1831), Vol. 1, 291, 471–3. Letters to Sinclair, 28 Feb. 1794 (extract) and 15 Apr. 1796.
6. *Gentleman's Magazine*, Vol. 102 pt. ii (Aug. 1832), 121–3. Letters to Percy, 19 Nov. 1785, 23 Jan. 1786, 18 Dec. 1786, 4 Sept. 1794.
7. Margaret Forbes, *Beattie and his Friends* (1904), 181, 199, 284–5. Abstracts, extracts of four letters to James Beattie, 1782–94.
8. *Report on the Laing Manuscripts*, Vol. 2 (1925), 578–9. Letter to Charles Dilly, 23 July 1795.
9. *Horace Walpole's Correspondence*, Vol. 16 (1951), 325–6. Letter to Walpole, 3 Feb. 1795.
10. *Correspondence of Thomas Percy & John Pinkerton*, ed. Wood (1985). JP's letters to Percy, 1778–94.

VII.   Manuscript letters by Pinkerton cited or consulted

Including originals, drafts, fragments, and copies; where no specific documentary reference is given, it is to be understood that copies were supplied by the library.

1. Bibliothèque Nationale: Joseph Van Praet, French MS 12773, f. 128 (12 [?Feb.] 1823).
2. Bodleian Library: Thomas Burgess, MS Douce d. 37, f. 56 (27 Jan. 1824); Cadell & Davies, MS Montagu d. 9, f. 146 (c. May, 1802), f. 148 (1 June 1802), f. 104 (1 Mar. 1811); Francis Douce, MS Douce d. 37, ff. 1–66 (letters, ?1789–1825); William Herbert, MS Montagu d. 5, f. 138 (28 Dec. 1787); Messrs. Perregaux (bankers, Paris), MS Montagu, d. 15, f. 106 (12 Aug. 1802).
3. Boston Public Library: Cadell & Davies (13 Feb. 1802).

4. British Library: Joseph Banks, Add. MS 31299, f. 18 (29 Jan. 1795), f. 22 (2 Feb. 1795); Cadell & Davies, Add. MS 34486, f. 11 (24 Apr. 1800), f. 14 (15 Apr. 1802); Edward Gibbon, Add. MS 34886, ff. 405–6 (2 Sept. 1793), ff. 416–17 (28 Oct. 1793); John B. Holroyd (Lord Sheffield), Add. MS 34887, f. 278 (28 July 1794), f. 279 (1 Aug. 1794); Macvey Napier, Add. MS 34612, ff. 172–3 (28 Jan. 1818), f. 234 (31 Oct. 1818), f. 303 (3 Nov. 1819); Thomas Warton, Add. MS 42561, f. 139 (2 Apr. 1784).

5. Edinburgh University Library: John Bradfute, MS La. II. 597/2 (5 Oct. 1794), 597/[14] (7 Aug. 1795); William Buchan, MS La. III. 379/669 (June 1787); Cadell & Davies, MS La. II. 597/[13] (14 Oct. 1799), 597/6 (19 Mar. 1809); Charles Dilly, MS La. II. 597/3 (9 May 1796); James Dodsley, MS La. II. 266 (6 Feb. 1778); David Steuart Erskine (Lord Buchan), MS La. II. 588 (11 Mar. 1792); James Gibson, MS La. II. 597/4 (6 Aug. 1799); Longman & Co., MS La. II. 597/7 (24 Feb. 1810); Grímur Thorkelín, MS La. III. 379/664–672 (letters, 1787–92).

6. Henry E. Huntington Library: Cadell & Davies (28 Jan. 1802); Charles O'Conor (13 Mar. 1786).

7. Historical Society of Pennsylvania: Joseph Banks (8 May 1798); Henry Brevoort (27 Aug. 1813); Cadell & Davies (7 Oct. 1799); Longman & Co. (8 Dec. 1804); Thomas Percy (14 June 1784).

8. Lewis Walpole Library, Yale University. MSS Vol. 82: J.B. Nichols (31 Mar. 1800, Vol. 3, copy); Longman & Co. (8 Dec. 1804, Vol. 3, draft). Misc. MSS: Earl Buchan (1 July [1788]); Egerton Brydges (5 Aug. 1814); Cadell & Davies (6 Mar. 1810, frag.).

9. Liverpool Public Libraries: William Roscoe (1 Aug., 20 Nov. 1814).

10. Maryland Historical Society: Jean G. Dentu (6 May 1810); Francis P. Plowden (28 May 1815); David Warden (letters, 1805–15).

11. National Library of Scotland: James Beattie, MS 1713, f. 4 (12 Aug. 1778), ff. 4–6 (18 Sept. 1778); John Bell, MS 583, f. 160 (26 Feb. 1782); William Blackwood, MS 4001, f. 196 (20 Apr. 1813), f. 198 (19 July 1813), f. 200 (24 July 1813); Cadell & Davies, MS 948, no. 7 (5 June 1801); MS 1709, f. 177 (n.d.); Archibald Constable, MS 2524, ff. 21–2 (19 May 1796), ff. 23–4 (6 Oct. 1796); Jean G. Dentu, MS 1810, f. 124 (2 June 1815); [?Fouché, Joseph], MS 1709, f. 214 (c. May, 1803); Longman & Co., MS. 1709, ff. 167–70 (c. 2 Mar. 1810); Thomas Percy, MS 1713, ff. 2–3, 6–10 (c. Apr. 1778–27 July 1779); Messrs. Perregaux (bankers, Paris), MS 2671, f. 125 (16 Jan. 1806); Harriet Elizabeth & Mary Margery Pinkerton, MS 1709, f. 120 (13 Jan. 1809); Henrietta Pinkerton, MS 1709, ff. 104, 106, 110, 112, 115, 117, 122 (20 Aug. 1805–20 Feb. 1809); William Roscoe, MS 2521, ff. 145–6 (19 Sept. 1814); Walter Scott, MS 3874, f. 138 (16 Apr. 1802), MS 3900, f. 284 (25 June 1825);

Andrew Stuart, MS 5401, f. 44 (4 Dec. 1796), f. 46 (6 Dec. 1796); John Young, MS 585, f. 44 (10 July 1784).

12. Natural History Museum, Botanical Library, Dawson Turner copies of Banks' correspondence: Joseph Banks, Vol. 12, ff. 11–12 (18 Jan. 1800), ff. 24–7 (28 Jan. 1800), ff. 51–3 (24 Mar. 1800).

13. New York Public Library, Monroe Papers: James Monroe (10 June 1806).

14. Pierpont Morgan Library: Cadell & Davies (4 Apr. 1800).

15. Sir Duncan Rice Library, University of Aberdeen: William Ogilvie, MS 914 (17 Apr. 1800)

16. Yale University Library: Thomas Astle (6 Dec. 1793, 5 Mar. 1795); Cadell & Davies (letters, 1804–10); George Nicol (26 June 1799); John Spottiswoode (Jan. 1805).

VIII.   Other manuscripts relating to Pinkerton, cited or consulted

1. Archives de la Seine et de la Ville de Paris: Acte de Décès 168/211 16 (JP's death certificate); file of marked graves in Paris.

2. Bodleian Library: Notes on Douce's copy of *Walpoliana* (1800), shelfmarked Douce, W. 103; MS Montagu d. 15, f. 104, Cadell & Davies to JP, 2 Mar. 1811; MS Douce d. 2, ff. 7–47, Douce's letters to JP, 1818–23; MS Douce d. 37, ff. 61–2, Mary Margery Pinkerton to Douce, 17 Mar. 1826, and reply; MS Eng. Misc. d. 244, f. 79, Thomas Percy to JP, 21 July 1792.

3. British Library: Add. MS 34486, ff. 16–23, "Idea of a new system of Geography," by JP, written Oct. 1799.

4. Edinburgh University Library: MS La. II. 647/62, William Bywater to JP, 7 Dec. 1785; MS La. II. 588, David Dalrymple to Lord Buchan, 10 May 1791; MS La. III. 625[1], "Reliques of Dissertation on the Goths and Enquiry into Scotish History," materials relating to the two works, written by JP prior to May 1788; MS La. III. 625[4]–[14], transcripts by JP from various historical documents, collected c. 1788–95, at the end [14] a "Chronicon Hebudum," a chronological sequence of events in the Hebrides, 795–1067.

5. General Register House, Edinburgh: CC8/8/125/1236–1238 (James Pinkerton's testament); Register of Deeds, Second Series, McKen. Office, Vol. 230, 873–4 (JP's Deed of Factory with William Buchan).

6. Historical Society of Pennsylvania: Cadell & Davies to JP, 9 Oct. 1799.

7. Lewis Walpole Library, Yale University: LWL MSS Vol. 82. A coll. titled "Letters addressed to John Pinkerton, Esq. from 1775 to 1815," bought by Dawson Turner from Longman's in June, 1826. It is arranged chronologically in 4 bound vols., with a separate printed title-page for

each. Vol. 1 covers 10 Jan. 1778–13 Feb. 1793; Vol. 2, 16 June 1793–24 Mar. 1800; Vol. 3, 25 Mar. 1800–11 Feb. 1808; Vol. 4, 10 Mar. 1808–31 Jan 1815 (with appendices, assorted dates). The letters are not numbered, and there is no pagination or numbering of folios. Identification must be by vol. and date. The letters are bound into the vols., but some are loose. Extractions have been made, e.g., of all letters from Horace Walpole. The letters are overwhelmingly addressed to JP; there are enclosures in them of various description, and other docs. A small number of letters, either copies, drafts, or originals, are by JP. The coll. has some printed matter, incl. prospectuses for *Vitae Antiquae Sanctorum* and *Iconographia Scotica*. Turner drew on this coll. for his edit. of JP's *Literary Correspondence*, editing items as he saw fit, e.g., by omitting what he thought non-"literary" material towards the end of letters or in postscripts, combining two letters into a single one, &c. Also from this coll: John Smith to JP, 17 Dec. 1812, 29 Nov. 1814; transcripts sent to PO'F by W.S. Lewis.

8. Longman & Co. archives, London (now in Reading University Library): "Cadell Purchase" papers, 11(a)–11(g). I have used the references from my research in Longman & Co. archives in the 1960s.

9. National Archives, Kew, London: Chancery documents, C13/113, 114.

10. National Library of Scotland: MSS 1709–21. The Pinkerton Papers, comprising 13 vols. of misc. literary matter presented to the Library by Mrs B.M.H. Riddell, JP's great granddaughter, in 1934. Much of the material consists of transcripts of books and MSS, jottings, outlines of literary projects, lists of portraits, notes on history, mineralogy, geography, etc. – a residue left in Henrietta Pinkerton's hands following the breakup of her marriage to JP in 1810, yet an important source for many aspects of his life. a. MS 1709. Letters and miscellaneous matter. The important items here are JP's letters to Henrietta, listed in VII (11), and her annotations on them, plus a few letters from JP, including one to his daughters. Among letter writers are: Adam de Cardonnel, Jean Dentu, R.P. Gillies, William Herbert, R.W. Heron, Thomas Holcroft, John Philip Kemble, Edmund Lodge, George Paton, Richard Southgate, and Andrew Stuart. Joseph Ritson's angry response to the *Critical Review*'s savaging of his *Pieces of Ancient Popular Poetry* is f. 44. b. MS 1710. Notes for the *Enquiry* and *History of Scotland* form the bulk of this MS; c. MS 1711. A section concerns JP's scientific writings and books of portraits. Chief items of interest: David Herd's notes on *Ancient Scotish Poems* (ff. 9–10), Percy's corrected proof-sheets of the second part of *Hardyknute* (ff. 11–18), JP's "Address

to the Eminent, the Learned, and the lovers of the early litterature and history of England, concerning an intended publication, to be intituled Rerum Anglicarum Scriptores" (ff. 154–9), and his proposed pamphlet "Centum Gravamina The Hundred Grievances of the People of England (ff. 246–50); d. MS 1712. This contains two texts of JP's play "The Heiress of Strathern" (ff. 1–119), the "General Argument" for an epic poem to be called "Ratho king of the Orkneys" (ff. 120–2), an unfinished text of his "Gothic Tale" (ff. 123–49), an outline of his proposed play "The British Princess, a Tragedy in Three Acts" (ff. 150–1), and the unfinished "An Easy Introduction to the Greek Language" (ff. 154–67); e. MS 1713. Notebook with drafts of letters to Percy and Beattie (listed in VII [11] above) and forged items for *Scottish Tragic Ballads* and *Select Scotish Ballads*; f. MS 1714. Notebook containing drafts of poems incl. in *Rimes* and *Tales in Verse*, with notes on English comedy and Thomas Gray's poems; g. MS 1715. Notebook featuring the corrected MS of *Rimes* and the MSS of *Two Dithyrambic Odes* and *Tales in Verse*; h. MS 1716. Extracts from and notes on MSS of early English and Scottish poetry in the British Museum library, with notes on scholarly books; i. MS 1717. Notebook containing JP's transcript of "The Geste of King Horn" from a Harleian MS, copied Jan. 1785; j. MS 1718. Notebook containing JP's notes on books connected with early Scottish and early European history; k. MS 1719. Notebook with sketches for poems, plays, and other literary work; l. MS 1720. Notebook containing mainly material for JP's *Enquiry* and *Dissertation on the Scythians or Goths*; m. MS 1721. Notebook containing private memoranda, 1777–9. MS 2670, ff. 174–5, Robert Ferguson to JP, 22 Nov. 1812.

11. National Library of Wales: MSS 13223C, Vols. II and III; 13224B, Vol. IV; William Coxe's letters to William Owen Pughe, 1802–3.
12. Natural History Museum, Botanical Library, Dawson Turner copies of Joseph Banks' corres: Vol. 12, ff. 13, 54, Banks to JP, 19 Jan, 26 Mar. 1800.
13. Patton, Lewis: Info. from William Godwin's diary.
14. Royal Literary Fund, London: Case No. 486.
15. Society of Antiquaries of Scotland, Edinburgh: MS minutes.
16. Somerset House, London: Wills for 1826, "Swabey," f. 284 (JP's will); marriage of one John Pinkerton (not JP) to Rachel Hudson, 12 July 1790, St. George's Church, Hanover Square (reference lost). See ch. 3, n. 85. Somerset House marriage records are now in the General Register Office, Southport, Merseyside.
17. Yale University Library: Cadell and Davies, arbitration bond re dispute with JP, 17 Mar. 1804 & receipt signed by JP, 12 Mar. 1806; George Nicol, letter to JP 28 June 1799.

IX.   Other works cited, or works previously listed with short titles

I've shortened some titles, adjusted capitalization, and occasionally added punctuation.

Adam, Alexander. *A Summary of Geography and History, Both Ancient and Modern.* 3rd ed. London: Cadell and Davies, 1802.

Addison, Joseph, Richard Steele, et al. *The Spectator.* 4 vols. London: J.M. Dent & Co., [1906].

Aikin, John. *Geographical Delineations.* London: J. Johnson, 1806.

Anderson, Alan Orr. *Early Sources of Scottish History: A.D. 500 to 1286.* 2 vols. Edinburgh and London: Oliver and Boyd, 1922.

– *Early Sources of Scottish History: A.D. 500 to 1286.* Edited by Marjorie Anderson. 2 vols. Stamford, Lincolnshire: Paul Watkins, 1990.

Anderson, Marjorie O. "Picts – the Name and the People." In *The Picts: a new look at old problems,* edited by Alan Small, 7–14. Dundee: Graham Hunter Foundation, 1987.

Andrews, J.H. *Maps in Those Days: Cartographic Methods before 1850.* Dublin: Four Courts Press, 2009.

Arnot, Hugo. *The History of Edinburgh.* Edinburgh: W. Creech; J. Murray, London, 1779.

Baker, J.N.L. *The History of Geography: Papers by J.N.L. Baker.* New York: Barnes & Noble, 1963.

– *A History of Geographical Discovery and Exploration.* New York: Cooper Square Publishers, 1967.

Barbour, John. *The Bruce.* Edited by John Jamieson. Edinburgh: Manners and Miller et al., 1820. (Vol. 1 of 2-vol. book; Vol. 2 *Wallace.*)

– *The Bruce.* Edited by W.W. Skeat. 2 vols. Edinburgh and London: Scottish Text Society, 1894.

Barker, J.R., "Cadell and Davies and the Liverpool Booksellers." *The Library,* 5th ser., 14 (Dec. 1959): 274–80.

Bawcutt, Priscilla. *Dunbar the Makar.* Oxford: Clarendon Press, 1992.

Beattie, James. *Dissertations Moral and Critical.* London: W. Strahan and T. Cadell; W. Creech, Edinburgh, 1783.

– *Letters.* 2 vols. London: John Sharpe, 1820.

– *Poetical Works.* Edited by Alexander Dyce. London: Bell and Daldy, 1866.

Beattie, William. *The Chepman and Myllar Prints.* Edinburgh: Edinburgh Bibliographical Society, 1950.

Bede. *A History of the English Church and People.* Translated by Leo Sherley-Price. Revised by R.E. Latham. London, Penguin Books, 1970.

Beveridge, Craig, "Childhood and Society in Eighteenth-Century Scotland." In *New Perspectives on the Politics and Culture of Early Modern Scotland*, edited by John Dwyer, Roger A. Mason, and Alexander Murdoch, 265–80. Edinburgh: John Donald Publishers Ltd., c. 1982.

Black, Jeremy. *Natural and Necessary Enemies: Anglo-French Relations in the Eighteenth Century*. Athens, Georgia: University of Georgia Press, 1986.

Blind Harry. *See* Henry the Minstrel.

Boardman, Steve. "Late Medieval Scotland and the Matter of Britain." In *Scottish History*, edited by Cowan and Finlay, 47–72.

Bosker, Aisso. *Literary Criticism in the Age of Johnson*. Groningen: J.B. Wolters, 1930.

Boswell, James. *Letters*. Edited by Chauncey Tinker. 2 vols. Oxford: Clarendon Press, 1924.

– *The Private Papers of James Boswell from Malahide Castle*. Edited by Geoffrey Scott and F.A. Pottle. 18 vols. New York: privately printed, 1928–34.

– *Boswell's Life of Johnson*. Edited by G.B. Hill. Revised by L.F. Powell. 6 vols. Oxford: Clarendon Press, 1934–50.

Bronson, Bertrand H. *Joseph Ritson: Scholar at Arms*. 2 vols. Berkeley: University of California Press, 1938.

Burns, Robert. *Letters*. Edited by J. De Lancey Ferguson. 2 vols. Oxford: Clarendon Press, 1931.

Burton, John Hill. *The History of Scotland*. 2nd ed. 8 vols. Edinburgh and London: William Blackwood and Sons, 1873.

Byerley, Thomas, and Joseph Robertson, eds. *The Percy Anecdotes*. 3 vols. London: T. Boys, 1821–3.

Cairns, Craig, ed. *The History of Scottish Literature*. 4 vols. Aberdeen: Aberdeen University Press, 1987–8.

Campbell, Alexander. *An Introduction to the History of Poetry in Scotland*. 2 pts. Edinburgh: Andrew Foulis, 1798–9.

Caw, James L. *Scottish Portraits*. 2 vols. Edinburgh: T.C. and E.C. Jack, 1903.

Chadwick, Nora K. *Celtic Britain*. London: Thames and Hudson, 1963.

Chalmers, George. *The Life of Thomas Ruddiman*. London: John Stockdale; William Laing, Edinburgh, 1794.

– *Caledonia: or, An Account, Historical and Topographic, of North Britain*. 3 vols. London: Cadell and Davies; Archibald Constable, Edinburgh, 1807–24.

Chapman, John H., and George J. Armytage, eds. *The Register Book of Marriages belonging to the Parish of St. George, Hanover Square*. 4 vols. London: Harleian Society, 1886–97.

Chapman, R.W., ed. *Johnson's Journey to the Western Islands of Scotland and Boswell's Journal of a Tour to the Hebrides with Samuel Johnson, LL.D.* London: Oxford University Press, 1965.

Child, Francis J. *The English and Scottish Popular Ballads.* 5 vols. Boston and New York: Houghton, Mifflin and Co., 1882–98.

Clark, Alexander F.B. *Boileau and the French Classical Critics in England (1660–1830).* Paris: Édouard Champion, 1925.

Clerke, Sir John. *Dissertatio de Monumentis quibusdam Romanis in Boreali Magnae Britanniae parte detectis.* Edinburgh: T. and W. Ruddiman, 1750.

Coleridge, S.T. *Selected Poetry and Prose.* Edited by Elisabeth Schneider. 2nd ed. San Francisco: Rinehart Press, 1971.

Colley, Linda. *Britons: Forging the Nation.* London: Pimlico, 2003.

Collins, W. Lucas. *Virgil.* Edinburgh and London: William Blackwood and Sons, 1890.

The Complaynt of Scotland. Written in 1548. Edited by John Leyden. Edinburgh: Archibald Constable, 1801.

Constable, Thomas. *Archibald Constable and his Literary Correspondents.* 3 vols. Edinburgh: Edmonston and Douglas, 1873.

Constantine, Mary-Ann. *The Truth Against the World: Iolo Morganwg and Romantic Forgery.* Cardiff: University of Wales Press, 2007.

Cowan, Edward J., and Richard J. Finlay, eds. *Scottish History: The Power of the Past.* Edinburgh: Edinburgh University Press, 2002.

Cowper, William. *Poetical Works.* Edited by Robert Southey. London: Henry G. Bohn, 1854.

Craigie, Sir William A., ed. *The Maitland Folio Manuscript.* 2 vols. Edinburgh and London: Scottish Text Society, 1919–27.

Crone, G.R. *Maps and their Makers: An Introduction to the History of Cartography.* London: Hutchinson University Library, 1968.

Crone, G.R., and R.A. Skelton. "English Collections of Voyages and Travels 1625–1846." In *Richard Hakluyt & His Successors: A volume issued to commemorate the centenary of the Hakluyt Society,* edited by Edward Lynam, 65–140. London: The Hakluyt Society, 1946.

Cummins, W.A. *The Age of the Picts.* Phoenix Mill, Gloucestershire: Alan Sutton Publishing Ltd., 1995.

Curley, Thomas M. *Samuel Johnson, the Ossian Fraud, and the Celtic Revival in Great Britain and Ireland.* Cambridge: Cambridge University Press, 2009.

Dalrymple, David (Lord Hailes). *Ancient Scottish Poems. Published from the MS. of George Bannatyne, MDLXVIII.* Edinburgh: John Balfour, 1770.

Dalyell, Sir John Graham. *Scotish Poems, of the Sixteenth Century.* Edinburgh: Archibald Constable; Vernor and Hood, London, 1801.

Davis, Bertram H. *Thomas Percy: A Scholar-Cleric in the Age of Johnson.* Philadelphia: University of Pennsylvania Press, 1989.

De Montluzin, Emily Lorraine. "Attributions of Authorship in the *British Critic* during the Editorial Regime of Robert Nares, 1793–1813." *Studies in Bibliography* 51 (1998): 241–58.

De Quincey, Thomas. *Collected Writings*. Edited by David Masson. 14 vols. London: A. and C. Black, 1896–7.

Dean, Dennis R. "The Word 'Geology'." *Annals of Science* 36 (1979): 35–43.

– *James Hutton and the History of Geology*. Ithaca and London: Cornell University Press, 1992.

Dempster, George. *Letters of George Dempster to Sir Adam Fergusson 1756–1813*. Edited by James Fergusson. London: Macmillan and Co., 1934.

Dentu, Jean Gabriel. *Moyen de Parvenir en Littérature, ou mémoire à consulter, sur une question de propriété littéraire, dans lequel on prove que le sieur Malte-Brun ... A copié littéralement une grande partie des Œuvres de M. Gossellin, ainsi que celles de MM. Lacroix, Walkenaer, Pinkerton, Puissant, etc., etc., et les a fait imprimer et débiter sous son nom*. Paris: n.p., 1811.

Dibdin, James C. *The Annals of the Edinburgh Stage*. Edinburgh: Richard Cameron, 1888.

Dickinson, R.E. *The Makers of Modern Geography*. New York and Washington: Frederick A. Praeger, 1969.

Dickinson, R.E., and O.J.R. Howarth. *The Making of Geography*. Westport, Connecticut: Greenwood Press, 1976.

Disraeli, Isaac. *The Illustrator Illustrated*. London: Edward Moxon, 1838.

Dobson, Austin. *Eighteenth Century Vignettes First Series*. London: Chatto and Windus, 1906.

Douglas, David C. *English Scholars 1660–1730*. 2nd ed. London: Eyre and Spottiswoode, 1951.

Douglas, Gawin. *Select Works of Gawin Douglass, Bishop of Dunkeld*. Perth: R. Morison and Son, 1787.

Duffy, Edward. *Rousseau in England: The Context for Shelley's Critique of the Enlightenment*. Berkeley: University of California Press, 1979.

Dunbar, William. *The Poems of William Dunbar*. Edited by James Kinsley. Oxford: Clarendon Press, 1979.

Eagles, Robin. *Francophilia in English Society, 1748–1815*. Houndmills, UK: Macmillan Press, 2000.

Edwards, A.S.G. "Editing Dunbar: The Tradition." In *William Dunbar, "The Nobill Poyet": Essays in Honour of Priscilla Bawcutt*, edited by Sally Mapstone, 51–68. East Linton, Lothian: Tuckwell Press, 2001.

Ellis, Peter Berresford. *Celt and Saxon: The Struggle for Britain AD 410–937*. London: Constable, 1993.

Eyre-Todd, George. *Scottish Poetry of the Eighteenth Century*. 2 vols. Glasgow: William Hodge and Co., 1896.

Ferguson, William. *Scotland: 1689 to the Present*. Edinburgh: Oliver & Boyd, 1968.

– *The Identity of the Scottish Nation: An Historic Quest*. Edinburgh: Edinburgh University Press, 1998.

Fjalldal, Magnús. "To Fall by Ambition – Grímur Thorkelín and his *Beowulf* Edition." *Neophilologus* 92 (2008): 321–32.

Forbes, Margaret. *Beattie and his Friends*. Westminster: Archibald Constable & Co., 1904.

Forbes, Sir William. *An Account of the Life and Writings of James Beattie, LL.D.* 2 vols. Edinburgh: Archibald Constable and Co. et al., 1806.

Fordun, John of. *John of Fordun's Chronicle of The Scottish Nation*. Edited by William Skene. Edinburgh: Edmonston and Douglas, 1872.

Fortescue, J.B. *The Manuscripts of J.B. Fortescue, Esq., preserved at Dromore*. 10 vols. London: Her Majesty's Stationery Office, 1892–1927.

French, Josephine, Valerie Scott, Mary Alice Lowenthal, eds., *Tooley's Dictionary of Mapmakers*. 4 vols. Tring, Hertfordshire: Map Collector Publications, 1999–2004.

Friedman, Albert B. *The Ballad Revival: Studies in the Influence of Popular on Sophisticated Literature*. Chicago: University of Chicago Press, 1961.

Garfield, Simon. *On the Map: A Mind-Expanding Exploration of the Way the World Looks*. New York: Gotham Books, 2013.

Gascoigne, John, "Joseph Banks, mapping and the geographies of natural knowledge." In *Georgian Geographies: Essays on Space, Place and Landscape in the Eighteenth Century*, edited by Miles Ogburn and Charles W.J. Withers, 151–73. Manchester and New York: Manchester University Press, 2004.

Gerald of Wales (Giraldus Cambrensis). *The Journey through Wales and The Description of Wales*. Translated by Lewis Thorpe. London: Penguin Books, 1978.

Gibbon, Edward. *Miscellaneous Works*. Edited by J.B. Holroyd (Lord Sheffield). 2 vols. London: A. Strahan, Cadell and Davies, 1796.

– *Miscellaneous Works*. 5 vols. London: John Murray, 1814.

– *The History of the Decline and Fall of the Roman Empire*. Edited by J.B. Bury. 7 vols. London: Methuen, 1896–1900.

– *Letters*. Edited by J.E. Norton. 3 vols. London: Cassell, 1956.

– *The English Essays of Edward Gibbon*. Edited by Patricia B. Craddock. Oxford: Clarendon Press, 1972.

Gillies, Robert P. *Recollections of Sir Walter Scott*. London: James Fraser, 1837.

– *Memoirs of a Literary Veteran*. 3 vols. London: Richard Bentley, 1851.

Glover, T.R. *Virgil*. 6th ed. London: Methuen & Co., 1930.

Grant, Francis J. *Register of Marriages of the City of Edinburgh, 1751–1800*. Edinburgh: Scottish Record Society, 1922.

Grant, James. *History of the Burgh Schools of Scotland*. London and Glasgow: William Collins, 1876.

– *Cassell's Old and New Edinburgh: Its History, its People, and its Places*. 3 vols. London, Paris, and New York: Cassell, Petter, Golpin and Co., 1884–7.

Gray, Thomas. *Correspondence*. Edited by Paget Toynbee and Leonard Whibley. 3 vols. Oxford: Clarendon Press, 1935.

Groom, Nick. *The Making of Percy's Reliques*. Oxford: Clarendon Press, 1999.

– *The Forger's Shadow: How Forgery Changed the Course of Literature*. London: Picador, 2002.

Hamer, Douglas. "The Bibliography of Sir David Lindsay (1490–1555)." *The Library*, 4th ser., 10 (June, 1929): 1–42.

Hardyknute, A Fragment. Edinburgh: printed by James Watson, 1719.

Hardyknute, A Fragment. Being the First Canto of an Epick Poem. London: Robert Dodsley, 1740.

Hardyknute. A Fragment of an Antient Scots Poem. Glasgow: Robert Foulis, 1745.

Harford, John S. *The Life of Thomas Burgess, D.D. … Late Lord Bishop of Salisbury*. London: Longman, Orme, Brown, Green, & Longmans, 1840.

Harvey, P.D.A. *The History of Topographical Maps: Symbols, Pictures and Surveys*. London: Thames and Hudson, 1980.

Hazen, A.T. *A Bibliography of Horace Walpole*. Folkestone and London: Dawsons of Pall Mall, 1973.

Hazlitt, William. *The Round Table: A Collection of Essays on Literature, Men, and Manners*. 2 vols. Edinburgh: Archibald Constable and Co.; Longman, Hurst, Rees, Orme, and Brown, London, 1817.

– *The Life of Thomas Holcroft*. Edited by Elbridge Colby. 2 vols. London: Constable and Co., 1925.

Henderson, Isabel. "The Problem of the Picts." In *Who are the Scots? and The Scottish Nation*, edited by Gordon Menzies, 20–35. Edinburgh: Edinburgh University Press, 2002.

Henry of Huntington. *Chronicle*. Translated and edited by Thomas Forester. London: Henry G. Bohn, 1853.

Henry the Minstrel. *The Metrical History of Sir William Wallace, Knight of Ellerslie*. 3 vols. Perth: R. Morison and Son, 1790.

Heron, Robert. *A New General History of Scotland*. 5 vols. Perth: R. Morison and Son; Vernor and Hood, London, 1794–9.

Hitchcock, Tim. *English Sexualities, 1700–1800*. Houndmills, UK: Macmillan Press Ltd., 1997.

Hodgkin, R.H. *A History of the Anglo-Saxons*. 2 vols. Oxford: Clarendon Press, 1935; 3rd ed., 1952.

Houston, R.A., and W.W.J. Knox, eds. *The New Penguin History of Scotland*. London: Penguin Press, 2001.

Hume, David. *The History of England*. 6 vols. Boston: Philips, Sampson and Co., 1850.

– *Selected Essays*. Edited by Stephen Copley and Andrew Edgar. Oxford: Oxford University Press, 1993.

Innes, Thomas. *A Critical Essay on the Ancient Inhabitants of the Northern Parts of Britain or Scotland*. Edited by George Grub. Edinburgh: William Paterson, 1879.

Irving, George Vere, and Alexander Murray. *The Upper Ward of Lanarkshire Described and Delineated*. 3 vols. Glasgow: Thomas Murray and Son, 1864.

Irving, Washington. *The Sketch Book of Geoffrey Crayon, Gent*. New York and Scarborough, Ont.: New American Library, c.1981.

Jackson, Kenneth. "The poem *A Eolcha Alban Uile*." *Celtica: Journal of the School of Celtic Studies* 3 (1956): 149–67.

Johnson, Samuel. Poems. Edited by E.L. McAdam, Jr, with George Milne. New Haven and London: Yale University Press, 1964.

– *The Rambler*. Edited by W.J. Bate and Albrecht B. Strauss. 3 vols. New Haven and London: Yale University Press, 1969.

– *A Voyage to Abyssinia (Translated from the French)*. Edited by Joel J. Gold. New Haven and London: Yale University Press, 1985.

Johnston, Arthur. *Enchanted Ground: The Study of Medieval Romance in the Eighteenth Century*. London: Athlone Press, 1964.

Johnstone, James, ed. *Antiquitates Celto-Normanicae*. Copenhagen: Aug. Frid. Stein, 1786.

Keates, J.S. *Understanding Maps* 2nd ed. Harlow, UK: Longman Group Ltd., 1996.

Kelly, Christopher. *A New and Complete System of Universal Geography*. Vol. 1. London: Thomas Kelly, 1819.

Kerr, Robert. *Memoirs of the Life, Writings & Correspondence of William Smellie*. 2 vols. Edinburgh: John Anderson, 1811.

Kersey, Mel. "Ballads, Britishness and *Hardyknute*, 1719–1859." *Scottish Studies Review* 5, 1 (2004): 40–56.

Ketton-Cremer, R.W. *Horace Walpole: A Biography*. London: Faber and Faber, 1946.

Kidd, Colin. *Subverting Scotland's Past: Scottish whig historians and the creation of an Anglo-British identity, 1689–c. 1830*. Cambridge: Cambridge University Press, 1993.

– "The Ideological Uses of the Picts, 1707–c.1990." In *Scottish History*, edited by Cowan and Finlay, 169–90.

King, Everard H. *James Beattie*. Boston: Twayne Publishers, 1977.

Kinsley, James, ed. *Scottish Poetry: A Critical Survey*. London: Cassell, 1955.

Knox, Vicesimus. *Winter Evenings: or, Lucubrations on Life and Letters*. 3 vols. London: Charles Dilly, 1788.

Law, Alexander. *Education in Edinburgh in the Eighteenth Century*. London: University of London Press Ltd., 1965.

Le Breton, Anna Letitia. *Memoir of Mrs. Barbauld, including Letters and Notices of her Family and Friends*. London: George Bell, and Sons, 1874.

Lewis, C.S. *The Discarded Image: An Introduction to Medieval and Renaissance Literature*. Cambridge: University Press, 1964.

Livingstone, David N., and Charles W.J. Withers, eds. *Geography and Enlightenment*. Chicago and London: University of Chicago Press, 1999.

Lloyd, Sir John Edward. *A History of Wales*. 3rd ed. 2 vols. London: Longmans, Green and Co., 1934.

Lockhart, J.G. *The Life of Robert Burns ... To which is added Thomas Carlyle's Review Essay*. Edited by J.M. Sloan. London: Hutchinson & Co., n.d.

Lynch, Jack. *Deception and Detection in Eighteenth-Century Britain*. Aldershot, UK: Ashgate, 2008.

Lyndsay, Sir David. *Poetical Works*. Edited by George Chalmers. 3 vols. London: Longman, Hurst, Rees, and Orme; Archibald Constable, Edinburgh, 1806.

Macaulay, T.B. *Lays of Ancient Rome*. London: G.P. Putnam's Sons, n.d.

Mackenzie, Agnes M. *An Historical Survey of Scottish Literature to 1714*. London: Alexander Maclehose, 1933.

Macpherson, James. *An Introduction to the History of Great Britain and Ireland*. Dublin: James Williams, 1771.

Macpherson, John. *Critical Dissertations on the Origin, Antiquities, Language, Government, Manners, and Religion, of the Ancient Caledonians*. London: T. Becket and P.A. De Hondt; J. Balfour, Edinburgh, 1768.

Maitland, William. *The History and Antiquities of Scotland, from the earliest account of time to the death of James I. Anno 1437*. 2 vols. London: A Millar, 1757.

Manning, Susan. "Antiquarianism, the Scottish Science of Man, and the emergence of modern disciplinarity." In *Scotland and the Borders of Romanticism*, edited by Leith Davis, Ian Duncan, and Janet Sorensen, 57–76. Cambridge: Cambridge University Press, 2004.

Mayhew, Robert. "William Guthrie's *Geographical Grammar*, the Scottish Enlightenment and the Politics of British Geography." *Scottish Geographical Journal* 115, 1 (1999): 19–34.

– *Enlightenment Geography: The Political Languages of British Geography, 1650–1850*. Houndmills, UK: Macmillan Press Ltd., 2000.

McClure, Ruth K. *Coram's Children: The London Foundling Hospital in the Eighteenth Century.* New Haven and London: Yale University Press, 1981.

McCordick, David. *Scottish Literature: An Anthology.* 2 vols. New York: P. Lang, 1996.

McCrone, David. "Tomorrow's Ancestors: Nationalism, Identity and History." In *Scottish History*, edited by Cowan and Finlay, 253–71.

McFarland, Thomas. *Romanticism and the Heritage of Rousseau.* Oxford: Clarendon Press, 1995.

Miles, David. *The Tribes of Britain.* London: Weidenfeld & Nicolson, 2006.

Minot, Laurence. *Poems on Interesting Events in the reign of King Edward III.* [Edited by Joseph Ritson.] London: T. Egerton, 1795.

Montgomerie, Alexander. *Poems.* Edited by James Cranstoun. Edinburgh and London: Scottish Text Society, 1887.

Murray, John. *A Comparative View of the Huttonian and Neptunian Systems of Geology.* Edinburgh: Ross and Blackwood, 1802.

Nangle, Benjamin C. *The Monthly Review First Series 1749–1789.* Oxford: Clarendon Press, 1934.

– *The Monthly Review Second Series 1790–1815.* Oxford: Clarendon Press, 1955.

Nichols, John. *Literary Anecdotes of the Eighteenth Century.* 9 vols. London: printed for the author, 1812–15.

Nichols, John, and J.B. Nichols. *Illustrations of the Literary History of the Eighteenth Century.* 8 vols. London: vols. 1–4 printed for the author; vols. 5–8 printed by and for J.B. Nichols and Son, 1817–58.

O'Conor, Charles. *The Letters of Charles O'Conor of Bellangare.* Edited by Catherine Ward and Robert E. Ward. 2 vols. Ann Arbor, Michigan: University Microfilms International, for the Irish American Cultural Institute, 1980.

O'Flaherty, P. "Johnson in the Hebrides: Philosopher Becalmed." *Studies in Burke and his Time* 13 (Fall, 1971): 1986–2001.

Paton, Henry. *Register of Marriages For the Parish of Edinburgh, 1701–1750.* Edinburgh: Scottish Record Society, 1908.

Payne, John. *Universal Geography, formed into a New and Entire System.* 2 vols. Dublin: printed by Zachariah Jackson, 1794.

Pearson, Jacqueline. *Women's Reading in Britain 1750–1835: A Dangerous Occupation.* Cambridge: Cambridge University Press, 1999.

Percy, Thomas. *Reliques of Ancient English Poetry.* 3 vols. London: J. Dodsley, 1765.

– *Bishop Percy's Folio Manuscript.* Edited by J.W. Hales and F.J. Furnivall. 3 vols. London: N. Trubner, 1867–8.

– *The Correspondence of Thomas Percy & Edmond Malone.* Edited by Arthur Tillotson. Baton Rouge, Louisiana: Louisiana State University Press, 1944.

Phillimore, W.P.W., et al., eds., *Hampshire Parish Registers: Marriages*. 16 vols. London: Phillimore and Co., 1899–1914.

Phillips, Mark Salber. *Society and Sentiment: Genres of Historical Writing in Britain, 1740–1820*. Princeton: Princeton University Press, 2000.

Phillipson, Nicholas. "The Scottish Enlightenment." In *The Enlightenment in National Context*, edited by Roy Porter and Mikuláš Teich, 19–40. Cambridge: Cambridge University Press, 1981.

Pinkerton, J.C. "Pinkerton, East Lothian." *Notes and Queries* 176 (3 June 1939): 388.

Pittock, Murray G.H. *Inventing and Resisting Britain: Cultural Identities in Britain and Ireland, 1685–1789*. New York: St Martin's Press, 1997.

– *Celtic Identity and the British Image*. Manchester and New York: Manchester University Press, 1999.

– *A New History of Scotland*. Phoenix Mill, Gloucestershire: Sutton Publishing, 2003.

– *Scottish and Irish Romanticism*. Oxford: Oxford University Press, 2008.

Pliny the Elder. *Natural History*. Translated by John Bostock and H.T. Riley. 6 vols. London: Henry G. Bohn, 1855–7.

Poems in the Scottish Dialect by Several Celebrated Poets. Glasgow: Robert Foulis, 1748.

Porter, Roy. *The Making of Geology: Earth Science in Britain 1660–1815*. Cambridge: Cambridge University Press, 1977.

Ramsay, Allan. *The Ever Green, Being a Collection of Scots Poems, Wrote by the Ingenious before 1600*. Edinburgh: printed by Thomas Ruddiman, 1724.

– *Poems*. Edited by George Chalmers. 2 vols. London: Cadell and Davies, 1800.

Ramsay, John. *Scotland and Scotsmen in the Eighteenth Century*. Edited by Alexander Allardyce. 2 vols. Edinburgh and London: William Blackwood, 1888.

Rees, Thomas, and John Britton. *Reminiscences of Literary London from 1779 to 1853*. London: Suckling and Galloway, 1896.

Rennell, James. *Memoir of a Map of Hindoostan; or the Mogul Empire*. 2nd ed. London: for the author, 1792.

– *The Geographical System of Herodotus, Examined and Explained*. London: for the author, 1800.

Ridley, Florence H. "Did Gawin Douglas write *King Hart*?" *Speculum* 34, 3 (1959): 402–12.

Ritson, Joseph. *Scotish Song*. 2 vols. London: J. Johnson and J. Egerton, 1794.

– *Robin Hood*. 2 vols. London: T. Egerton and J. Johnson, 1795.

– *Letters from Joseph Ritson, Esq., to Mr. George Paton. To which is added, a critique by John Pinkerton, Esq. upon Ritson's Scotish Songs*. Edinburgh: John Stevenson, 1829.

– *Letters*. Edited by Harris Nicolas. 2 vols. London: William Pickering, 1833.

Robberds, J.W. *A Memoir of the Life and Writings of the late William Taylor*. 2 vols. London: John Murray, 1843.

Roberts, Barré Charles. *Letters and Miscellaneous Papers*. London: n.p., 1814.

Roberts, W. Rhys. "The Quotation from *Genesis* in the *De Sublimitate* (IX. 9)." *Classical Review* 11, 9 (1897): 431–6.

Rogers, Pat. *Johnson and Boswell: The Transit of Caledonia*. Oxford: Clarendon Press, 1995.

Roper, Derek. *Reviewing before the Edinburgh 1788–1802*. London: Methuen, 1978.

Scott, Walter. *The Antiquary*. London: Marcus Ward, n.d.

– *The History of Scotland*. 2 vols. London: Longman, Rees, Orme, Brown, and Green, and John Taylor, 1830.

– *Poetical Works*. Edited by J.G. Lockhart. 12 vols. Edinburgh: Robert Cadell; Houlston and Stoneman, London, 1851.

– *Letters*. Edited by H.J.C. Grierson. 12 vols. London: Constable and Co., 1932–7.

Shakespeare, William. *Plays*. Edited by Samuel Johnson and George Steevens. 2nd ed. 10 vols. London: C. Bathurst et al., 1778.

– *Plays*. Edited by Samuel Johnson and George Steevens. Revised by Isaac Reed. 15 vols. London: T. Longman et al., 1793.

Shearer, Thomas, and Arthur Tillotson. "Percy's Relations with Cadell and Davies." *The Library*, 4th ser., 15 (Sept., 1934): 224–36.

Sher, Richard B. *The Enlightenment & the Book: Scottish Authors & Their Publishers in Eighteenth-Century Britain, Ireland, & America*. Chicago and London: University of Chicago Press, 2006.

Sherbo, Arthur. "Isaac Reed and the *European Magazine*." *Studies in Bibliography* 37 (1984): 210–27.

Shields, Juliet. "Situating Scotland in Eighteenth-Century Studies." *Literature Compass* 9, 2 (2012): 140–50.

Sibbald, Sir Robert. *The History, Ancient and Modern, of the Sheriffdoms of Fife and Kinross*. Edinburgh: for the author, 1710.

Skene, William F. *Celtic Scotland: A History of Ancient Alban*. 2nd ed. 3 vols. Edinburgh: David Douglas, 1886–90.

Snyder, Christopher A. *The Britons*. Malden, Mass.: Blackwell Publishing, 2003.

Somerville, Thomas. *My Own Life and Times 1741–1814*. Edinburgh: Edmonston and Douglas, n.d.

Southey, Charles, ed. *The Life and Correspondence of Robert Southey*. 6 vols. Longman, Green Brown, and Longmans, 1849–50.

Stone, Lawrence. *The Family, Sex and Marriage In England 1508–1800*. London: Weidenfeld and Nicolson, 1977.

Strong, John. *A History of Secondary Education in Scotland*. Oxford: Clarendon Press, 1909.

Strout, Alan Lang. "Some Unpublished Letters of John Gibson Lockhart to John Wilson Croker." *Notes and Queries* 187 (23 Sept. 1944): 134–7.

Supplement to the fourth, fifth, and sixth editions of the Encyclopædia Britannica. Edited by Macvey Napier. 6 vols. Edinburgh: Archibald Constable & Co.; Hurst, Robinson & Co., London, 1824.

Sweet, Rosemary. *Antiquaries: The Discovery of the Past in Eighteenth-Century Britain*. London and New York: Hambledon and London, 2004.

Swift, Simon. *Romanticism, Literature and Philosophy: Expressive Rationality in Rousseau, Kant, Wollstonecraft and Contemporary Theory*. London: Continuum, 2006.

Tacitus. *Agricola; Germania*. Translated by Harold Mattingly. Revised by J.B. Rives. London: Penguin Books, 2009.

Taylor, Griffith, ed. *Geography in the Twentieth Century*. New York: Philosophical Library; Methuen, London, 1953.

The Thre Prestis of Peblis: how thai tald thar talis. Edited by T.D. Robb. Edinburgh and London: Scottish Text Society, 1920.

Tooley, R.V., Charles Bricker [text], and Gerald Roe Crone. *A History of Cartography: 2500 Years of Maps and Mapmakers*. London: Thames and Hudson, 1969.

Transactions of the Literary and Antiquarian Society of Perth. Vol. 1. Perth: for the Society, 1827.

Trevor-Roper, Hugh. *The Invention of Scotland: Myth and History*. New Haven: Yale University Press, 2008.

Trumbach, Randolph. "Erotic Fantasy and Male Libertinism in Enlightenment England." In *The Invention of Pornography: Obscenity and the Origins of Modernity, 1500–1800*, edited by Lynn Hunt, 253–82. New York: Zone Books, 1993.

Turner, Sharon. *The History of the Anglo-Saxons, from their first appearance above the Elbe, to the death of Egbert*. London: Cadell and Davies, 1799.

– *The History of the Anglo-Saxons, from the earliest period to the Norman Conquest*. 7th edition. London: Longman, Brown, Green, and Longmans, 1852.

Wainwright, F.T., ed. *The Problem of the Picts*. Edinburgh: Nelson, 1955.

Walker, Joseph Cooper. *Historical Memoir on Italian Tragedy, From the Earliest Period to the Present Time*. London: E. Harding, 1799.

– *An Historical and Critical Essay on the Revival of the Drama in Italy*. Edinburgh: Mundell and Son; Longman, Hurst, Rees, and Orme, London, 1805.

Wallace, James. *The History of the Kingdom of Scotland*. Dublin: for the author, 1724.

Walpole, Horace. *Horace Walpole's Correspondence with Thomas Chatterton, Michael Lort, John Pinkerton, John Fenn and Mrs Fenn, William Bewley, Nathaniel Hillier*. Edited by W.S. Lewis and A. Dayle Wallace. New Haven: Yale University Press, 1975; Vol. 16 of the 48-vol. *Horace Walpole's Correspondence* (1937–83).

Warton, Thomas. *The Union: or Select Scots and English Poems*. Edinburgh: Archibald Munro & David Murray, 1753.

– *Observations on the Faerie Queene of Spenser*. London: R. and J. Dodsley; J. Fletcher, Oxford, 1754.

– *The History of English Poetry, from the Close of the Eleventh to the Commencement of the Eighteenth Century*. 3 vols. London: Dodsley et al., 1774–81.

Wellens, O. "The 'Critical Review': New Light on its Last Phase." *Revue Belge de Philosophie et d'Histoire* 56, 3 (1978): 678–94.

Whitaker, John. *The Genuine History of the Britons Asserted*. London: Dodsley et al., 1772.

Wilson, Kathleeen. *The Island Race: Englishness, Empire and Gender in the Eighteenth Century*. London and New York: Routledge, 2003.

Wordsworth, William. *Poetical Works*. Edited by Thomas Hutchinson. London: Henry Frowde, 1903.

Wyntoun, Androw of. *Đ Orygynale Cronykil of Scotland*. Edited by David Macpherson. 2 vols. London: Thomas Egerton; William Laing, Edinburgh, 1795.

– *The Orygnale Cronykil Of Scotland*. Edited by David Laing. 3 vols. Edinburgh: Edmonston and Douglas, 1872–9.

Young, G.M. *Gibbon*. 2nd ed. London: Rupert Hart-Davis, 1948.

Zachs, William. *Without Regard to Good Manners: A Biography of Gilbert Stuart 1743–1786*. Edinburgh: Edinburgh University Press, 1992.

– *The First John Murray and the Late Eighteenth-Century London Book Trade*. Oxford: Oxford University Press, for The British Academy, 1998.

Index

Lightning Source UK Ltd.
Milton Keynes UK
UKHW041642221222
414329UK00013B/127/J

9 781442 649286